The Movement of Molecules across Cell Membranes

THEORETICAL AND EXPERIMENTAL BIOLOGY

An International Series of Monographs

CONSULTING EDITOR

J. F. Danielli

State University of New York at Buffalo, Buffalo, New York

Volume 1 J. L. Cloudsley-Thompson, Rhythmic Activity in Animal Physiology and Behaviour. 1961

Volume 2 W. P. Rogers, The Nature of Parasitism: The Relationship of Some Metazoan Parasites to Their Hosts. 1962

Volume 3 C. B. Williams, Patterns in the Balance of Nature. 1964

Volume 4 D. J. Triggle, Chemical Aspects of the Autonomic Nervous System. 1965

Volume 5 R. T. Treager, Physical Functions of Skin. 1966

Volume 6 W. D. Stein, The Movement of Molecules across Cell Membranes. 1967

The Movement of Molecules across Cell Membranes

W. D. STEIN

LECTURER IN BIOPHYSICAL CHEMISTRY AT THE VICTORIA UNIVERSITY OF MANCHESTER, MANCHESTER, ENGLAND

AP

ACADEMIC PRESS New York San Francisco London 1967

A Subsidiary of Harcourt Brace Jovanovich, Publishers

ACADEMIC PRESS, INC.
111 Fifth Avenue, New York, New York 10003

United Kingdom Edition published by
ACADEMIC PRESS, INC. (LONDON) LTD.
24/28 Oval Road, London NW1

LIBRARY OF CONGRESS CATALOG CARD NUMBER: 66–30105

PRINTED IN THE UNITED STATES OF AMERICA

To my father, Professor P. Stein,
and to the memory of my late father-in-law,
Rabbi Moshe Baruch Morgenstern,
wise men and scholars.

Preface

This monograph is an attempt to analyze the molecular basis of the movement of substances across the cell membrane. By summarizing pertinent current knowledge on the structure and composition of the cell membrane and on the kinetics of movement across the membrane, one can begin to build a consistent physical picture of such membranes. One can attempt, for any particular molecular species, to assess whether movement occurs through the bulk of the membrane or through specific limited regions of the membrane, or whether it is brought about by the action of specific membrane components. With the pathway of movement more or less established, one can begin to analyze the molecular events which occur during this movement and to discuss the factors controlling the rate of this movement. This work is concerned with these problems.

While the book will be of interest primarily to research students and research workers in the field of membrane studies and in cognate fields, that is, to biophysicists, biochemists, and physiologists, it is hoped that the level of presentation is such that those who intend to make use of their knowledge of the properties of cell membranes—for example, pharmacologists and research workers in the clinical fields and in agriculture—will also find the book of value, not only for the general picture of the membrane which is presented but also for the information on membrane properties which is collated. Although this monograph was not a result of a lecture course, it may prove suitable for use in one.

The battered condition of my copy of "The Permeability of Natural Membranes," by Davson and Danielli, testifies to the frequency with which that book has been consulted during the fifteen years it has been on my bookshelf. During this period a number of excellent monographs covering much the same ground and containing newer material have been published, yet it is to "Davson and Danielli" that I most often turn for information or for clarification of a problem of the cell membrane. Among the possible reasons for the preeminence of this work are the clarity of expression and argument used by the authors and the consistent development of the consequences of a physically plausible model for the cell membrane—that of the "paucimolecular membrane"—which lends a unity and profundity to their treatment. In addition, one turns to "Davson and Danielli" to consult the tables of permeability values and other membrane parameters with which the book abounds. Yet much has changed in this field since the second edition was published in 1952, and it is now clear that one waits in vain for a third edition.

It should be said at once that the present work cannot hope to reach the distinction of that earlier work; the author is neither a Davson nor a Danielli. But it is hoped that some of the spirit of their work will be found here. In particular, what has been attempted is the exploration of the degree to which the Davson–Danielli model of the cell membrane is still valid as an explanation of membrane transport phenomena and the extension of this model to take account of the newer studies.

Any model of the cell membrane must accord with the available data on the structure and composition of the membrane so far as this is known, and in Chapter 1 we consider the current picture of the anatomy of the cell membrane. Two major theoretical advances made in the years since the Davson–Danielli model was proposed can be used in a modern analysis of this model. The more recent of these advances is the application by Kedem and Katchalsky of the theory of irreversible thermodynamics to membrane transport, a technique which illuminates much of the current discussion on movement across cell membranes. This theory is introduced in Chapter 2 and is referred to again on numerous occasions in later chapters. An earlier theoretical advance was the Eyring analysis of diffusion as a rate process, and the application of this viewpoint to diffusion into and across the cell membrane occupies Chapter 3 of this monograph. The existence of pores in the red cell membrane is also considered in Chapter 3. Chapter 4 treats the kinetic consequences of the analysis of the carrier model (that is, the assumption that specific membrane components are concerned in the movement of their substrates across the membrane) as applied to the simplest system, that of the equilibrating facilitated diffusion systems. Chapter 5 discusses the mechanism by which cells concentrate certain nonelectrolytes and the revolution in our understanding of this mechanism recently brought about by the analysis of ion and nonelectrolyte interactions. In Chapter 6, the concentrative ability of cells is further considered and a distinction is made between primary and secondary concentrative abilities, namely, those linked directly to an energy input and those being

linked only at one remove which feed on a preexisting concentration gradient of some penetrating molecule. Chapter 7 is concerned with the water balance of cells and tissues and is the most physiological of the chapters. I believe that Chapter 8 represents a somewhat new endeavor, the collating of much of the available material on the properties of the "carriers," some of these properties being solely inferential but some having a more direct physical and chemical basis. In the final chapter some possible mechanisms for the mode of action of the specialized transfer systems are considered in the light of the available evidence.

Many of the arguments presented rest on a mathematical analysis of the postulated physical models. Wherever possible the attempt has been made to keep this analysis simple by introducing successively the complexities of the model, rather than by using the more difficult, but inherently sounder, technique of beginning with the most general model. Similarly, the intuitive physical basis of the models used has been stressed so that the mathematical analysis can be used as an effective tool, rather than serve to overwhelm the less sophisticated reader.

The emphasis on *molecules* in the title not only stresses that this work is most concerned with analysis at the molecular level but that the movement of charged particles, of ions, has been given less detailed treatment. In part, this is because other recent sources have given adequate coverage of the available information and also because my own work has been concerned with the movement of nonelectrolytes. Yet it is hoped that most of the points concerning ion movements which illuminate the general principles of movement across cell membranes have been adequately covered.

One might perhaps hope that this monograph will be the last of its type which will need to be written. The current exciting progress in the isolation of specific carrier molecules from cell membranes must surely lead in the near future to the classification of membrane transport as a branch of enzymology. It is in the belief that the techniques of the enzymologist will lead us to a detailed picture of the physical basis of membrane transport that this book has been written at this stage, for it is the phenomena detailed in the following pages which the new generation of membrane enzymologists will have to account for and explain. It is hoped that the bringing together of this information will help them in their task.

It is a pleasure to acknowledge the help of my friends and colleagues in various aspects of the preparation of this book. In particular, I should like to thank Mrs. Molly Levine for reading and commenting on the entire manuscript and Dr. S. Shall for his comments on much of the first draft. Miss Ingrid Sandberg prepared a number of the original figures. I must thank, also, the authors and publishers who so generously allowed me to reproduce figures already published. My wife, Chana, refuses to let me thank her for her contributions, which included the preparation of the Subject Index, of the remaining original figures, the typing of the entire manuscript in its three drafts, and the proof-reading—apart from putting up with the conventional difficulties of being an author's wife.

I should like, finally, to record the debt I owe to my teachers: Professor Joseph Gillman, formerly of the University of Witwatersrand, Professor J. F. Danielli, F.R.S., formerly of King's College, London, and Professor H. N. Christensen, of the University of Michigan. Any virtues that this book may possess must largely arise from the stimulus and encouragement that I received from these distinguished men.

W. D. Stein

June, 1967

Contents

Preface vii

Glossary of Symbols xv

CHAPTER 1

The Anatomy of the Plasma Membrane

1.1 Some Considerations of Methodology 1
1.2 Structure and Composition of the Cell Membrane 2
1.3 The Structure of Myelin Forms—A Model System for the Cell Membrane 12
1.4 Nerve Myelin 19
1.5 Physical Stability of the Bimolecular Lipid Leaflet 22
1.6 Surface Tension of the Cell 31
1.7 Conclusions 34

CHAPTER 2

General Aspects of Diffusion across Membranes

2.1 Permeability and Diffusion Coefficients 36
2.2 Irreversible Thermodynamics of Membrane Processes 40
2.3 The Reflection Coefficient σ 45
2.4 The Relation between the Conventional and the Phenomenological Coefficients 47
2.5 The Physical Interpretation of the Phenomenological Coefficients 48
2.6 Methods of Measuring the Coefficients of Membrane Permeability 52
2.7 The Diffusion and Distribution of Ions 59
2.8 Active and Passive Transport 62

CHAPTER 3

The Molecular Basis of Diffusion across Cell Membranes

3.1 Introduction 65
3.2 Diffusion in Liquids as Movement within a Lattice 66
3.3 A Lattice Model for Diffusion into and across the Cell Membrane 72
3.4 Molecular Significance of the Parameters $\Delta F^{\ddagger}$, $\Delta H^{\ddagger}$, $\Delta S^{\ddagger}$, and $PM^{1/2}{}_{max}$ 85
3.5 The Movement of Ions 90

3.6 An Alternative Model—Pores in the Cell Membrane 106
3.7 The Permeability of Water 120
3.8 Conclusions 124

CHAPTER 4

Facilitated Diffusion—the Kinetic Analysis

4.1 Introduction 126
4.2 Criteria for Identifying Facilitated Diffusion Systems 127
4.3 Distribution of the Facilitated Diffusion Systems 128
4.4 A Preliminary Kinetic Analysis of Facilitated Diffusion 129
4.5 The Mobile Carrier Hypothesis 148
4.6 The Direct Determination of the Parameters K_m and V_{max}, Using the Flux Equations 157
4.7 The Dimerizer Hypothesis for Facilitated Diffusion 176

CHAPTER 5

The Coupling of Active Transport and Facilitated Diffusion

5.1 The Development of the View That Sodium Transport and Metabolite Transport Are Coupled 177
5.2 Experimental Tests of the Co-transport Model 183
5.3 The Kinetics of Co-transport 192
5.4 The Distribution Ratios at the Steady State 198
5.5 A Tertiary Active Transport System 203

CHAPTER 6

The Primary Active Transport Systems

6.1 Criteria for Distinguishing between Primary and Secondary Active Transport Systems 207
6.2 Indications That Certain Active Transport Systems May Not Be Linked to Other Driving Fluxes 209
6.3 The Linkage of Active Transport to Chemical Reactions 215
6.4 Kinetics of the Primary Transport Systems 221
6.5 Primary Active Transport Systems Uncoupled to the Energy Input 228
6.6 The Kinetics of "Pump" and "Leak" Systems 231
6.7 Ion Pumping by Epithelial Cell Layers 235

CHAPTER 7

The Movement of Water

7.1 The Volume of a Cell at the Steady State 242
7.2 Why a Linked Sodium/Potassium Pump 251
7.3 Transport across Epithelial Cell Layers 253

CHAPTER 8

Molecular Properties of the Transport Systems

8.1 Information Derived from Studies of the Specificity of Transport Systems 266
8.2 Molecular Significance of the Action of Drugs on Transport 281
8.3 Inhibition by Chemical Reagents 289
8.4 Attempts at the Isolation of Transport Systems from Cell Membranes 295
8.5 Data Suggesting That a Number of Transport Systems May Be Bivalent toward Substrates or Inhibitors 308

CHAPTER 9

Possible Mechanisms for Mediated Transfer

9.1 What the Models Have to Account For: A Summary of the Properties of the Mediated Transfer Systems 309
9.2 The Criteria of Acceptability for Model Systems of Transport 311
9.3 Models for Mediated Transfer 314
9.4 The Future of Transport Studies 321

References 324

Author Index 353

Subject Index 361

Glossary of Symbols

A	activation energy for diffusion
A_i	concentration of nondiffusible intracellular matter
A_s	area of pores available to solute
A_w	area of pores available to solvent water
A_{I}, A_{II}	electrochemical potential of solute on side I or II of membrane
B	concentration of permeable anion
C_i	internal solute concentration
C_s	concentration of solute or permeant
$\overline{C}_s$	mean of solute concentrations on 2 sides of membrane
ΔC_s	concentration difference across a membrane
C_{I}, C_{II}	concentration of permeant on side I or II of membrane
D	diffusion coefficient
D_0	diffusion coefficient at the absolute zero of temperature (computed)
E	concentration of free carrier
E_D	electrical potential across the membrane
E_{I}, E_{II}	electrical potential on side I or II of membrane
E_m	maximum electrical potential within membrane
F	Faraday constant
$\Delta F^{\ddagger}$	free energy for formation of the transition state
G	concentration of permeant G
$\Delta H^{\ddagger}$	enthalpy change for the formation of transition state
I	concentration of inhibitor
J_D	exchange flow of solute and solvent
J_s	flux of solute
J_v	total flow of solute and solvent
$J_{\mathrm{I}\rightarrow\mathrm{II}}$	undirectional flux, in direction I to II
$J_{\mathrm{II}\rightarrow\mathrm{I}}$	undirectional flux, in direction II to I
K	distribution coefficient for solute between membrane and aqueous phase
K (with subscript)	refers in general to equilibrium constants, as defined in Sections 4.5 and 5.3
K_i	dissociation constant for inhibitor-carrier complex

K_m	substrate concentration for half-maximal unidirectional flux, "Michaelis constant," defined in Eq. (4.1)
K_s	dissociation constant of carrier-substrate complex, defined in Eq. (4.11)
$K^‡$	equilibrium constant for the formation of the transition state
L_D	coefficient of exchange flow, defined in Eq. (2.5)
L_{Dp}	ultrafiltration coefficient, Eq. (2.6)
L_p	pressure-filtration coefficient, Eq. (2.4)
L_{pD}	osmotic coefficient, Eq. (2.6)
M	molecular weight (Chapter 3)
M	concentration of permeable cation (Chapter 7)
N	number of hydrogen bonds that permeant makes with water (Chapter 3)
N	concentration of second permeable cation (Chapter 7)
P	permeability coefficient
P_s	permeability coefficient of solute
P_w	permeability coefficient of water
Q	free energy of formation of a single hydrogen bond between permeant group and water
Q_{10}	ratio of given parameters for a 10°C temperature rise
R	the gas constant
R_{ii}	coefficient of permeability of i, as a function of the concentration of i (Eq. 2.33)
R_{ij}	coefficient of coupling of the flows of species i and j (Eq. 2.33)
R_{ir}	coefficient of coupling between the flow of species i and the progress of a chemical reaction r (Eq. 2.33)
S	concentration of permeant S
$\Delta S^‡$	entropy change for the formation of the transition state
T	absolute temperature
V	volume of cell
$\overline{V}_s$	partial molar volume of solute
$\overline{V}_w$	partial molar volume of solvent
$V_{\max}$	maximal rate of unidirectional flux, defined in Eq. (4.1)
X_i	amount of intracellular matter in model cell
e (subscript)	concentration at external face of membrane
f_{sm}	a frictional coefficient (drag exerted by membrane on solute)
f_{ss}	a frictional coefficient (drag exerted by solute on solute)

f_{sw}	a frictional coefficient (mutual drag of solvent and solute)
f_{wm}	a frictional coefficient (drag exerted by membrane on solvent)
f_{ww}	a frictional coefficient (mutual drag of solvent molecules)
h	Planck's constant
i (subscript)	referring to the ith component (Chapter 2 and Section 6.1)
i (subscript)	internal concentration of permeant
j (subscript)	referring to the jth component
k	Boltzmann's constant
k (with subscript)	refers in general to rate constants. (for k_1 to k_5 see Fig. 4.10; for k_6 to k_8 see Section 5.3)
k	transmission coefficient
l	leak rate constant, in Eq. (7.2) and following
m (subscript)	referring to the membrane
n	number of pores per unit area of membrane (Chapter 3)
n	rate constant for leakage of cation N (Chapter 7)
Δp	pressure difference across membrane
p	pump rate constant, in Eq. (7.2) and following
r	radius of aqueous channels in pore model of membrane structure
r	radius of diffusing molecule, in Eq. (3.2)
r (subscript)	referring to the rth chemical reaction
$\mathbf{r}$	ratio of rate constants for transfer across the membrane of loaded and free carrier (Chapter 4)
R	distribution ratio for mobile ions (Chapter 2)
$\mathbf{s}$	radius of solute molecule
s (subscript)	referring to the solute (permeant)
$\mathbf{w}$	radius of water molecule
w (subscript)	referring to the solvent (or water)
Δx	thickness of membrane
α	activity
Δ	an increment or difference between some parameter in two states
η	viscosity of liquid
$\lambda_1\lambda_2\lambda_3$	lattice parameters on the lattice model for diffusion (Fig. 3.1)
$\Delta\mu_i$	electrochemical gradient of species i
ν^+, ν^-	number of cations or anions respectively in a salt
σ	reflection coefficient, defined in Eq. (2.12)
ϕ_w	volume fraction of water in membrane
ω	the solute permeability coefficient, at zero volume flow, defined in Eq. (2.14)

CHAPTER 1

The Anatomy of the Plasma Membrane

1.1 Some Considerations of Methodology

This book is concerned with the problem of how molecules and ions move across cell membranes. To handle this problem effectively, we will need to understand in some detail the structure of cell membranes, what molecules the membrane is composed of, and how these molecules are arranged. In the present chapter we shall consider these problems but shall find that they have no easy solution. One reason for this is that, as Ponder (1961) has pointed out, the very definition of the term "cell membrane" is a matter of contention. In fact, we cell biologists use the term "cell membrane" or "plasma membrane"—we shall use these terms interchangeably—in at least three quite different senses. In the anatomical sense, the cell membrane is the external limiting region of the cell, visible occasionally as a darkly staining region in the light microscope and with more certainty in the electron microscope as a layer (or pair of layers) of osmiophilic material. In the biochemical sense, the cell membrane is a "fraction" of the cell prepared by the now classical techniques of selective disintegration of the whole cell, followed by differential centrifugation. A preparation is obtained which can be analyzed chemically and which can, by electron microscopy, be compared with the "cell membrane" seen in the whole cell. Finally, in the physiological sense, the "cell membrane" is a hypothetical structure invented to explain certain data on the "permeability of cells" (that is, on the rate of entry into these cells of certain substances) and to explain other data on the distribution of metabolites and other molecules between the cell and the fluid in which the cell is immersed. Such data often suggest the presence of a "permeability barrier" between the cell and its environment.

Now, as Ponder (1961) and Dervichian (1955) have pointed out, it has been generally *assumed* that these three definitions of the cell membrane refer to the same entity, but this is a matter which is as yet far from being *proved*. For it is clear and widely accepted that the "cell membrane" of the biochemists may include (by absorption—physical and mechanical) many substances which are not present in the "cell membrane" of the anatomist and may have lost many components (possibly crucial to the architecture of the membrane) during even the gentlest separation and washing procedures. We shall see, too, that so little is it necessary to equate the anatomists' and physiologists' "membrane" that a number of schools of physiologists indeed maintain that the existence of a "cell membrane" barrier need not be postulated at all, while those physiologists who do feel the need for a "cell membrane" cannot with certainty exclude the possibility that this permeability barrier may be external or internal to the anatomical "membrane."

We shall see that the conclusions drawn by many students of the membrane on the basis of anatomical, physiological, and biochemical studies allow them to form a consistent picture of the membrane at the molecular level—but it is often in the nature of such studies to reinforce one another. We shall discuss these and other studies in the following chapters, attempting to distinguish carefully in each case between the experimental findings and the conceptual picture on which the study is based, or which emerges from the given study. In this chapter we shall confine ourselves strictly to a consideration of the anatomical and biochemical pictures of the cell membrane and in the remainder of the book consider the physiological analyses.

1.2 Structure and Composition of the Cell Membrane

A. Electron Microscopy

One of the most generally valid findings of current electron microscopy (J. D. Robertson, 1964) is the demonstration of the presence, at the limiting surface of almost all cell types studied, of a layer of material some 75 Å wide which binds (and is thus made visible by) electron-dense compounds such as osmium tetroxide, potassium permanganate, and uranyl acetate [see Figs. 1.1a, 1.1b, 1.2a, and 1.2b for some representative examples, and J. D. Robertson (1964) for many others]. The use especially of potassium permanganate as a stain has shown clearly that this limiting layer is in fact composite, being made up of two dark (that is, heavy-metal adsorbing) layers each about 25 Å wide and separated by a light layer (one which does not take up the electron-dense stain) also some 25 Å wide. Thus at the interface between two cells (Figs. 1.1b, 1.2a, and a portion of 1.2b) four dense layers (two contributed by each cell) and three light layers (one from each cell and from the space between the two cells) can be seen. Each set of a *pair* of dense layers separated by a light layer is referred to as a "unit membrane." Similar unit membranes have been found as a general feature of many intracellular organelles. Thus the endoplasmic reticulum is a system of such membranes arranged in concentric shells, while the mitochondrion is composed similarly of a more or less complex unit membrane; but there is less agreement in this case that the unit membrane is the dominant structure (Sjöstrand, 1963). Unit membranes in layers form the basis of the architecture of the chloroplast, of the retinal rod, and of the myelin sheath that surrounds medullated nerve fibers (see Section 1.5). (For mitochondrial membranes, see Fig. 1.19) Electron microscope studies may use freeze-etching techniques in which tissues are frozen at −180°C and fractured at that temperature. The fracture faces then are etched by vacuum sublimation at low temperatures. It would seem that for a variety of plant materials, including onion root tip and yeast cells, the plane of fracture is between the two halves of the unit membrane (Branton, 1966). Thus the fracturing process appears to split the membrane and expose an internal membrane face. These studies, which do not depend on the use of fixatives to produce the image, provide strong support for the validity of the unit membrane concept in living tissues.

The structure of the unit membrane, as seen in the electron microscope, is evidently that of a highly ordered lamellar object. We might inquire at once as to the nature of the molecular organization that forms this supramolecular structure. Does the order visible in the electron microscope reflect a corresponding segregation at the molecular level of the chemical species that form the membrane? To answer this question we must identify the molecular species that comprise the cell membrane.

B. Chemical Composition

Most chemical analyses of the cell membrane have been performed on fractions isolated from mammalian red blood cells. These cells are readily available in large quantities as a pure cell type and can be induced by a variety of simple techniques (such as their suspension in media of low osmolarity, the repeated freezing and thawing of cells, and their subjection to the action of detergents) to lose the greater part of their hemoglobin, the major internal component. This is the process of "hemolysis." Light and electron microscopy confirm that these ghost preparations are indeed the cell membranes (Ponder, 1955), and in a number of cases (Teorell, 1952; Stein, 1956; LeFevre, 1961b)

Fig. 1.1(a)

Fig. 1.1. Cell membranes in profile. (a) Portion of a human red blood cell—fixed with permanganate and sectioned—showing the unit membrane structure

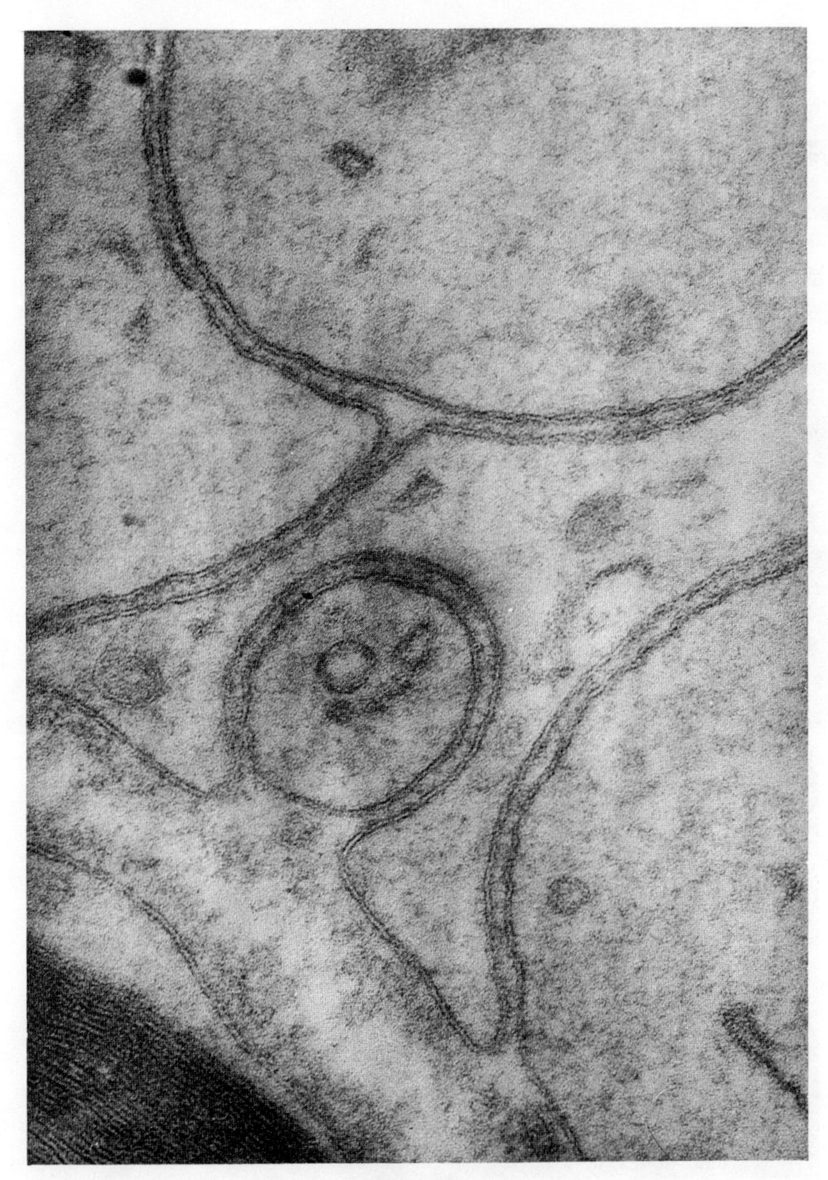

Fig. 1.1(b)

bounding the cell. Magnification, 280,000 ×. (b) Unit membranes in a mature unmyelinated mouse sciatic fiber at high magnification, 155,000 ×. (Taken with kind permission from J. D. Robertson, 1964.)

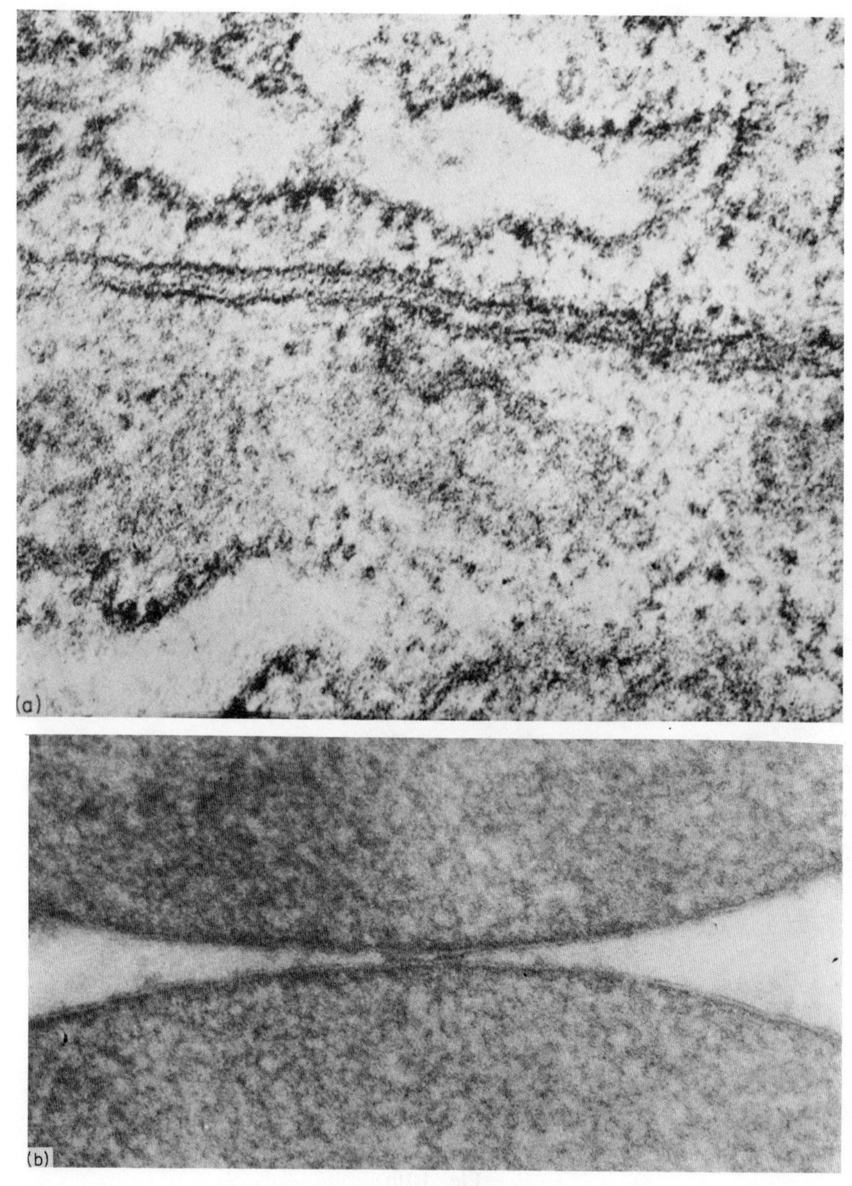

Fig. 1.2. Further cell membranes. (a) Cell boundary between two exocrine pancreas cells showing two (asymmetric) unit membranes separated by a narrow interspace. Osmium fixed and stained with uranyl acetate. Magnification, × 180,000. (Taken with kind permission from Sjöstrand, 1962.) (b) Two rat erythrocytes from liver tissue, fixed in acrolein, stained with uranyl acetate. Dense areas indicative of uranyl ions bound to protein. (Taken with kind permission from Swift, 1962.)

such membranes isolated by gentle hemolytic procedures have been shown to retain to a surprising degree their capacity to act as a permeability barrier. The material remaining after hemolysis, termed variously the "ghosts" or "stroma" or "post-hemolytic residue," can be washed by centrifugation, isolated, and analyzed chemically.

With a number of possible methods available for hemolysis and with a variety of washing procedures, it is obvious that a spectrum of membrane fractions can be derived, the chemical composition of which can be varied—perhaps not at will, but certainly over a large range. The repeated washing of red cells in either high or low ionic strength media has been reported to result in a loss of lipid as well as nonlipid material, from whole cells (Lovelock, 1955) and from ghosts (Ponder, 1955), a finding which might, in part, account for the variations in the composition of the membrane as these have been reported by different workers (Table 1.1). Recently, however, a systematic study of the retention of hemoglobin by stroma when this was prepared in a variety of media of different pH and ionic strength has been carried out by Dodge *et al.* (1963), a study which greatly clarifies the understanding of the basic protein and lipid content of the membrane. These workers have found conditions (hemolysis in 120 volumes of 20 m*M* phosphate buffer at pH 7.65, 4°C) where less than 0.1% of hemoglobin is retained by the stroma after a single hemolysis step and in which the lipid content of the obtained stroma is essentially equal to that of the intact cells. Similarly, ghosts prepared by a procedure of "successive hemolyses" in saline media of reduced osmolarity (Weed *et al.*, 1963) retain their entire content of lipid, while retaining less than 0.5% of the hemoglobin. If stroma is prepared at pH 6.15, however, substantial proportions of hemoglobin and significant amounts of nonhemoglobin protein are bound to the stroma (C. D. Mitchell *et al.*, 1965). Furthermore, hemoglobin-free stroma prepared at pH 7.65 will, if then taken to pH 6.15, combine with hemoglobin to form a tightly bound complex, while stroma prepared at pH 6.15 and containing hemoglobin will lose this if taken to pH 7.65. The retention of hemoglobin is thus pH-dependent and is reversible. It would appear, therefore, from these studies that hemoglobin is not a structural component of the membrane, under physiological conditions, but that the conditions of preparation of the ghosts might be, on occasion, such as to induce the binding of hemoglobin and presumably other proteins. Some of the studies recorded in Table 1.1 were performed on ghosts prepared at pH values of 6.0 and below and are thus likely to be subject to error.

It can be confidently stated, however, that all preparations studied contain both lipid and protein in large amounts but that polysaccharides,

TABLE 1.1

CHEMICAL COMPOSITION OF HUMAN ERYTHROCYTE MEMBRANES[a]

					% Total phospholipid as		
References	Total protein	Total lipid	Total cholesterol	Total phospholipid	Lecithin	"Cephalin"	Sphingomyelin + lysophosphatide
Erickson *et al.* (1938)	17.7	3.99	1.20	2.27		60	
Parpart and Dziemian (1940)	—	3.94	—	—			
Reed *et al.* (1960)	—	4.95	1.13	2.88	30	40	22
de Gier and van Deenen (1961)	—	—	—	—	38	25	37
Dodge *et al.* (1963)	6.0	5.24	1.42	3.15	27	27 (PE)[b]	22
Weed *et al.* (1963)	5.61	4.95	1.08	3.15	29	36	25
Ways and Hanahan (1964)	—	4.76	1.17	—	30	41	24
van Gastel *et al.* (1964)	—	4.80	1.13	2.73			

[a] All values recorded as $10^{13} \times g$ substance per red cell.

[b] Here PE stands for phosphatidylethanolamine.

and nucleic acids (if present at all) are present in only small amounts. The mucopolysaccharides, although only minor components, play a major role in determining both the immunological behavior of the erythrocytes and the electrical potential of the cell surface. Presumably the outermost layer of the cell is a loose mucopolysaccharide-containing region bearing a number of strongly negatively charged groups (Seaman and Heard, 1960; Heard and Seaman, 1960) of which sialomucopeptides may well be the dominant constituents (Cook *et al.*, 1960).

The lipid present is largely cholesterol (Fig. 1.3) and phospholipid in roughly equal molar amounts, the bulk of the phospholipid being lecithin and "cephalin" (phosphatidylserine and phosphatidylethanolamine) (Fig. 1.3). The precise details of the lipid composition of the erythrocyte membrane are species-dependent and attempts have been made to correlate the lipid composition of membranes isolated from the red cells of different animal species with the physiological properties of such membranes (van Deenen and de Gier, 1964).

The weight of protein is some one to two times that of the combined lipids. A number of different protein fractions have been identified and some beginnings have been made on the study of their physical properties. One component isolated by Moskowitz and Calvin (1952), termed "reticulin" or "stromin," is a lipoprotein appearing as a rod-shaped particle 5–11 μ long and 0.3–1.3 μ wide in the electron microscope. When dried and extracted with ether, this lipoprotein gives *elinin,* which on extracting with alcohol gives the lipid-free *stromatin.* A protein having enzymic activity that has been identified in red cell ghosts is the sodium/potassium-activated adenosine triphosphatase which we shall discuss at length in Chapter 8. Another membrane-bound enzyme is acetylcholinesterase (3.1.1.7), a protein which has been the subject of a number of careful investigations. Thus C. D. Mitchell *et al.* (1965) chose five representative enzymes present in whole human red cells, namely, carbonic anhydrase (4.2.1.1), adenosine deaminase (3.5.4.4), aldolase (4.1.2.7), glyceraldehyde phosphate dehydrogenase (1.2.1.12), and acetylcholinesterase, in order to discover by analyses of stroma which, if any, of these enzymes is an integral part of the cell membrane. Carbonic anhydrase and adenosine deaminase were never found bound to stroma. By a correct choice of the pH and ionic strength during hemolysis, stroma could be prepared very largely free of both aldolase and glyceraldehyde phosphate dehydrogenase, but acetylcholinesterase was almost wholly retained by the stroma at all values of pH and ionic strength used. Thus, acetylcholinesterase would seem to be a real constituent of the membrane. Aldolase and glyceraldehyde phosphate dehydrogenase can, however, be bound by electrostatic linkages and may

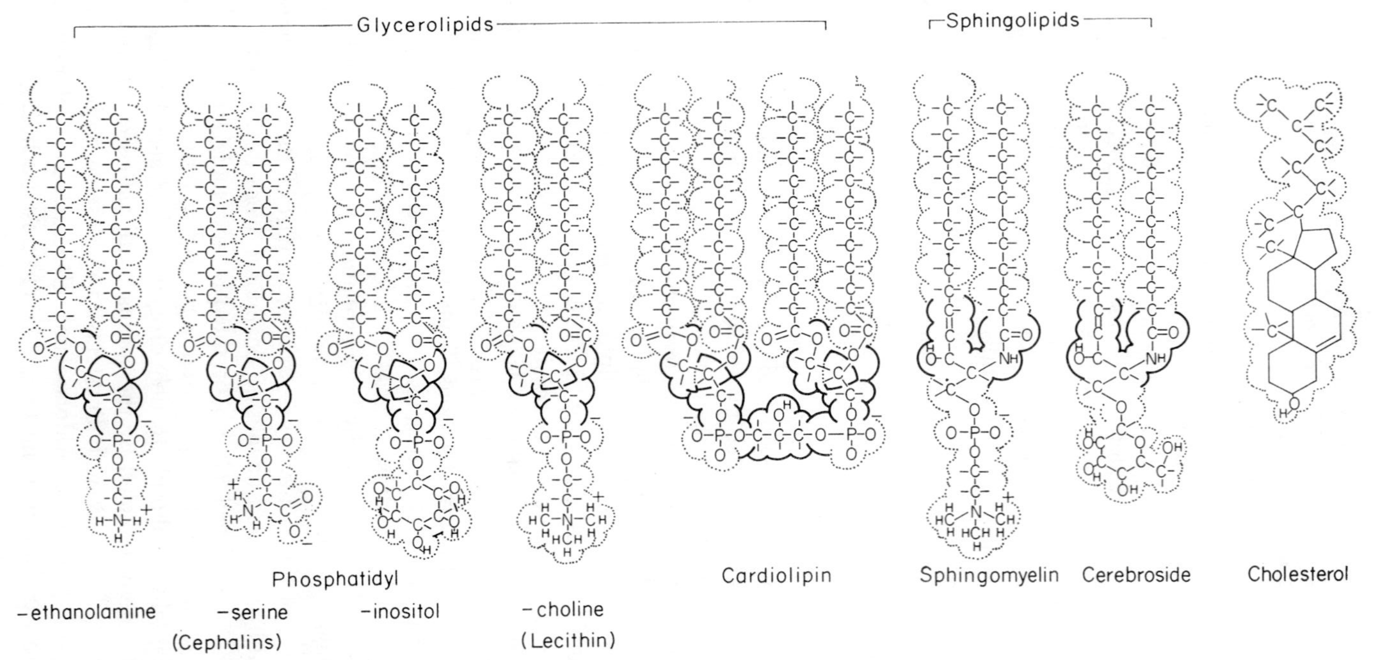

Fig. 1.3. Chemical formulas of structurally important lipids showing also the approximate spatial outlines of the molecular configuration. (Taken with kind permission from Finean, 1961.)

or may not be attached to the membrane under physiological conditions, while carbonic anhydrase and adenosine deaminase must be present in the cell water. It is the difficulty in solubilizing the membrane proteins that has hampered their further study. Acetylcholinesterase has been solubilized and then purified 250-fold from erythrocyte membranes (J. A. Cohen and Warringa, 1953) but only after treatment of the stroma with butanol, a finding suggestive of the role of lipid-protein interactions in the association of acetylcholinesterase with the membrane. Ohnishi (1962) has reported the extraction of actin- and myosin-like proteins from erythrocyte ghosts, while in the author's laboratory a study has begun of the proteins liberated from human erythrocyte stroma when a suspension of stroma in water is treated with ice-cold butanol, a method introduced for ox erythrocytes by Maddy (1964, 1966). The proteins of hemoglobin-free stroma can be very largely solubilized by this procedure, yielding a major component which forms an optically clear solution after high speed centrifugation and which behaves in our hands as a homogeneous, large-molecular-weight (7×10^5) substance on gel filtration. The relation between this protein and elinin or stromatin is yet to be established. For the ox erythrocyte stroma, Maddy (1966) finds that the major protein is a sialoprotein of molecular weight of some 3×10^5. There is evidence that this protein is itself composed of smaller subunits.

The applicability of the *n*-butanol procedure to the solubilization of red cell stroma has been reported also by Poulik and Lauf (1965) who have fractionated the solubilized proteins first by gel filtration and then by gel electrophoresis. Gel electrophoresis shows that the major protein-containing peak found by gel filtration is composed of at least seven subfractions. Schniedermann (1965), in a parallel study, has reported that treatment of red cell stroma with the nonionic detergent Triton X-100 solubilizes the major portion of the membrane and subsequent gel electrophoresis demonstrates the presence of some twenty components. [In this connection, Bonsall and Hunt (1966) have shown that the acetylcholinesterase activity associated with the cell stroma is lost after treatment of the wet material with *n*-butanol but is retained in the Triton extract.] Finally, a third solubilization procedure has been developed by Bobinski (1966) in the author's laboratory. Treatment of red cell stroma with a molar solution of sodium iodide is also effective in solubilization of the stroma. From such an iodide extract, five major ultraviolet absorbing fractions can be separated by chromatography on diethylaminoethyl-cellulose columns. A detailed study of the interrelationships of the different fractions separated by the different procedures will clearly be necessary before the full significance of these findings can be evaluated. Such a detailed study has been begun by Hanahan and his colleagues. Thus C. D. Mitchell and Hanahan (1966) have shown that the extraction of hemoglobin-free stroma with strongly hypertonic solutions of sodium chloride affords an excellent solubilization of the membrane proteins. These authors have analyzed the phospholipid and total protein contents of the fractions isolated from such saline extracts and have assayed these fractions for acetylcholinesterase activity. Acetylcholinesterase, which remained firmly bound when the stroma was treated with hypotonic buffers (see above), was extracted into hypertonic saline as a lipoprotein. When the stroma were treated by ultrasonic irradiation in a 10% solution of butanol, a lipoprotein was liberated containing 68 to 80% of the original lipid of the stroma and only 9 to 20% of the stromal protein (T. E. Morgan and Hanahan, 1966). The protein of this lipoprotein could be released by extraction with ethanol/diethyl ether and was found to have a molecular weight of 163,000.

In recent years a good deal of hard thought has been applied to the problem of obtaining membrane fractions from cells other than the erythrocyte. In a number of cases useful preparations have been obtained, and in some cases analytical data on chemical composition are available. These are collected in Table 1.2. The data of Tables 1.1 and 1.2 concur in suggesting that cholesterol, phospholipid, and protein are the major components of the cell membrane (as this is defined by the isolation procedures of the biochemist). We must consider now the details of how these components are involved in the structure of the membrane. Do the darkly staining layers of the unit membrane contain the protein molecules? Do they contain the lipid? Or are both species present in both regions of the structure revealed by electron microscopy? Electron microscopy by itself has not yielded the answer to these questions and we must consider first how the study of model systems has helped in the interpretation of the electron micrographs.

1.3 The Structure of Myelin Forms—A Model System for the Cell Membrane

Many phospholipids swell when placed in contact with water and form the so-called myelin figures (Fig. 1.4), which are visible as long tubular structures under the light microscope. Similar structures are produced when nerve fibers are wetted, the phospholipid in the nerve fiber sheath being the agent active in producing these figures. (The alkali salts of oleic acid also produce similar structures on wetting, but phospholipids give the most striking effects.) Myelin forms have been studied with the

TABLE 1.2

CHEMICAL COMPOSITION OF ISOLATED CELL MEMBRANES

Species or tissue	References	Protein (%)	Lipid (%)	Other constituents or comments
Microorganisms				
Mycoplasmas sp.	Razin *et al.* (1963)	47–60	35–37	Only 10% lipid is cholesterol; also present, 4–7% carbohydrate
Micrococcus lysodeiktius	Salton and Freer (1965)	63–65	15–26	Circa 2% RNA
Sarcina lutea	Salton and Freer (1965)	53–61	20–27	Circa 5% RNA
Bacillus lichenformis	Salton and Freer (1965)	75	28	Circa 0.8% RNA
Staphylococcus aureus	P. Mitchell and Moyle (1956)	41	22.5	
Bacillus megaterium	Yudkin (1966)	70	25	Lipid largely cephalin
Amoeba proteus	Wolpert *et al.* (1965)	25	32	15% polysaccharide
Mammalian cells				
Muscle, rat	Kono and Colowick (1961)	65	15	Lipid largely phospholipid
Liver, rat	Emmelot *et al.* (1964)	85	10	Half phospholipid is lecithin; cholesterol is ⅓ total lipid
Nerve myelin, ox brain	Autilio *et al.* (1964)	18–23	73–78	As liver. Over 90% protein is proteolipid, chloroform-methanol soluble

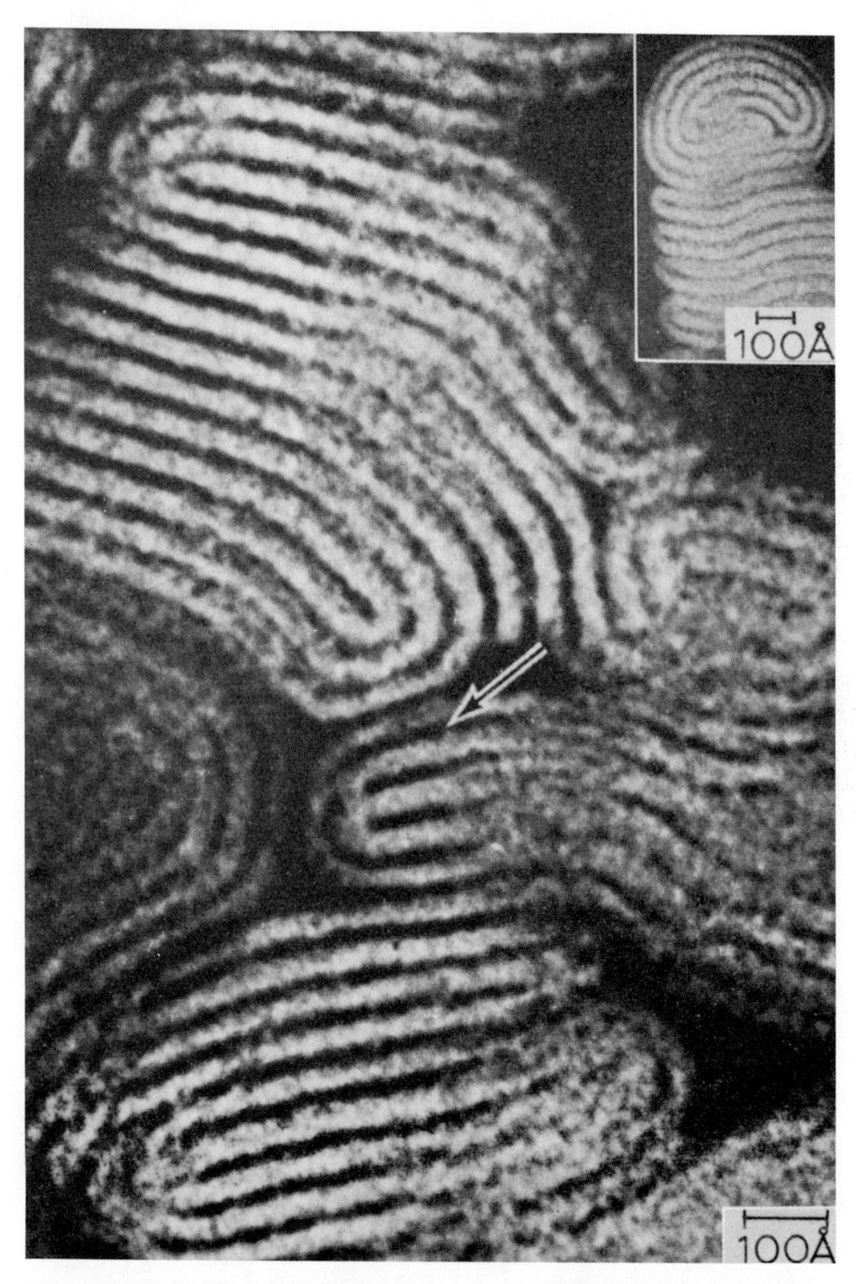

Fig. 1.4. Electron micrographs of lecithin micelles (myelin forms) embedded in phosphotungstic acid. The alternation of dark (arrow) and light regions reflects a corresponding structural organization at the molecular level of phosphotungstate-

electron microscope, by X-ray crystallography and by optical birefringence techniques. Their structure is now well understood and we shall see that it has an important bearing on the problem of membrane structure.

Figure 1.4, taken from Fernández-Morán (1962), shows the myelin forms produced by adding lecithin to water, the forms being embedded for electron microscopy in buffered phosphotungstic acid. We have here a regular array of dense lines (10–20 Å wide) separated by light bands (25–30 Å), the center-to-center distance between dark bands being 45–50 Å in the specimen depicted and up to 80 Å in partially hydrated specimens. (Compare this figure with Figs. 1.1 and 1.2.) Myelin forms can adsorb protein to form structures (Fig. 1.5) in which the dense bands are more dense than in pure lipid forms but in which the characteristic array of alternating light and dark lines of the myelin form is preserved. The light layers of the structure seen in Fig. 1.5 are 20–25 Å wide and should be compared with Figs. 1.1 and 1.2. Clearly, phospholipids alone (in contact with water) will form an array of structures

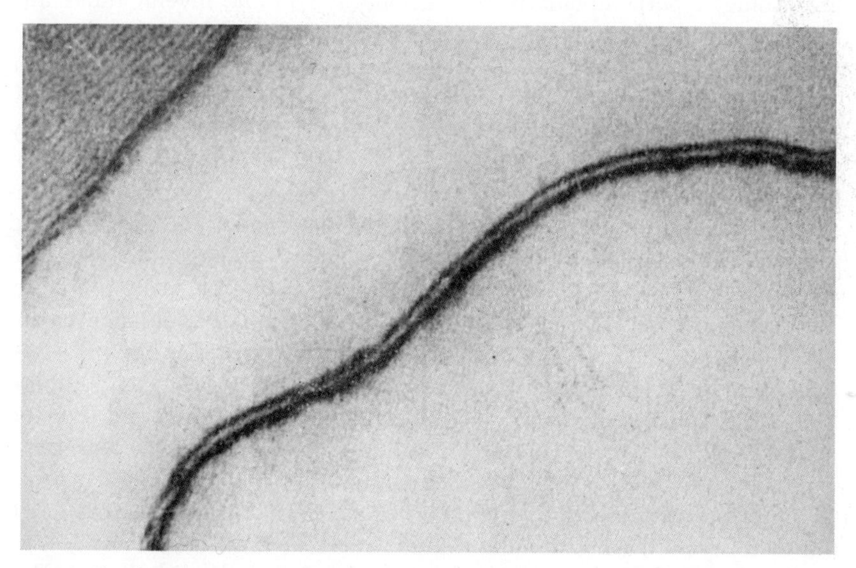

Fig. 1.5. Electron micrograph of a lamellar structure found in lipid-protein-water preparations. In the upper left-hand corner, a palisade of such structures can be seen. Compare with Figs. 1.1 and 1.2. Magnification, 400,000 ×. (Taken with kind permission from Stoeckenius, 1962.)

---

binding components. Magnification, 1,250,000 ×; insert, magnification 550,000 ×. (Taken with kind permission from Fernández-Morán, 1962.)

resembling unit membranes. The addition of protein intensifies the already dark layers, not the light layer. We may tentatively conclude that the protein segregates together with part of the lipid in the dense area and away from the remainder of the lipid which forms the light layer.

X-ray crystallographic analyses of the myelin forms indicate a regularly repeating unit, some 60 Å in width in the driest specimens, the repeat distance increasing to as much as 150 Å in heavily hydrated preparations. This analysis confirms the electron microscopic observations of, for example, Fig. 1.4, and is additionally valuable in that it refers to wet unfixed specimens of the myelin forms. The fixing of the specimens for electron microscopy does not apparently *produce* the layered structure—it is present in the wet specimens. Observations of the myelin forms with polarized light reveals that the forms are optically birefringent (Fig. 1.18)—they refract light to a different extent depending on whether the light is polarized at right angles to, or parallel with, the long axis of the myelin form. The sign of the birefringence is negative with respect to the long axis of the myelin tubes, suggesting that the lipid molecules are arranged perpendicularly to the long axis of the myelin tubes and hence perpendicularly to the concentric light and dark layers of the tube walls of Fig. 1.4. [For an excellent discussion of these matters, see Frey-Wyssling (1953).] The dimensions of the layer-to-layer distance (40–50 Å) are those predicted for a *pair* of phospholipid molecules (the fatty acid chains are between 20 and 25 Å long when extended), and the total picture is thus consistent with the myelin forms being composed of concentric shells of radially oriented bimolecular leaflets of phospholipid interspersed with layers of water. (This is best understood by a consideration of Fig. 1.6.) Now a phospholipid molecule is amphiphilic, being composed of a large hydrophobic region, the fatty acid side chains (Fig. 1.3), together with a hydrophilic phosphate ester grouping. It is, therefore, to be expected that the hydrophilic groups of the phospholipid will be preferentially situated in the aqueous interface and that the hydrophobic fatty acid chains will interlock with one another as these are drawn in Fig. 1.6. Such a model would account for the regular array of light and dark layers of Fig. 1.4, but the question still remains: Are the dark layers the hydrophilic-charged faces of the bimolecular leaflet, or is it the ethylenic double bonds of the unsaturated lipid chains (Fig. 1.3) that take up the stain? Either interpretation is consistent with the facts as stated thus far.

Once again direct analysis of the situation has not yielded a solution of this problem but a solution has come from observations of another phospholipid system (Stoeckenius, 1962). Phospholipid-water systems of very low water content show, at 37°C, a type of structure (Fig. 1.7) which is very different from that of the myelin form—a hexagonal array of light and dark areas, rather than the concentric lamellae of myelin. X-ray analysis of these forms reveals that they consist of cylindrical tubes of water surrounded by a lesser amount of phospholipid, the cylinders being stacked in the hexagonal array. The electron micrograph of Fig. 1.7 is a cross section of such an array, the cylinders showing up as small black dots with the osmic stain. Clearly, it is the aqueous phase that takes up the stain and thus presumably the charged regions of the phospholipid—although the phospholipids here contain unsaturated double bonds, which might perhaps have been expected to stain. Conversely, in certain soap-water systems, X-ray analysis reveals the structure to be an array of cylinders of soap surrounded by water; here the electron micrograph shows a honeycomb pattern of light dots surrounded by dense walls, again the hydrophilic region taking up the electron-dense stain. This is a clear suggestion that in the similarly treated myelin forms

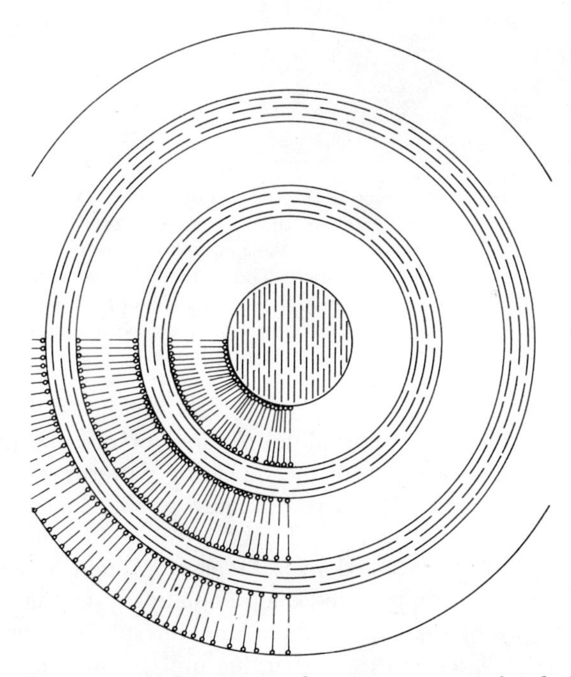

Fig. 1.6. Diagrammatic representation of a cross section of a hydrated myelin form. The smallest circles represent the hydrophilic heads of the phospholipid molecules depicted as interacting with the aqueous regions, the concentric hatched circles. The nonpolar tails of the phospholipid molecules in the array are directed away from the aqueous phases and are radii of the cylinder forming the myelin form. The diameter of each double layer of lipid molecules will be 40–50 Å.

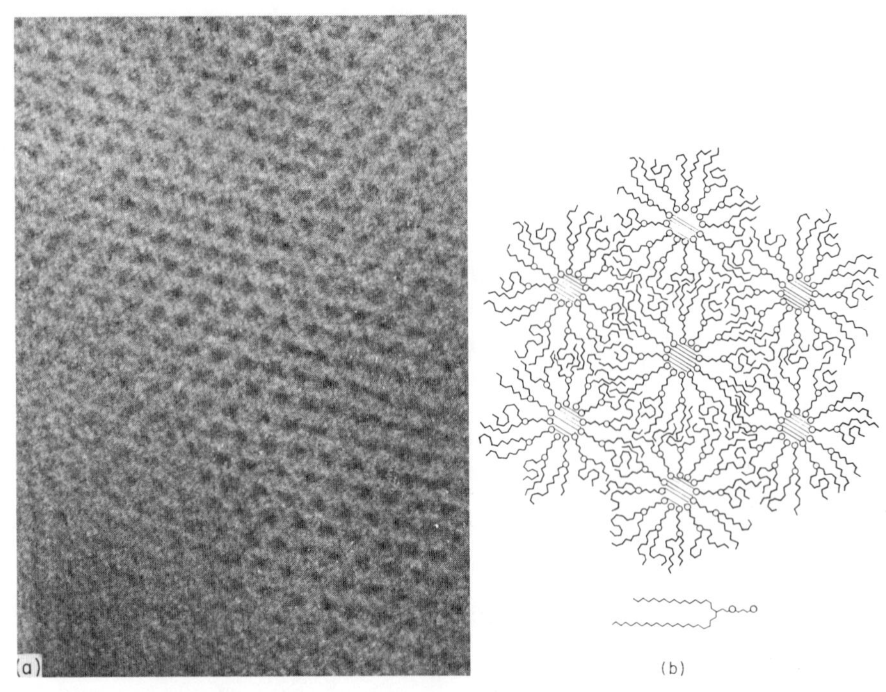

Fig. 1.7. (a) Electron micrograph of a cross section of the hexagonal lattice formed in a concentrated phospholipid/water mixture. Fixed with osmium tetroxide. Magnification, 930,000 ×. The dense spots are the hydrophilic groups. (b) The proposed molecular architecture of such a lattice. The hatched areas correspond to cross sections of the water cylinders forming the array. These are surrounded by the polar groups of the lipid molecules, and these again, by the nonpolar chains. (Taken with kind permission from Stoeckenius, 1962.)

of Fig. 1.4 it is again the hydrophilic portions of the phospholipid that segregate into the dark layers of the repeating patterns, the lipid chains taking up the stain very poorly. With regard to Fig. 1.5, we can deduce that the region stained is composed once again of the charged portions of the phospholipid, reinforced by a layer of protein at each aqueous interface in a manner represented diagrammatically in Fig. 1.8.

Thus the study of the phospholipid and phospholipid-protein model systems shows that structures similar to "unit membranes" form spontaneously, when phospholipids are in contact with water, and the study suggests further that these model membranes have the structure of Fig. 1.8. Returning now to consider the cell membrane of Figs. 1.1 and 1.2, the analogy in structure and in composition between the true and model systems is so close as to suggest very strongly that the cell membrane, too, has the structure of Fig. 1.8. Certainly one can say that the components of the cell membrane (Section 1.2) in the presence of water will spontaneously form structures such as Fig. 1.5. If the cell membrane does not have the structure of Fig. 1.8, there must be some compelling constraint which prevents this.

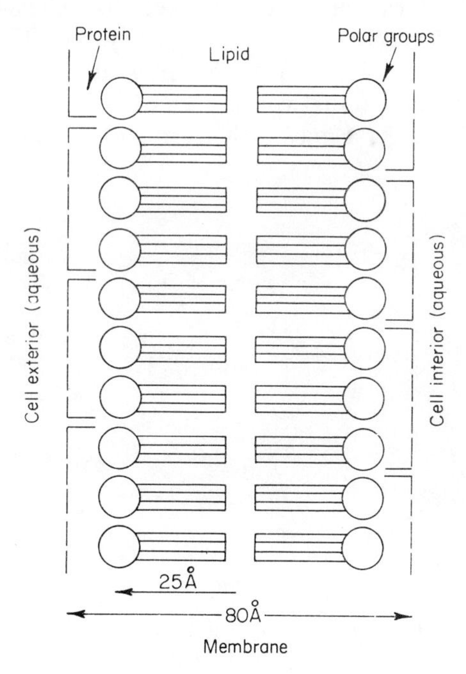

Fig. 1.8. The "paucimolecular" model of cell membrane structure. (After Davson and Danielli, 1952.)

1.4 Nerve Myelin

The point reached thus far in the present argument is that by virtue of the similarity in composition and electron microscopic structure of the cell membrane (Figs. 1.1 and 1.2) and the myelin forms (Figs. 1.4 and 1.5), the cell membrane may well have a molecular structure such as that of Fig. 1.8. But in at least one case we can go beyond such an argument by analogy, to a more firmly based interpretation. The case in question is that of the nerve myelin sheath or neurolemma, a set of unit membranes that surround the axon of the nerve fiber in a series of concentric lamellae.

It is now clear (Geren and Schmitt, 1955) that this structure arises during the development of the nerve tissue as a result of the action of an associated cell, the Schwann cell, which wraps itself around the nerve in a manner illustrated in Fig. 1.9. The myelin sheath is, therefore, made up of a set of cell membranes arranged in pairs, the two halves of a pair coming together as the cytoplasm is lost from between the membranes—each membrane itself being a double-layered unit membrane (Fig. 1.10). The myelin sheath forms an excellent object for X-ray crystallographic analysis (Finean, 1961), displaying a regularly repeating array of concentric units of spacing 150–180 Å (in different species). These rather large units are divided into two very similar halves, each half being therefore some 75–90 Å, a good value for the unit membrane. Optical

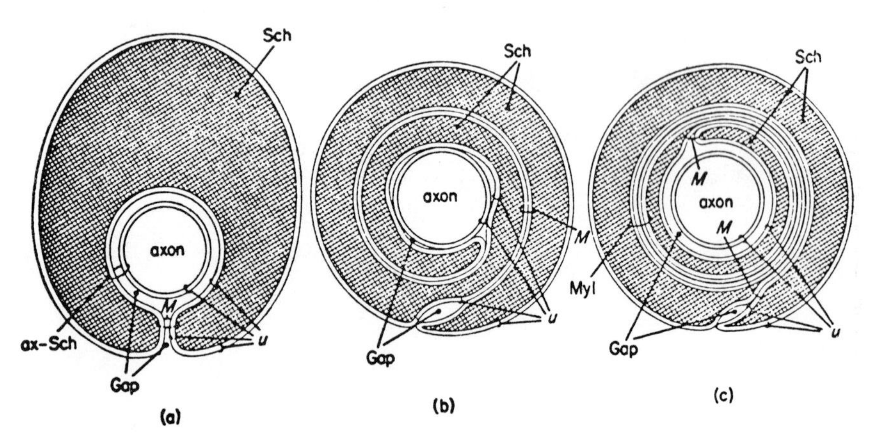

Fig. 1.9. Diagrams illustrating the successive stages (a, b, c) of the development of nerve myelin. A small axon is shown enveloped by a Schwann cell (Sch) which winds around the axon, the unit membranes (*u*) of the Schwann cell coming in contact to form the myelin layer (Myl). (Taken with kind permission from J. D. Robertson, 1960.)

birefringence studies show unequivocally, as with the myelin forms, that the lipid components of the sheath are arranged anisotropically—having a different structure along different axes—with the long axes of the molecules lying along the radii of the concentric shells of unit membranes (Fig. 1.6). From the dimensions of the X-ray spacings (75–90 Å) each layer can be assumed to be composed of at least two sheets of lipid molecules, each sheet being 20–25 Å thick. Other arguments support the presence of protein in the layers between the lipid layers depicted in Fig. 1.10, and a structure combined of two sheets of lipid molecules together with adsorbed protein adequately accounts for the dimensions of the X-ray. That the hydrophilic regions of the phospholipid molecules in nerve myelin are the sites of deposition of the electron-dense stains

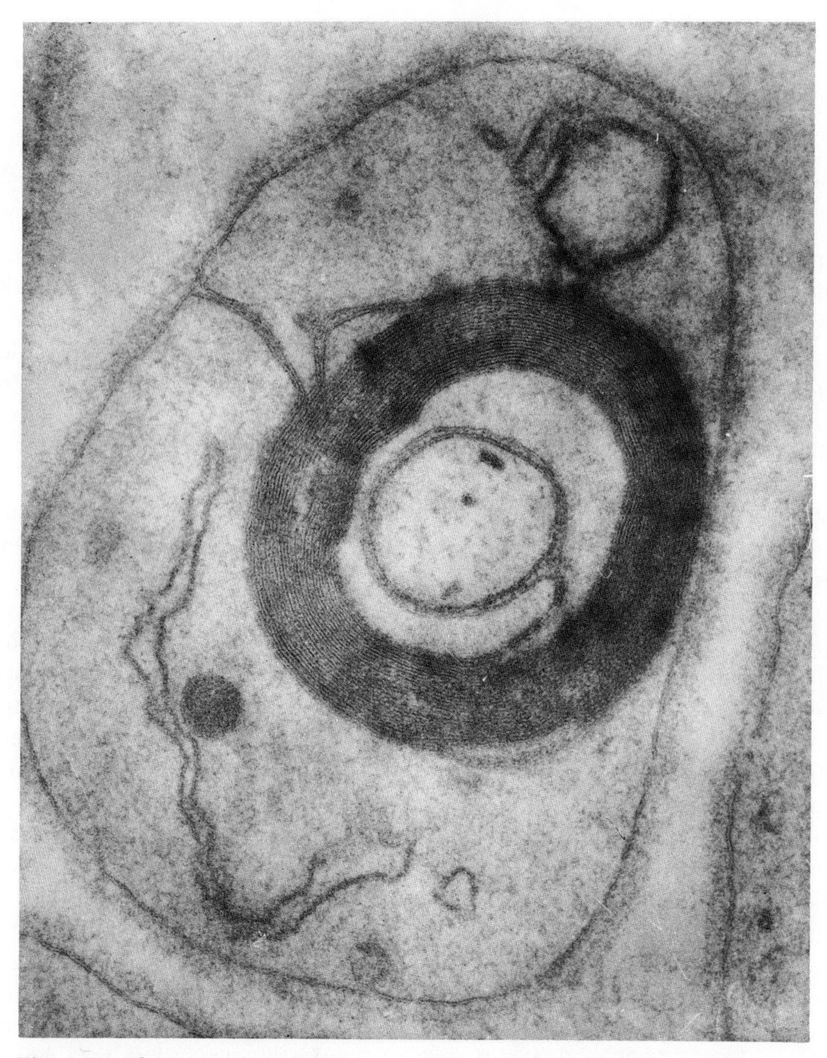

Fig. 1.10. Electron micrograph of a myelinating fiber from developing mouse sciatic nerve (compare Fig. 1.9c). Magnification, 90,000 ×. (Taken with kind permission from J. D. Robertson, 1960.)

has been shown by Finean (1962) in a parallel study of X-ray diffraction and electron microscopy of stained and unstained preparations of nerve myelin.

Thus the structure shown in Fig. 1.8 proves to be valid for the true cell membranes of the Schwann cell surrounding the nerve axon, and the suggestion that Fig. 1.8 represents the structure of cell membranes generally can be defended now by induction from one specific cell membrane type to the generality, as well as by the analogy with the myelin forms.

1.5 Physical Stability of the Bimolecular Lipid Leaflet

The study of the myelin forms (Section 1.3) suggests that the bimolecular lipid leaflet is the most stable structure for a dilute phospholipid/water mixture, that is, is the form into which such a mixture will spontaneously organize itself. The probable importance of such forms for the structure of the cell membrane (Section 1.4) makes it desirable to consider in greater detail the physical chemistry of such systems. We shall follow here the treatment given in the excellent articles of Haydon and Taylor (1963) and Green and Fleischer (1964).

We have emphasized above that the salient feature of the chemical structure of the phospholipids is that these are amphiphilic, being made up of a strongly hydrophobic portion, the two long-chain fatty ester residues and a strongly hydrophilic portion—the partially esterified phosphate residue (Fig. 1.3). This hydrophilic portion is fully ionized at physiological pH but carries no *net* charge in the case of lecithin (phosphatidylcholine), a slight negative charge for phosphatidylethanolamine, and a full negative charge for both phosphatidylserine and inositol. The amphiphilic nature of the phospholipids indicates that they are intensely surface-active, and will thus be concentrated at a polar-nonpolar interface (that is, the water-air and the water-hydrocarbon interface), with the charged hydrophilic portion in the aqueous phase and the hydrophobic chains inserted into the nonpolar phase. In this way both the tendency of the hydrophilic portion to be surrounded by water and that of the hydrophobic region to be out of the aqueous phase can be simultaneously satisfied. Lecithin will spread on a surface of clean water to form a monomolecular film, if the available water surface is sufficient. This film can then be compressed by a clean glass slide and the molecules will pack together until, at the point of closest packing, each molecule in the surface occupies only 96 Å (Dervichian, 1958). At a greater degree of compression (that is, if further attempts are made to decrease the area available for each molecule), the monomolecular film will buckle to form a many-layered structure. In the case of a more concentrated phospholipid/water mixture, the water-air interface will soon become saturated with phospholipid so that excess phospholipid remains within the aqueous phase. The tendency of the hydrophobic portions of the molecule to withdraw from the aqueous phase can now only be satisfied by the mutual aggregation of hydrocarbon chains from different molecules, to form a "microphase," nonpolar in nature. This association will continue until all the phospholipid is organized in such a fashion that the hydrocarbon chains are oriented away from the water while the charged portions of the molecule remain embedded in the aqueous phase. This can be most *economically* achieved in the form of the bimolecular lipid leaflet (Fig. 1.8 with the protein absent) in which all hydrocarbon chains are surrounded by other hydrocarbon chains and all charged groups are in the aqueous phase.

Other configurations are, of course, possible—we have seen one such form in Fig. 1.7—but are not in general as energetically favorable as is the bimolecular leaflet. Thus, for example, polar group-polar group interactions will be favored as well by the bimolecular leaflet as by a globular micelle (Fig. 1.11), but such a micelle will not maximize the nonpolar-nonpolar contacts. Conversely, a large three-dimensional aggregate of lipid will satisfy the nonpolar-nonpolar interactions but will not maximize the interaction of water with the charged phosphatidyl residues. The bimolecular leaflet is clearly a compromise between the two forms of Figs. 1.7 and 1.11 and will be the form in which both types of mutual interaction are satisfied to the greatest possible extent.

It is inconceivable, however, that the planar bimolecular leaflet will exist in solution as a set of rigid flat planes, for the planes have a marked tendency to curvature due to the wedge-shaped form of the head group of the phospholipids.

The particular form of bimolecular leaflet that phospholipids take up

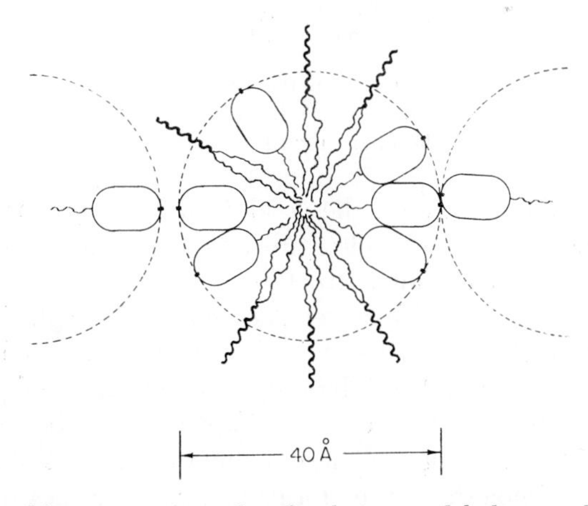

Fig. 1.11. Possible arrangement of molecules in a globular micelle composed of cholesterol (ovoid structures) and lecithin (**Y**-shaped structures). The hydrophobic regions of these molecules pack within the interior, hydrophilic groups are at the surface of the globule. (Taken with kind permission from Lucy and Glauert, 1964.)

in water has been studied experimentally by Fleischer and by Fernández-Morán and Green (review by Green and Fleischer, 1964). If some phospholipid dissolved in an organic solvent is stirred into distilled water and the mixture then vigorously dialyzed against water, an optically clear solution will be obtained in which the phospholipid exists in the form of *micelles* of uniform size, having a molecular weight (as determined by ultracentrifugation) of some two to three million and hence containing two to three thousand phospholipid molecules. These micelles appear as two-dimensional spirals when examined by electron microscopy (Fig. 1.4), Fig. 1.12 being an interpretation of the electron micrograph

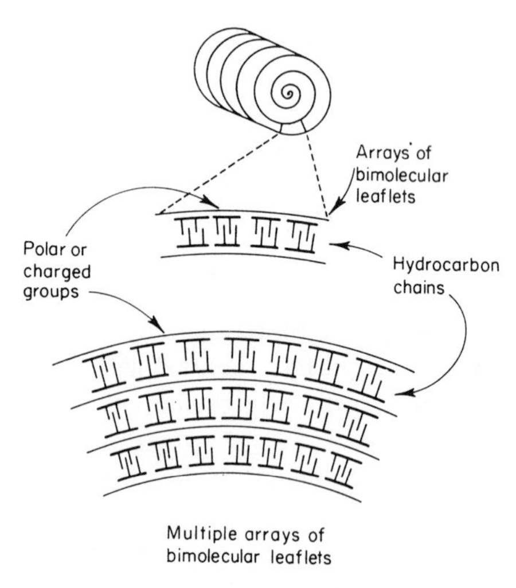

Fig. 1.12. A molecular interpretation of the electron micrograph of Fig. 1.4. Note the postulated spiral form of the micelle in this interpretation. (Taken with kind permission from Green and Fleischer, 1964.)

of Fig. 1.4. In these micelles the lipid is still in the form of the bimolecular leaflet, but these leaflets are now rolled upon one another to form the stable micellar leaflet structure.

Cholesterol in contact with water does not swell in the same manner as do the phospholipids. The sole hydroxyl group of cholesterol is far less polar than the phosphatidyl residue of the phospholipids and two such hydroxyl groups can conceivably form a bridge between two cholesterol molecules forming a dimeric complex which is stable within a nonpolar phase. Cholesterol thus forms globular micelles of nonuniform size in water or else crystallizes out of the aqueous phase. Nevertheless, cholesterol can be incorporated into myelin forms and bimolecular micelles in the presence of phospholipid and water. If a mixture of lecithin and cholesterol is spread at the air/water interface and the monolayer then compressed by a glass slide (Fig. 1.13), the area occupied by each lecithin molecule is now a function of the amount of cholesterol in the mixture and at a ratio of 3 cholesterol molecules to 1 lecithin molecule, the effective area occupied by each of the lecithin molecules is reduced to 50 square Å, rather than the value of 96 square Å found for pure lecithin. At the inverse ratio of 3 lecithin molecules to 1

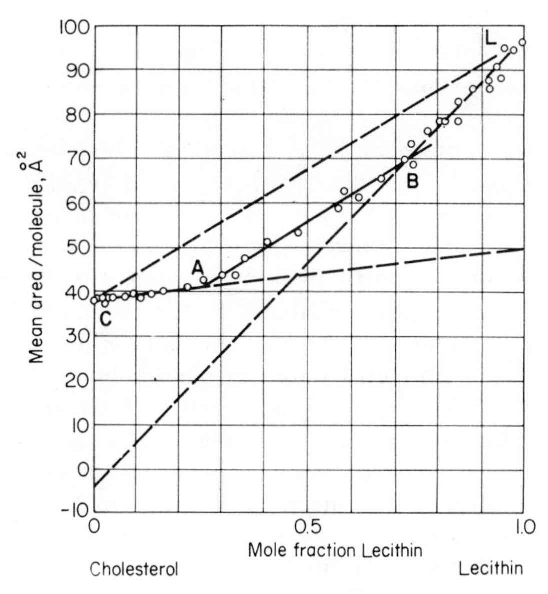

Fig. 1.13. The variation of the mean area per molecule (ordinate) as a function of composition (abscissa) for monolayers spread from mixtures of cholesterol and egg lecithin. The circles give the experimental points, the dashed line *CL* is the prediction were the molecular areas of cholesterol (*C*) and lecithin (*L*) to be simply additive. (Taken with kind permission from Dervichian, 1958.)

of cholesterol, the area occupied by the lecithin molecule is 82 square Å. The sharp breaks in the curve of Fig. 1.13 suggest that the two critical ratios of cholesterol to lecithin define critical points for the formation of particular forms of cholesterol-lecithin aggregates. In fact, the ratios found are those to be expected if the cholesterol-lecithin film is composed of a hexagonal array of molecules—a two-dimensional crystal similar in basic form to the structures discussed in Section 1.3 for water/phospholipid and water/soap mixtures. Thus lecithin and cholesterol can pack together in mixed films to form molecular aggregates with a packing that differs from those of the pure components. When cholesterol is incorporated into myelin forms containing phospholipids, the mixed layers in the myelin form have a smaller width than those made from pure lecithin. The length of the extended cholesterol molecule (17–20 Å) is less than the 22–25 Å of the phospholipid side chain. Thus in maximizing hydrocarbon/hydrocarbon interactions in the mixed layers, the lecithin side chains apparently bend backward over themselves (like a walking stick) to pack together with the cholesterol. In the case of a bimolecular leaflet made from lecithin and cholesterol, as the concentration of cholesterol in the bilayer is increased, a point will be reached at which the bimolecular layer is no longer the most stable form and conversion into globular micelles, the stable form for pure cholesterol, may follow.

Although in a dilute phospholipid solution, a bimolecular leaflet will be the most stable form, yet as we have seen (Fig. 1.7) in very concentrated solutions—when the amount of aqueous phase is the limiting factor—a type of inverted myelin form results, the hexagonal array of cylindrical tubes of water surrounded by lecithin. This form ensures that the limited amount of water available is used in the most efficient manner and interacts with the maximum possible number of phosphatidyl groups. A number of other structures found in phospholipid/cholesterol/water solutions have been investigated, particularly with the aid of the electron microscope (Lucy and Glauert, 1964) (Figs. 1.14 and 1.15). These structures include lamellae, hollow tubes, helices, and hexagonal arrays, all apparently built up of a fundamental globular subunit 40–50 Å in diameter. Globular micelles of this size are not, however, likely to be the stable forms for dilute cholesterol/lecithin mixtures when the sterol and phospholipid are present in comparable concentration. As we have seen, the transformation of such a globular array into bimolecular leaflets must lead to an increase in the hydrocarbon/hydrocarbon contacts and to a decrease in hydrocarbon-aqueous interaction—an energetically favorable transition. But obviously these forms are stable in certain hydration and concentration situations, and in a number of circumstances in the cell as well as in model systems the globular micelle will clearly be a more stable form.

Consider, for example, Fig. 1.16, taken from J. D. Robertson (1964). This is a scale drawing of a close-packed array of phospholipid molecules forming the cross section of a spherical vesicle. At the inner rim of the bimolecular leaflet, the heads of the phospholipid are in closest possible contact (approximately 10 Å) yet at the outer rim, the separation (20 Å) is such that interaction between neighboring groups is beginning to be unfavorable. This structure, then, encloses a vesicle of the smallest diameter that it is possible to build from a bimolecular assembly. The

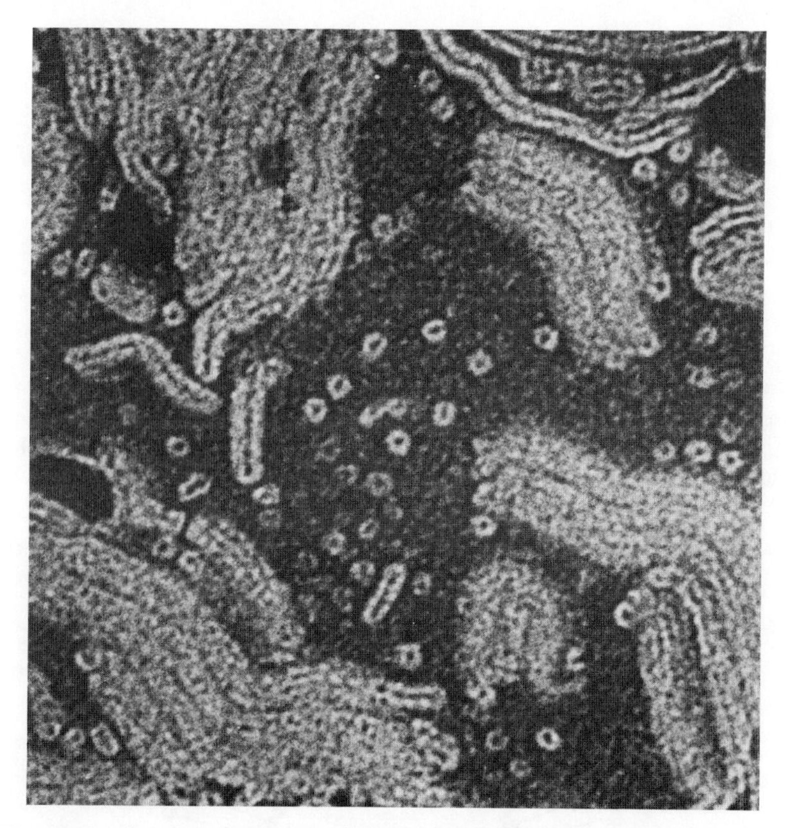

Fig. 1.14. Electron micrograph of lecithin/cholesterol/water mixtures. A negatively stained (phosphotungstate) preparation. Clearly shown are the profiles and end view of what appear to be hollow tubes filled with phosphotungstate. Magnification, 300,000 ×. (Taken with kind permission from Lucy and Glauert, 1964.)

outer diameter of the structure is 300 Å, and the inner vesicle is 150 Å in diameter. Robertson points out that unit-membrane bounded vesicles do seem to have a lower limit of size of about this order of magnitude. Yet as Fig. 1.4 indicates, myelin forms are often found with curvatures greater than a consideration of Fig. 1.16 would allow.

At the front edge of an advancing myelin form, then, the hydrocarbon chains can no longer pack perpendicularly to the plane of the phosphatidyl-aqueous interface in the body of the structure, but rather polar-nonpolar contacts may well be minimized by some configuration such as that shown in Fig. 1.17. This suggestion is confirmed by Fig. 1.18 which shows that the ends of such a myelin form display optical birefringence of opposite sign to the main body of the form.

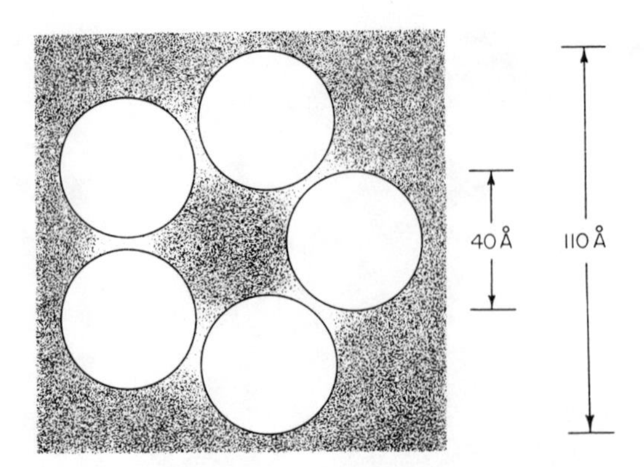

Fig. 1.15. An interpretation of the structures seen in end-on view in Fig. 1.14. The circles are cross sections of globular micelles, as in Fig. 1.11. Such globules could also be arranged in a helical array to give the picture of Fig. 1.14. (Taken with kind permission from Lucy and Glauert, 1964.)

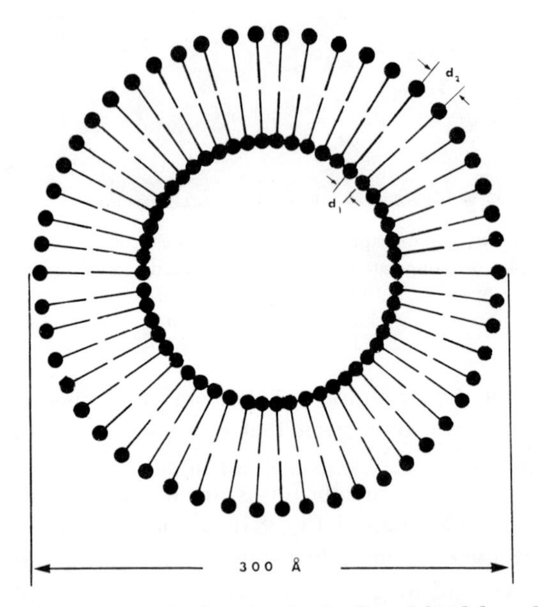

Fig. 1.16. A scale drawing of a bimolecular leaflet of lipid bounding a spherical vesicle. Each lipid molecule has a polar head of diameter 10 Å. The bimolecular leaflet has a thickness of 75 Å. The closest packing distance d_1 is 10 Å; d_2 is 20 Å. The diameter of the vesicle is thereby limited to the value shown of 300 Å. Thus vesicles of less than 300 Å in diameter cannot be formed on the bimolecular leaflet model. (Taken with kind permission from J. D. Robertson, 1964.)

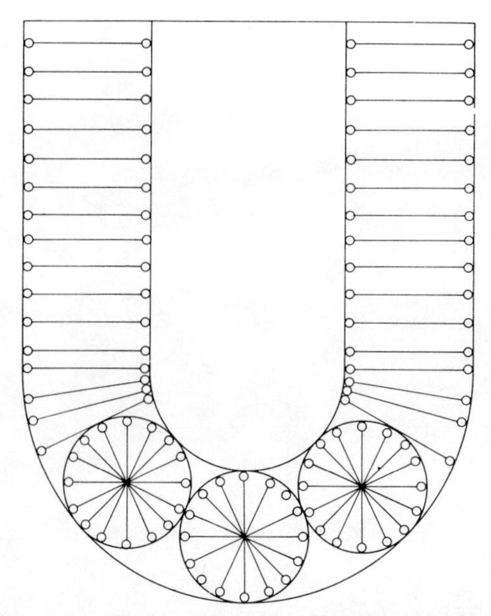

Fig. 1.17. Diagrammatic representation of the front of an advancing myelin figure. The curvature here is too great to allow a stable bimolecular leaflet (compare Fig. 1.16). Globular micelles, as depicted, might thus be expected to be the most stable form in this region.

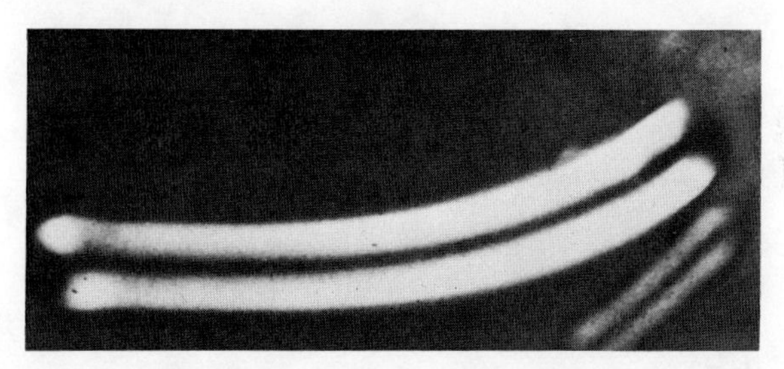

Fig. 1.18. A myelin form, prepared from the alcohol-ether extract of cat nerve, seen in polarized light. The light area indicates that the birefringence of the myelin form with respect to its long axis is negative, but note the characteristic change in sign of the birefringence–appearance of dark areas–at the ends of the myelin form (compare Fig. 1.17). (Taken with kind permission from J. D. Robertson, 1960.)

There will be many situations—consider, for example, the globular particles forming the membrane of the mitochondrion as these appear in the electron micrographs of Sjöstrand (for example, Fig. 1.19)—where cell organelles or components of organelles contain considerable amounts of lipid, yet are smaller than the 300-Å diameter structures of Fig. 1.16.

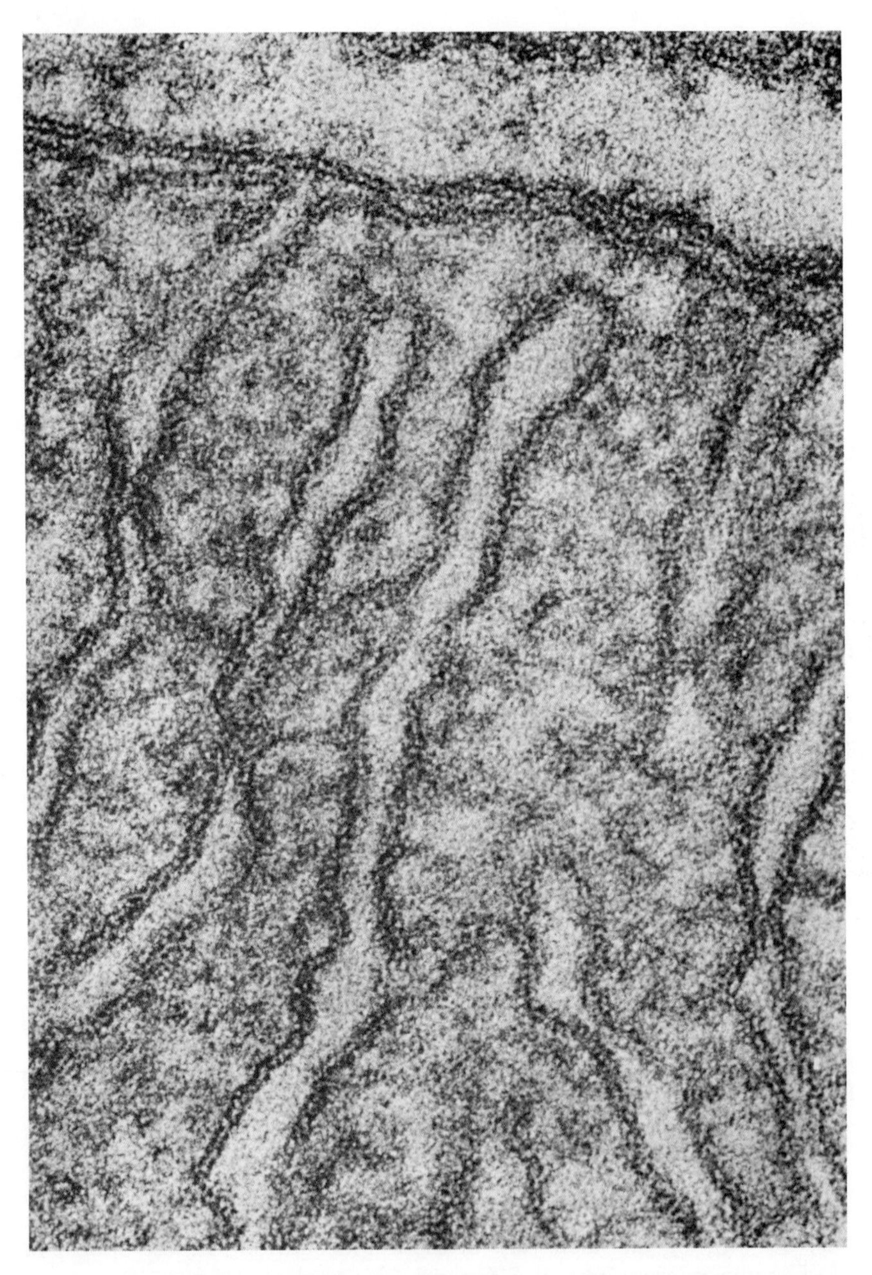

Fig. 1.19. Globular particles of the mitochondrial membrane. A high magnification (400,000 ×) view of part of a mitochondrion of mouse kidney. Osmium fixed. Note how the "unit membrane" here appears to be made up of a globular substructure, the heavy lines of osmium-binding material being regularly interrupted by light areas. (Taken with kind permission from Sjöstrand, 1963.)

Here also a globular micelle may well be the structural basis of such particulate units. Finally, Lucy and Glauert (1964) point out that the close packing of globular micelles leads to the formation of spaces about 5 Å in diameter between the globules, a structural feature which may yet be found to be important in the architecture of the cell membrane, where—as the discussion of Chapter 3 indicates—it is a widely held view that such pores are required to account for certain permeability phenomena. [There is good evidence from studies of freeze-etched membrane faces (Branton, 1966) that globular particles of some 85 Å in diameter are embedded within the unit membrane bilayer. Different tissues possess a more or less extensive complement of these globular particles.]

The configuration taken up by a lipid–water system will always be that which will result in a minimum free energy in the system considered—a configuration which will generally be that of the bimolecular leaflet in which polar-polar contacts and nonpolar-nonpolar contacts are maximized while polar-nonpolar interactions are at a minimum. But in the biological situation of the cell membrane our ignorance of detail is such that we cannot be sure that constraints do not exist (imposed, for example, by an underlying lattice or framework of protein) which impose on some part of the configuration a form of lesser inherent stability than the bimolecular lipid layer and, in such circumstances, the globular micelles of Figs. 1.11, 1.14, and 1.15 may well be important. For the membranes of the axolemma (the myelin sheath), the structure that has been most thoroughly studied by physical techniques, the X-ray crystallographic data (Finean, 1962) are sufficient to rule out the possibility that a quantitatively significant amount of the membrane is composed of globular micelles. Even in this case, however, such micelles, even when present in an amount which would escape identification by crystallography, might be significant from the point of view of physiological function.

1.6 Surface Tension of the Cell

Any liquid surface (and this includes liquid/liquid interfaces) behaves as if a skin of molecules existed at the surface of the liquid, the skin resisting any attempt at an enlargement of the surface area. This tension in the surface results from the fact that the molecules in the superficial layer are more strongly attracted by molecules in the interior than they are attracted by the molecules in the air above the surface (or by the molecules of the overlying liquid phase). The resulting inward pull on the molecules in the surface resists any attempt to extend the surface area. The water/air surface has a tension of some 76 dynes cm^{-1} at 25°C, while water/hydrocarbon interfaces have lower tensions—from 10 to 50 dynes cm^{-1}. [A useful tabulation of the data is given by Adamson (1960a).]

Inasmuch as the cell interior is a different phase from the external medium in which the cell is immersed, it is possible to talk of the surface tension at the cell surface and various methods of measuring this quantity have been devised (see, for example, Frey-Wyssling, 1953). The values obtained range from 0.1 to 2.0 dynes cm^{-1}, values far lower than those found for the water/air interface and at least an order of magnitude below those for the water/hydrocarbon interface. Thus, the view we are developing here—that the cell surface is composed of a hydrocarbon phase—is apparently not borne out by the surface tension data. A cell and an oil droplet of the same size will have very different surface tensions. Danielli (see especially, Davson and Danielli, 1952) first considered this problem. He showed that the data demonstrating a very great difference between the surface tension of cells and of oil droplets could in fact be reconciled. In model experiments, protein adsorbed onto oil droplets could be shown to cause a reduction of surface tension to a value within the range found for cells. Danielli postulated that the hydrocarbon surface which he assumed to be the basis of the cell membrane could perhaps be coated by a layer of adsorbed protein, considerations which led to the hypothesis of the cell membrane as a "paucimolecular" aggregate of lipid and protein arranged in layers as in Fig. 1.8.

[This brilliant hypothesis has been the basis of almost all work in the exploration of membrane structure and function during the last thirty years, and it is most gratifying to see how the most recent technical advances of electron microscopy and X-ray crystallographic analysis have confirmed the major part of this view.]

But, as Haydon and Taylor (1963) point out, it is not to be expected that values for the *bulk* surface tension of a hydrocarbon-water interface be applicable to the bimolecular lipid layer. If we realize that the bimolecular leaflet is perfectly symmetrical about the plane containing the nonpolar ends of the hydrocarbon chains, it will be apparent that any excess pull of molecules at one polar-nonpolar interface will be exactly compensated for by an oppositely directed pull at the other interface. Putting this another way, we have in the bimolecular lipid leaflet a water-hydrocarbon interface of tension, say, 50 dynes cm^{-1}, directed upward (producing a curvature toward the water phase) sitting on top of a hydrocarbon-water interface of tension again 50 dynes cm^{-1}, directed downward (producing a curvature in the opposite direction). The net surface tension is thus zero—or at most reflects the slight difference in composition of the cellular and extracellular phases. Thus Danielli's reasons for postulating the existence of a layer of protein exterior to each surface of the bimolecular lipid layer are invalid, but the presence of such layers of protein has been amply confirmed as we have seen by recent electron microscopic and X-ray crystallographic investigations. The elasticity of the cell surface, where this has been measured (Frey-Wyssling, 1953) is rather higher than would be expected for the essentially fluid, inelastic film that is the bimolecular lipid leaflet, and this increase in elasticity could well arise from the presence of structural proteins (or other molecules) at the cell surface.

Danielli assumed that the first layer of protein adsorbed on each surface of the lipid leaflet was unrolled in the manner that a protein unrolls at a lipid-water interface. But since the surface of the lipid leaflet is not in fact hydrocarbon, being formed by the charged phosphatidyl residues of the phospholipid, any protein complexed with the unit membrane need not unroll. That protein does interact with such bimolecular lipid leaflets is apparent from the electron microscopic and X-ray crystallographic studies of myelin forms, produced in the presence of protein (for example, Fig. 1.5).

Proteins and phospholipids can interact either by electrostatic interaction—between charged groups on the protein and oppositely charged groups on the phospholipid—or by hydrophobic group–hydrophobic group interaction, when the hydrophobic side chains of the proteins would provide the acceptor sites. In a penetrating survey of lipid-protein interactions, Green and Fleischer (1964) provided criteria by which the electrostatic interactions and hydrophobic interactions can be distinguished and provided carefully studied examples where each kind of interaction is dominant in a particular situation. Thus the complex formed between cytochrome c (a protein component of the mitochondrion) and micelles of acidic phospholipids is largely electrostatic in origin. As the ionic strength of the medium surrounding the complex is increased to a critical value, the charge-charge interactions are suppressed to a point where the complex is no longer stable. Yet at low ionic strength, the protein-micelle complex can be extracted into heptane while, in an aqueous medium, the protein will sediment together with the micelles on ultracentrifugation. Acidic or neutral proteins do not interact with the micelles but the basic protamines, histones, and ribonuclease will form stable complexes. A completely different situation is found for the structural protein of the mitochondrion which interacts with all phospholipids (acidic or neutral) to form complexes containing 20 to 25% by weight of lipid. Such complexes, once formed, are stable even at high ionic strength—so that here hydrophobic bonding is paramount in complex formation. Similarly, for the enzyme β-hydroxybutyric dehydrogenase

(1.1.1.30), an enzyme which requires lecithin as a cofactor for its activity, the complex between protein and phospholipid is hydrophobic in character.

It is clear that the association of protein with the lipid bilayers will have a profound effect on the properties of the bilayer, as well as having an effect on the properties of the protein. In particular, in studies on lipid monolayers, it has been quite clearly shown (see Adamson, 1960b) that proteins can penetrate such monolayers, in the same way as cholesterol was found to penetrate a lecithin monolayer in the experiment depicted in Fig. 1.13. It is thus very likely that proteins possessing the correct spatial orientation of their polar and nonpolar side chains might be able to coexist within lipid bilayers to form a protein plug extending through the plane of the bilayer. There is no direct experimental evidence for such a structure, however, either from studies with model systems or from electron microscopic or X-ray crystallographic investigations of cell membranes. Indeed, these physical techniques strongly suggest that such imperfections in the otherwise homogeneous lipid bilayer are not quantitatively significant. Studies of the optical rotatory dispersion (ORD) spectra of erythrocyte ghosts (Lenard and Singer, 1966) and of membranes prepared from ascites tumor cells (Wallach and Zahler, 1966) suggest that about one third of the membrane protein is in the α-helical form and very little in the β form suggested in the original proposal of Danielli [see Davson and Danielli (1952) and Fig. 1.8]. The shift (in comparison with simple aqueous systems) of these ORD spectra toward longer wavelengths, observed in all membrane preparations studied thus far (Lenard and Singer, 1966), suggests that the α-helices exist in some special environment, very probably arising from the interaction of these helices with the hydrophobic groups of the membrane lipid (Wallach and Zahler, 1966). These interactions could be accounted for if the α-helical regions extend *through* the membrane in the form of the plugs considered above.

1.7 Conclusions

We may summarize the argument thus far as follows:

1. Nearly all cell membranes so far examined show in the electron microscope the "unit membrane" structure of two dark (osmiophilic) layers separated by a light layer.

2. Cell membranes isolated and analyzed biochemically consist for the most part of cholesterol, phospholipid, and protein.

3. Model studies show that the osmiophilic regions of these compo-

nents would be the charged portions of the phospholipid molecules and of the proteins. The hydrocarbon regions would not stain heavily.

4. A particular cell membrane, that of the nerve myelin sheath (formed by the membrane of the surrounding Schwann cell) has been shown by X-ray crystallography to be composed of a bimolecular layer of lipid molecules with the hydrocarbon chains forming an orderly array, directly perpendicular to the plane containing the charged phosphatidyl groups and the proteins.

5. The components of cell membranes (cholesterol, phospholipid, and protein) in model experiments spontaneously form arrays of lipid leaflets, which can be shown by electron microscopy and crystallography to be bimolecular.

6. Physicochemical considerations suggest that the bimolecular leaflet is indeed the most stable form for such systems.

7. The dimensions of the "unit membrane" and its staining properties are completely consistent with the bimolecular lipid layer model of Fig. 1.8.

Put this way, the force of the argument for the validity of the bimolecular lipid layer model can be seen, but it is obvious that there are still some gaps in the logic. In particular, arguments on the behavior of model systems must always be suspect while the full details of the biological situation remain obscure. It must also be emphasized that, while the cell membrane appears as a simple double-layered system at the present level of resolution of the electron microscope, improvements in resolution, the use of thinner sections and of more specific stains, and the development of techniques for the study of hydrated specimens are likely to reveal complexities within this organization.

Nevertheless we shall assume that, as a first approximation, the arguments of this section hold, and we shall attempt to see just how far a physiological analysis of the properties of the cell membrane is consistent with the simple model of Fig. 1.8. A number of the more complex models that have been suggested as the basis of physiological studies are discussed in later chapters.

CHAPTER 2

General Aspects of Diffusion across Membranes

In this chapter we shall develop certain theoretical principles underlying the passage of molecules and ions across cell membranes and obtain a number of results which we shall use in later chapters. If these results only, and not their derivation, are of interest to the reader, this chapter may perhaps be taken as read. We include also a brief section on experimental methods.

2.1 Permeability and Diffusion Coefficients

The term "permeability" has the clear connotation of a rate. We can define a permeability coefficient P as the number of molecules of the penetrating species (the permeant) crossing in unit time, unit area of the cell membrane specified, when a unit concentration difference is applied across the membrane. Clearly, both the permeant and the cell membrane must be specified if the remark "substance X has permeability P" is to have any real meaning. The permeability coefficient is a directly measurable quantity, but its molecular significance is complex—it depends both on the diffusivity of the molecule in question through the membrane barrier and also on the thickness of that barrier. If the thickness of the cell membrane and, in particular, the thickness of that part of the membrane which is the major barrier to diffusion were known, then we could transform the measured concentration *difference* applied across the membrane into a concentration *gradient*. The permeability coefficient would thereby be converted into the diffusion coefficient D for transfer across the membrane. In general, the thickness of the permeability barrier is not known; but if we assume that this barrier is the bimolecular lipid leaflet discussed in Chapter 1, having a thickness of some 50 Å, then $P = D/50$ Å.

From the definition of P it will be seen that it has the dimensions

$ML^{-2}\ T^{-1}\ M^{-1}\cdot L^{3}$ or LT^{-1}; that is, it has the dimensions of a velocity. If, in the definition of P, the unit of concentration is expressed as the number of molecules (or gram molecules) per cubic centimeter while the area of the cell membrane is expressed in square centimeters and the transfer rate as a number of molecules (or gram molecules) per second, the unit of permeability appears as the centimeter per second. The diffusion coefficient has the dimension $L^2\ T^{-1}$ and its cgs unit is the square centimeter per second. Expressed in these units, for a 50 Å (5×10^{-7} cm) thick membrane

$$P \equiv D/5 \times 10^{-7}$$

or

$$D \text{ cm}^2 \text{ sec}^{-1} \equiv 5 \times 10^{-7}\ P \text{ cm sec}^{-1}$$

P is often reported in other units. For example, the unit μmole/μ^2/gram molecule/liter/second has components all of a convenient size for reporting the experimental measurements, while the unit centimeter/hour is often used in work on plant cells. The unit millimole/minute/millimolar/gram dry weight cells is much used in reporting data on cells of unknown surface area. Finally, permeabilities are often reported as "half-times"—the time taken for the concentration of permeant within the cell to rise (or fall) to a point midway between the initial and the final value relevant to that study, or as "hemolysis times"—the time taken for a certain percentage of erythrocytes to burst when these are added to a solution of the permeant.

It would be preferable if all data were reported in cgs units (cm sec^{-1}) as then a direct comparison between data obtained in different studies would be possible. Tables 2.1 and 2.2 have been drawn up to enable the most commonly used units to be interconverted, where this is possible from the data given in any particular study. Most of the data collected in this book have been converted into cgs units using Tables 2.1 and 2.2.

A comparison between values of P determined by different procedures or in different cells is valid only if the number of molecules crossing unit membrane area in unit time is directly proportional to the applied concentration difference—as implied in the definition of P. This is equivalent to the assumption that for these membranes Fick's first law of diffusion (Fick, 1855) strictly applies. (In addition, we must assume that the thickness of the barrier to diffusion is not affected by the prevailing permeant concentrations.) Fick's law is expressed formally as

$$\frac{dn}{dt} = -DA\frac{dc}{dx} \tag{2.1}$$

TABLE 2.1

INTERCONVERSION OF UNITS FOR SOLUTE PERMEABILITY COEFFICIENTS P_s [a,b]

	cm sec^{-1}	cm hr^{-1}	gm mole/μ^2/ gm mole/liter/sec	gm mole/μ^2/ gm mole/liter/min	mmole/min/ mM/gm dry wt.	$1/t_{1/2}$ (sec^{-1})	$1/t_{\text{hemolysis}}$ (sec^{-1})[c]
cm sec^{-1}	1	3.6×10^{3}	10^{-11}	6×10^{-10}	$\frac{1}{F} \times \frac{A}{V} \times 6 \times 10^{2}$	$\frac{1}{C} \times \frac{A}{V} \times 1.45 \times 10^{4}$	$\frac{1}{H} \times 10^{4}$
cm hr^{-1}	2.78×10^{-4}	1	2.78×10^{-15}	1.67×10^{-13}	$\frac{1}{F} \times \frac{A}{V} \times 1.67 \times 10^{-1}$	$\frac{1}{C} \times \frac{A}{V} \times 4.03$	$\frac{1}{H} \times 2.78$
gm mole/μ^2/ gm mole/ liter/sec	10^{11}	3.6×10^{14}	1	60	$\frac{1}{F} \times \frac{A}{V} \times 6 \times 10^{13}$	$\frac{1}{C} \times \frac{A}{V} \times 1.45 \times 10^{15}$	$\frac{1}{H} \times 10^{15}$
gm mole/μ^2/ gm mole/ liter/min	1.67×10^{9}	6×10^{12}	1.67×10^{-2}	1	$\frac{1}{F} \times \frac{A}{V} \times 10^{12}$	$\frac{1}{C} \times \frac{A}{V} \times 2.42 \times 10^{13}$	$\frac{1}{H} \times 1.67 \times 10^{13}$
mmole/min/ mM/gm dry wt	$F \times \frac{V}{A} \times 1.67 \times 10^{-3}$	$F \times \frac{V}{A} \times 6$	$F \times \frac{V}{A} \times 1.67 \times 10^{-14}$	$F \times \frac{V}{A} \times 10^{-12}$	1	$\frac{F}{C} \times 24.2$	$\frac{F}{H} \times \frac{V}{A} \times 16.7$ or $\frac{F}{R^2 - 1} \times 33.3$
$1/t_{1/2}$ (sec^{-1})	$C \times \frac{V}{A} \times 6.9 \times 10^{-5}$	$C \times \frac{V}{A} \times 2.48 \times 10^{-1}$	$C \times \frac{V}{A} \times 6.9 \times 10^{-16}$	$C \times \frac{V}{A} \times 4.15 \times 10^{-14}$	$\frac{C}{F} \times 4.15 \times 10^{-2}$	1	$\frac{C}{H} \times \frac{V}{A} \times 6.9 \times 10^{-1}$ or $\frac{C}{R^2 - 1} \times 1.38$
$1/t_{\text{hemolysis}}$ (sec^{-1})[c]	$H \times 10^{-4}$	$H \times 3.6 \times 10^{-1}$	$H \times 10^{-15}$	$H \times 6 \times 10^{-14}$	$\frac{H}{F} \times \frac{A}{V} \times 6 \times 10^{-2}$ or $\frac{R^2 - 1}{F} \times 3 \times 10^{-2}$	$\frac{H}{C} \times \frac{A}{V} \times 1.45$ or $\frac{R^2 - 1}{C} \times 7.25 \times 10^{-1}$	1

[a] If the value of the permeability coefficient is *given* in a unit recorded in the extreme left-hand column of the table, find the *desired* unit in the upper row. At the junction of row and column is the factor by which the given value must be multiplied to convert it into the new unit. Thus a permeability coefficient of, say, 1.2×10^{-15} gm mole/μ^2/gm mole/liter/sec must be multiplied by 3.6×10^{14} to convert it to cm hr^{-1}. It is thus 4.3×10^{-2} cm/hr.

[b] A is the area of the cell studied in square micra. C is a dimensionless term—an integration constant—which depends on the concentration of solute and on the particular details of the procedure used to measure $t_{1/2}$. C has the value [1 + concentration of permeant in isotonic units] if the equations of Stein and Danielli (1956) are required and the value [0.28 + concentration in isotonic units] if the procedure of Stein (1962b) is followed. F is the volume of cells associated with unit dry weight of cells. It is given by the fraction of the cell that is the dry weight divided by the specific gravity of the cells. H is the hemolysis constant of Jacobs (1952). It is given by $\frac{1}{2}(V/A)(R^2 - 1)$ where R is defined below. It is also (Jacobs, 1952) equal to the product $t_w \times L_p \times I \times 10^4$, where t_w is the time of hemolysis in pure water in seconds, L_p is the osmotic permeability coefficient of water in cm^4 osmole^{-1} sec^{-1} (see Table 2.2), and I is the initial internal concentration of impermeable solute in the cell in gm mole/cm^3. R is the fragility of the cell, given by the ratio of the volume at the hemolysis point to the initial volume of the cell. V is the initial volume of the cell in cubic micra.

[c] Hemolysis times some 20× that for the hemolysis time in pure water will give accurate values of P_s by these formulas. For shorter times, the hemolysis time in pure water should be subtracted from the hemolysis time in the solute (see also discussion in Section 2.4).

TABLE 2.2

INTERCONVERSION OF UNITS FOR OSMOTIC (HYDRAULIC) WATER PERMEABILITY COEFFICIENTS L_p AT 25° AND 0°C[a,b]

	cm sec^{-1}	cm^4(os)mole^{-1} sec^{-1}	cm^3 cm^{-2} (cm H_2O)$^{-1}$ sec^{-1}	μsec^{-1}	$\mu^3\mu^{-2}$ (atm)$^{-1}$ sec^{-1}	$\mu^3\mu^{-2}$ (atm)$^{-1}$ min^{-1}	cm^3(cell)$^{-1}$ (cm H_2O)$^{-1}$ sec^{-1}	cm^4 (os)mole^{-1} hr^{-1}
cm sec^{-1}	1	18.1 (18)	7.19×10^{-7} (7.78)	10^4	7.41 (8.04)	4.45×10^2 (4.82)	$A \times 7.19 \times 10^{-15}$ (7.78)	6.51×10^4 (6.48)
cm^4(os)mole^{-1} sec^{-1}	5.52×10^{-2} (5.56)	1	3.97×10^{-8} (4.31)	5.52×10^2 (5.56)	4.10×10^{-1} (4.46)	24.6 (26.8)	$A \times 3.97 \times 10^{-16}$ (4.31)	3.6×10^3
cm^3 cm^{-2} (cm H_2O)$^{-1}$ sec^{-1}	1.39×10^6 (1.29)	2.52×10^7 (2.32)	1	1.39×10^{10} (1.29)	1.03×10^7	6.19×10^8	$A \times 10^{-8}$	9.07×10^{10} (8.35)
μsec^{-1}	10^{-4}	1.81×10^{-3} (1.8)	7.19×10^{11} (7.78)	1	7.41×10^{-4} (8.04)	4.45×10^{-2} (4.82)	$A \times 7.19 \times 10^{-19}$ (7.78)	6.51 (6.48)
$\mu^3\mu^{-2}$(atm)$^{-1}$ sec^{-1}	1.35×10^{-1} (1.24)	2.44 (2.24)	9.68×10^{-8}	1.35×10^3 (1.24)	1	60	$A \times 9.68 \times 10^{-16}$	8.79×10^3 (8.06)
$\mu^3\mu^{-2}$(atm)$^{-1}$ min^{-1}	2.24×10^{-3} (2.07)	4.07×10^{-2} (3.73)	1.62×10^{-9}	2.24×10 (2.07)	1.67×10^{-2}	1	$A \times 1.62 \times 10^{-17}$	1.45×10^2 (1.34)
cm^3(cell)$^{-1}$ (cmH_2O)$^{-1}$ sec^{-1}	$1/A \times 1.39 \times 10^{14}$ (1.29)	$1/A \times 2.52 \times 10^{15}$ (2.32)	$1/A \times 10^8$	$1/A \times 1.39 \times 10^{18}$ (1.29)	$1/A \times 1.03 \times 10^{15}$	$1/A \times 6.19 \times 10^{16}$	1	$1/A \times 9.07 \times 10^{18}$ (8.35)
cm^4(os)mole^{-1} hr^{-1}	1.54×10^{-5} (1.55)	2.78×10^{-4}	1.10×10^{-11} (1.20)	1.54×10^{-1} (1.55)	1.14×10^{-4} (1.24)	6.85×10^{-3} (7.45)	$A \times 1.10 \times 10^{19}$ (1.20)	1

[a] Symbols and method of use as in Table 2.1.

[b] Values in parentheses indicate numerical parameters at 0°C. For temperatures between 0° and 25°C, a linear interpolation will be valid.

where dn/dt is the number of molecules dn crossing area A in the interface in time dt when a concentration difference dc is applied over a distance dx. In terms of the permeability constant P, Fick's law is

$$\frac{dn}{dt} = -PA\,dc \tag{2.2}$$

showing clearly the relation between P and D. Equations (2.1) and (2.2), however, often do not apply to diffusion of permeants across cell membranes. Chapters 4, 5, and 6 include data on many such deviant systems.

One more definition that we shall need is that of "flux." The "flux" of permeant is given by the total number of permeant molecules crossing unit area of the membrane in unit time. The flux, as expressed by Eq. (2.2), is determined by the *net* movement of permeant molecules across the membrane, that is, by the excess of, say, inward movement over outward movement. Where, as will happen repeatedly in Chapters 4, 5, and 6, we shall have to refer separately to the movements in each direction, inward or outward, we shall refer to the inward flux or outward flux across the membrane. The flux across the membrane is, by Eq. (2.2), proportional both to the concentration of permeant on the side from which the flux is directed and to the permeability coefficient P. A flux has the dimensions $ML^{-2}T^{-1}$.

A naive concept of the permeability coefficients P_s for a solute S might suggest that provided the solute, the cell membrane, the temperature and the solution compositions were defined, only one value of P_s would be able to be measured, that is, that P_s was invariant. But a deeper analysis, pioneered for cell membranes by Kedem and Katchalsky (1958, 1961), shows that the rate and direction of movement of *solvent,* if this occurs simultaneously with the movement of solute, may greatly affect P_s. For, if solute and solvent interact within the membrane, solute and solvent molecules will exert a drag on one another, movement of solvent affecting that of solute and vice versa. It turns out that a minimum of three parameters are required to characterize fully the behavior of a membrane to a particular solute. One parameter, closely related to the term P_s, will characterize the interaction between solute and membrane. Another parameter describes the solvent-membrane interactions—it is related to the permeability P_w of the membrane toward water. The third parameter is required to characterize the solute-solvent interaction and has no direct analogy among the classical permeability coefficients. It is clear that no statement regarding the permeability of a given membrane to a particular solute is valid unless the role played by the solvent is understood, that is, unless the three determining parameters are available. We shall proceed to consider these three parameters directly. The observed effects are quantitatively accounted for by the theoretical framework of the thermodynamics of irreversible processes and we must now briefly consider some of the methods and results of this theory, insofar as they concern diffusion across the cell membrane.

2.2 Irreversible Thermodynamics of Membrane Processes

Classical thermodynamics teaches us that if a reversible process is carried out in a cycle, no production of entropy—of disorder—will result. A reversible process, in this regard, is one which takes place infinitely slowly, the forward movement being always opposed by a restraining force only just insufficient to oppose movement, and in such a fashion that an infinitesimal increase in this restraining force is able to reverse the direction of movement. The results of classical thermodynamics concern such reversible processes and describe how the entropy production and the heat and free energy changes in one half of a reversible cycle are related. But most natural processes are irreversible and, in particular, diffusion and permeation are, in this thermodynamic sense, irreversible. We continually study systems in which movement occurs across an interface—on one side of which is a definite and large concentration of permeant molecules, while on the other side is a smaller or even zero concentration. Only at equilibrium when the process which we are studying—diffusion—is completed, do the criteria for a reversible process apply.

Irreversible thermodynamics concerns itself with such irreversible processes and in particular with the extent and rate of entropy production during the operation of an irreversible process. For it is the case that such a process, being irreversible, will produce entropy when carried out in a cycle. Even if one is able to restore all possible variables to their original position (that is, to reverse the operation of diffusion, concentrating the permeant once again on the original side of the interface), it can be shown that the universe as a whole will have increased in disorder by the end of this cycle [that is, entropy will have been produced (De Groot, 1959)]. Irreversible thermodynamics deals of necessity with rates—for if the process occurs infinitely slowly, one of the conditions for reversibility applies and the entropy production will be zero. We will want to relate the rate of entropy production to the rate of the irreversible process.

Any process, be it a translocation of matter, of heat, of energy or of volume (see below), can be considered as resulting from the operation of certain *forces* which produce a net *flux* of the quantity considered (matter, heat, etc.). We call the force that produces a particular flux the conjugate force of that flux. The amount of flux produced (at low rates of flux) is related to the amount of the conjugate force by an appropriate phenomenological coefficient (diffusion coefficient, coefficient of heat conductivity, etc.). It turns out that, provided certain conditions are satisfied, the rate of entropy production is given by the sum of the products of these forces and fluxes, a very natural result (for proof of this statement see De Groot, 1959).

For the process of permeation across membranes, Kedem and Katchalsky (1958) choose the following appropriate net fluxes:

(a) The total flow of volume—solute together with solvent—across the membrane (symbolized by J_v).

(b) The *relative* velocity of solute versus solvent (symbolized by J_D). This can also be termed the exchange flow of solute and solvent. Then the respective conjugate forces are (1) the pressure difference across the membrane Δp, and (2) a term $RT\,\Delta C_s$, related to the concentration difference of solute ΔC_s. This description refers to the simultaneous passage of a single solute s and of water w across the membrane. The rate of entropy production can be shown (Kedem and Katchalsky, 1958) to be given by the sum of the products of these forces and fluxes, that is,

$$\text{Rate of entropy production} = J_v\,\Delta p + J_D RT\,\Delta C_s \tag{2.3}$$

The phenomenological coefficient relating the flow of volume to its conjugate force, the pressure difference causing this volume flow, is the pressure-filtration coefficient L_p. Thus

$$J_v = L_p\,\Delta p \qquad \text{(at zero concentration difference)} \tag{2.4}$$

Likewise, the relative velocity of solute and solvent J_D is proportional to its conjugate force, the concentration difference of solute $RT\,\Delta C_s$, the coefficient of proportionality being symbolized by L_D. Thus,

$$J_D = L_D\,RT\,\Delta C_s \qquad \text{at zero pressure difference} \tag{2.5}$$

The term L_p is clearly related to the permeability coefficient for water P_w, while L_D is analogous to the solute permeability P_s, but the relationship between these terms is not a simple one, as we shall see.

Fundamental to the development of the theory of irreversible processes has been the concept of the *cross-coefficients.* If, in the case of membrane permeation, a pressure difference Δp is applied across the membrane, we have seen that a net flow of volume J_v results, determined by the pressure-filtration coefficient L_p. But if the membrane is semipermeable (that is, more permeable to water than to the solute), the applied pressure difference will force relatively more water than solute across the membrane. The extruded fluid will therefore have a different composition from the fluid prior to its extrusion, and there will thus be a change in the relative velocity of solute versus solvent as a consequence of the applied pressure difference. The flux in relative velocity J_D will thus be determined also by the applied pressure difference Δp, the relation between the flux and this force being given by a cross-coefficient which we shall write as L_{Dp}. Since a change in the composition of the extruded fluid with pressure is the process of ultrafiltration, L_{Dp} is the ultrafiltration coefficient. Similarly, if a concentration difference is applied across the membrane a net flow of volume will take place. This is the phenomenon of *osmosis* and the appropriate cross-coefficient, here symbolized L_{pD}—the flow of volume per unit concentration difference—is the osmotic coefficient. (The nonmathematical reader is likely to find the symbolism used here—that conventionally used—confusing. He might with profit rewrite the equations of this and the next section, using the symbol L_o for L_{pD} and L_f for L_{Dp}.) These cross-coefficients obviously relate to the solute-solvent interactions, which we are hoping to be able to quantitate.

Similar cross-coefficients are characteristic of all irreversible processes. Another example—and the earliest studied—of these cross-phenomena is concerned with the behavior of bimetallic strips. On the one hand, heat is evolved when an electric current flows across a junction of two metals (the Peltier effect); on the other hand, an electric current will be produced if the junctions between strips of different metals are kept at a different temperature. We have in this situation, as well as these cross-phenomena, the following two direct phenomena:

(1) A flux of electric current which occurs as a result of the conjugate electromotive force, here the electrical potential difference applied across the metals. The phenomenological coefficient in this case is the electrical conductivity, given by Ohm's law.
(2) A flux of heat which flows through the metal as a result of its conjugate force, the temperature gradient. The phenomenological coefficient here is the thermal conductivity of the metal.

In addition, we have the following cross-phenomena:

(3) The flux of current as a result of the temperature gradient force.
(4) The flux of heat as a result of the electromotive force.

These cross fluxes and forces are related by appropriate cross-coefficients and it has been found experimentally that if these cross-coefficients are expressed in the correct units, they are under all conditions identical. These striking experimental observations are predicted by Onsager's development of the theory of irreversible processes, in the form of a very general theorem: "If the proper choice is made of the fluxes and forces, then the cross-coefficients will be equal" (see De Groot, 1959). The "proper choice of fluxes and forces" turns out to be the one in which, when each force is multiplied by its conjugate flux and then these force-flux products are summed, the grand summation is exactly equal to the rate of entropy production in the irreversible process considered. Kedem and Katchalsky's choice of fluxes and forces in the case of membrane diffusion [Eq. (2.3) above] ensures that the conditions for the application of Onsager's theorem are satisfied.

When we consider, therefore, both the direct- and the cross-phenomena, we see that the total volume flow arises from the combined effects of the pressure difference Δp and the concentration difference ΔC_s. Similarly, the relative velocity of solute and solvent is likewise composed of two such parts, each characterized by the appropriate phenomenological coefficient. The formal statement of these relations is the following set of equations which specifically takes into account the mutual effects of solute and solvent flows.

$$\begin{aligned} J_v &= L_p\,\Delta p + L_{pD}\,RT\,\Delta C_s \\ J_D &= L_{Dp}\,\Delta p + L_D\,RT\,\Delta C_s \end{aligned} \tag{2.6}$$

Now, since the choice of forces and fluxes has been correctly made [Eq. (2.3)], Onsager's theorem applies.

$$L_{pD} = L_{Dp} \tag{2.7}$$

Thus, what we have termed the ultrafiltration coefficient and the osmotic coefficient will always be equal to each other if expressed in the correct units. Equations (2.4) and (2.5) follow from Eq. (2.6) if the appropriate substitutions are made.

Equation (2.6) shows that the permeability of a membrane to a given solute in the presence of water can be characterized by the four phenomenological coefficients L_p, L_D, L_{pD}, and L_{Dp}, while Eq. (2.7) shows that two of these coefficients, L_{pD} and L_{Dp}, are identical. We are then left with only three coefficients which, as we saw above, are the minimum number required to characterize fully the behavior of a membrane to a particular solute-solvent system. We must now show that these coefficients are indeed those required for the understanding of membrane permeability. To show this we must first introduce one more term, the reflection coefficient σ.

2.3 The Reflection Coefficient σ

For those cases where the solute is completely impermeable to the given membrane, that is, for the case of the ideal semipermeable membrane, fewer coefficients are required to characterize the membrane, and it is easy to prove that indeed only one phenomenological coefficient is necessary. In this case the flow of volume J_v is given exactly by the flow of water alone—there is no flow of solute—while the relative flow of solute versus solvent is also given by the flow of water alone—but now with a negative sign for, if the flow of water is positive, the relative flow of solute versus solvent is in the opposite direction. We have, then,

$$J_v = -J_D \tag{2.8}$$

and substituting in Eq. (2.6), the result is Eq. (2.9):

$$(L_p + L_{Dp})\ \Delta p + (L_D + L_{pD})\ RT\ \Delta C_s = 0 \tag{2.9}$$

Equation (2.9) can be satisfied for all values of Δp and ΔC_s only if the terms within the parentheses are always zero. Therefore,

$$L_p = -L_{Dp} \qquad \text{and} \qquad L_D = -L_{pD} \tag{2.10}$$

and since $L_{pD} = L_{Dp}$ [Eq. (2.7)], we have, therefore,

$$L_p = L_D = -L_{Dp} = -L_{pD} \tag{2.11}$$

Thus a single phenomenological coefficient suffices to characterize an ideal semipermeable membrane. Flow across such a membrane will be either hydrostatic, given by the pressure gradient Δp and the phenomenological coefficients L_p $(= -L_{Dp})$, or osmotic, given by the osmotic gradient $RT\ \Delta C_s$ and the identical coefficients L_D $(= -L_{pD})$. Thus the rates of hydrostatic flow and of osmotic flow are necessarily equal. This has often been denied in the past by a number of physiologists (Chinard, 1952; E. J. Harris, 1956). Durbin (1960) has, however, confirmed the predictions of Eq. (2.11) by direct measurements of the osmotic and hydrostatic flows on synthetic cellulose membranes (Table 2.3).

For a "leaky" membrane, Eq. (2.11) will not hold and the movement of the solute must be taken into account. In order to describe the relative rates of solvent and solute permeabilities, Staverman (1952) has defined the *reflection coefficient* σ, which is the ratio of $-L_{pD}$ to L_p, that is, the ratio of the osmotic coefficient (or the ultrafiltration coefficient) to the pressure-filtration coefficient (see Eq. 2.12).

$$\sigma = -L_{pD}/L_p \tag{2.12}$$

TABLE 2.3

OSMOTIC (L_{pD}) AND HYDROSTATIC (L_p) WATER FLOW ACROSS PERMEABLE CELLULAR MEMBRANES[a]

Water flow produced by gradient of	Net volume flow (μl M^{-1} min^{-1})			
	Dialysis tubing	Cellophane	Wet gel	Solute radius (Å)
D_2O	0.06	—	0.084	1.9
Urea	0.6	0.6	1.5	2.7
Glucose	5.1	4.2	5.8	4.4
Sucrose	9.2	7.0	10.4	5.3
Raffinose	11	8.5	13	6.1
Inulin	19	41	84	12
Bovine serum albumen	25.5	98	270	37
Hydrostatic pressure	25	95	370	
Derived pore size (Å)	23	41	82	

[a] Data taken from Durbin, 1960.

For an ideal semipermeable membrane σ is, from Eq. (2.11), equal to unity. In a coarse, nonselective membrane, there can be no ultrafiltration and σ is then zero. The Staverman factor σ is thus a measure of the *semipermeability* of the membrane to the given solute. A very useful way of looking at σ is to write it as follows:

$$\sigma = \frac{-L_{Dp}}{L_p} = \frac{-J_D}{J_v} \qquad (\Delta C_s = 0)$$

from Eq. (2.6), when it is clear that σ is the ratio of the flow of water relative to solute ($-J_D$) compared with the total volume flow (J_v) under the influence of a pressure gradient alone. If the membrane does not distinguish between solute and solvent there is no relative flow and σ is zero. If the membrane selects absolutely, $-J_D = J_v$ and $\sigma = 1$.

If the osmotic flow of volume occurring across a membrane in the absence of a pressure difference [$L_{pD}\, RT\, \Delta C_s$ from Eq. (2.6)] is compared with the volume flow occurring across this membrane when the concentration difference is zero [$L_p\, \Delta p$ from Eq. (2.6)], σ can be measured by a direct experiment. Table 2.3 records some data obtained in this way for various solutes by Durbin (1960); σ can also be obtained by finding what ratio of pressure difference to applied solute concentration difference will produce zero volume flow, using Eq. (2.6) with $J_v = 0$. This is the method used for human red blood cells by Goldstein and Solomon (see Section 2.6).

For the ideal semipermeable membrane, the rate of osmotic flow is given by van't Hoff's law as $L_p\ RT\ \Delta C_s$. For any real system this result will not hold exactly but rather the expression $\sigma L_p\ RT\ \Delta C_s$ [from Eq. (2.12)] will apply, where σ varies from 0 through 1.

2.4 The Relation between the Conventional and the Phenomenological Coefficients

We must now try to relate the phenomenological coefficients to the conventional parameters describing membrane permeability. L_p need cause little difficulty. It is merely the conventional coefficient describing the rate of volume flow of water across the membrane under the influence of a pressure gradient provided we specify [Eq. (2.6)] the absence of a concentration gradient of permeable solute. Likewise, the significance of the coefficients L_{pD} and L_{Dp} is easily understood from Eq. (2.6). The coefficient L_D, however, requires translation into the conventional form for, as it stands, it describes the exchange flow of solute versus solvent, a quantity not directly measured. (For the ultrafiltration flow, given by $L_{Dp}\,\Delta p$, exchange flow is, however, a natural measure.) What is most conveniently measured is the rate of flow of solute across the membrane, which we can symbolize as J_s, the flux of solute. In terms of our previous fluxes J_v and J_D, J_s is given by Eq. 2.13:

$$J_s = (J_v + J_D)C_s \tag{2.13}$$

since $J_v + J_D$ gives the total flow of that part of the volume of the solution that is occupied by the solute, and the product of this volume and C_s, the volume concentration, gives the total number of moles of solute that cross the membrane. Using Eqs. (2.6), (2.7), and (2.12), one can readily obtain Eq. (2.14):

$$J_s = (1 - \sigma)J_v C_s + (L_D - \sigma^2 L_p)C_s\ RT\ \Delta C_s \tag{2.14}$$

In the particular case of zero volume flow ($J_v = 0$), J_s is directly proportional to ΔC_s.

The coefficient of proportionality is $(L_D - \sigma^2 L_p)C_s\ RT$, which is of the same form as a conventional permeability constant and hence is directly measurable. The term $(L_D - \sigma^2 L_p)C_s$ can be replaced by the symbol ω. Thus ω, σ, and L_p are three independent parameters all of which are obtainable experimentally and from which L_p, L_D, and $L_{pD} = L_{Dp}$ can be derived; ω, σ, and L_p thus suffice to characterize the membrane.

In the past, permeability coefficients for various cells have often been determined by measuring the volume changes which occur as water enters cells osmotically following the entrance of the permeant (see

Section 2.6). The irreversible thermodynamic treatment shows clearly that this procedure suffers from three serious disadvantages. First, since the volume of the cell is continually changing during the course of the experiment, solute may be entering both by diffusional flow and by being dragged along by the water molecules, that is, by bulk flow. Hence a combination of the parameters ω and σ will be measured rather than either alone. Second, in the course of the derivation of the equations used to obtain the apparent permeability coefficients from the cell volume changes [see, for instance, among numerous other cases, the author's paper (Stein, 1962b)], the incorrect assumption is made that the permeant exerts its full osmotic effect across the cell membrane. Instead, an osmotic effect given by σ times the concentration of permeant will be found. The assumption that $\sigma = 1$, as made by the author and many earlier researchers, implies that the membrane has no permeability to the given permeant; yet it is precisely the permeability of this solute that is being measured. Finally, as was well known to earlier workers—and indeed allowed for in their studies—an error even in the apparent permeability constant will result if the permeant travels rapidly across the membrane (but if this rate is far from that of water—some one-hundred-fold less—little or no error results here). The author has done some preliminary calculations (Stein, 1964c) which show that the first two sources of error mentioned above determine that the parameter measured in such studies of cell volume changes is the ratio of ωRT to σ^2; that is, the permeability coefficient P_s as measured is close to $\omega RT/\sigma^2$ if the permeant does not move too rapidly across the membrane. Since values of σ reported for a number of the more slowly penetrating nonelectrolytes (Table 2.5) vary between 0.5 and 1, the early studies have overestimated the diffusional solute permeability coefficient by up to fourfold. For most of the slower permeants (where both ω and σ are small) the error is closer to 1½- to 2-fold.

2.5 The Physical Interpretation of the Phenomenological Coefficients

We have discussed in the previous section the relation between conventional permeability parameters and the phenomenological coefficients ω, σ, and L_p. In a further development of their analysis Kedem and Katchalsky (1961) have shown how it is possible to "translate" the phenomenological coefficients into "frictional coefficients" having a very readily interpretable physical meaning.

One can express the resistance to flow that the membrane offers to the solvent in terms of the frictional coefficient f_{wm}, the drag that the membrane m exerts on the solvent w. Similarly, there is a frictional coefficient

f_{sm} describing the interaction between solutes and membrane and a coefficient f_{sw} expressing the mutual drag of solute and solvent on each other. Kedem and Katchalsky (1961) and also Ginzburg and Katchalsky (1963) have been able to show that the frictional coefficients are related to the phenomenological coefficients by the expressions

$$\sigma = 1 - \frac{\omega \overline{V}_s}{L_p} - \frac{K f_{sw}}{\phi_w (f_{sw} + f_{sm})} \tag{2.15}$$

$$\omega = \frac{\phi_w}{\Delta x (f_{sw} + f_{sm})} \tag{2.16}$$

$$L_p = \frac{\phi_w}{\Delta x [(f_{wm}/\overline{V}_w) + f_{sm}(1 - \sigma)\, \overline{C}_s]} \tag{2.17}$$

where in addition to the terms defined earlier $\overline{V}_s$ is the partial molar volume of the solute, $\overline{V}_w$ that of the solvent, $\overline{C}_s$ the mean of the solute concentrations on the two sides of the membrane, K is the distribution coefficient for the partitioning of solute between the aqueous phase and the membrane, ϕ_w is the volume fraction of water in the membrane, and Δx is the thickness of the membrane. One may therefore characterize a solute-solvent-membrane system by the three phenomenological coefficients L_p, L_D, and L_{pD} $(= L_{Dp})$, or by these three parameters transformed into ω, σ, and L_p, or by the three frictional coefficients f_{sw}, f_{sm}, and f_{wm}. All three of these formalisms are equivalent, but in certain circumstances one may be more convenient to use than another. [Vaidhyanathan and Perkins (1964) have derived equations equivalent to (2.15) and (2.16) from statistical mechanical considerations.]

Of the terms in Eq. (2.15)—the most useful equation for biological material—ω, σ, L_p, and $\overline{V}_s$ can be found directly. Putting Eq. (2.15) into the form of Eq. (2.18) it is clear that the *difference* between $1 - \sigma$ and

$$(1 - \sigma) - \frac{\omega \overline{V}_s}{L_p} = \frac{K f_{sw}}{\phi_w (f_{sw} + f_{sm})} \tag{2.18}$$

the term $\omega \overline{V}_s / L_p$ will indicate whether f_{sw} is significant or not. The physical significance of Eqs. (2.15) and (2.18) is apparent from the following argument: We have seen in Section 2.3 that the term σ is given by $-J_D/J_v$ when $\Delta C_s = 0$, that is, it is the ratio of the flow of *water relative to solute* as compared with the *total* volume flow, under the influence of a pressure gradient alone. Now, a simple transformation of this expression yields the result that $1 - \sigma$ is given by $J_s/C_s J_v$, when $\Delta C_s = 0$. The expression J_s/C_s (the rate of flow of solute in moles per unit time, divided by the concentration of solute in moles per unit volume) will be seen to give the rate of flow of that portion of the volume that is occupied by the solute. Thus, $1 - \sigma$ is the ratio of the relative rates of flow of the volume occupied by solute to the flow of total volume, under the influence of a pressure gradient alone. The term $\omega\overline{V}_s$ is the rate of flow of solute (again expressed as the volume occupied by the solute, since $\overline{V}_s$ is the volume occupied by one mole of the solute) under the influence of a chemical gradient alone. The denominators in the terms $1 - \sigma$, and $\omega\overline{V}_s/L_p$ are identical, being the total volume flow under the influence of a pressure gradient. Thus if $1 - \sigma$ and $\omega\overline{V}_s/L_p$ are identical, the rate of flow of the solute is the same whether the flow is caused by a chemical gradient *or* by a pressure gradient. If, on the other hand, $1 - \sigma$ is greater than $\omega\overline{V}_s/L_p$, more solute will flow when a pressure gradient is imposed (that is, when a resulting flow of solvent takes place), indicating that the bulk flow of water drags along a significant amount of solute.

If it turns out, however, that the solute and solvent interact only to a very small degree during their passage across the membrane [that is, if the two sides of Eq. (2.18) are close to zero], it can be suggested that the solute and water cross the membrane by different routes, the solute perhaps through the lipid and the water through aqueous channels. If, to account for a high degree of solute-solvent interaction, the contrary assumption is made (that is, that the penetration of both solvent and solute takes place by diffusion through aqueous pores in the cell membrane), it can be shown (Kedem and Katchalsky, 1961) that the right-hand term of Eq. (2.15)—the term containing the frictional coefficient—can be replaced by the term A_s/A_w where A_s is the area of the pores available to the solute and A_w is the area available to the solvent water. We shall see in Section 3.6 how this expression has been used to derive an estimate of the radius of the pores postulated by some authors as being present in the cell membrane.

Equations (2.15), (2.16), and (2.17) can be solved for the three unknown frictional coefficients f_{sw}, f_{sm}, and f_{wm} if the phenomenological coefficients ω, σ, and L_p have been determined and if K, ϕ_w, and Δx are known. Ginzburg and Katchalsky (1963) give the solution of these simultaneous equations and describe measurements of the three phenomenological coefficients for two synthetic membranes, one being a cellophane dialysis tubing and the other a wet gel. From these data the frictional coefficients collected in Table 2.4 have been derived, and it appears that for these materials the coefficient f_{sm} is at least an order of magnitude smaller than f_{sw}, while f_{wm} is two orders of magnitude smaller again than f_{sm}. The main resistance to transport here is, therefore, the friction between solute and solvent. This result is perhaps to be expected for such highly water-swollen membranes. Corresponding determinations

TABLE 2.4

FRICTIONAL COEFFICIENT[a] FOR FREE DIFFUSION IN WATER AND FOR DIFFUSION IN DIALYSIS TUBING AND A WET GEL MEMBRANE[b,c]

		Diffusion in:						
		Free water	Dialysis tubing			Wet gel		
Solute	$10^5 \times$ conc. (mole cm^{-3})	$10^{-16} \times f_{sw}^{\circ}$	$10^{-16} \times f_{sw}$	$10^{-16} \times f_{sm}$	$10^{-13} \times f_{wm}$	$10^{-16} \times f_{sw}$	$10^{-16} \times f_{sm}$	$10^{-13} \times f_{wm}$
Sucrose	2.5	0.487	3.25	0.65	8.55	1.12	0.066	1.72
	10.0	0.521	3.57	0.57	8.95	1.14	0.068	1.79
Glucose	1.25	0.368	1.79	0.31	8.35			
	10.0	0.380	1.97	0.21	8.65	0.78	0.031	1.72
Urea	50	0.172	0.66	0.065	8.30	0.282	0.0046	1.68
HTO	—	0.101	0.328			0.114		

[a] In dynes sec $mole^{-1}$ cm^{-1}.

[b] Data taken from Ginzburg and Katchalsky, 1963.

[c] The values of f_{sw} are an order of magnitude higher than the free diffusion coefficient f_{sw}°. Both f_{sw} and f_{sw}° increase with concentration, their ratio being, however, invariant. f_{sm} is an order of magnitude lower than f_{sw}. f_{wm} is two orders of magnitude lower again than f_{sm}. Hence, the main resistance to transport here is the friction between solute and solvent within the membrane pores.

of the frictional coefficients for biological materials are still lacking since in general the values of ϕ_w, the water content of the membrane, Δx, its thickness, and K, the partition coefficient for the solute with the membrane, are unknown. From Eqs. (2.16) and (2.17) it is apparent that the ratio between ω and L_p is independent of the unknowns ϕ_w, K, and Δx. At low concentration C_s, this ratio becomes

$$\frac{\omega}{L_p} = \frac{f_{wm}}{f_{sw} + f_{sm}} \times \frac{1}{\bar{V}_w} \tag{2.19}$$

As would be expected if the friction between water and membrane (f_{wm}) is high as compared with that between solute and membrane (f_{sm}), a relatively higher flow of solute will be permitted to occur; that is, the ratio ω/L_p will be large. Equation (2.19) clearly illuminates the relation between phenomenological and frictional coefficients.

2.6 Methods of Measuring the Coefficients of Membrane Permeability

The older methods of measuring cell permeabilities have been collected in Davson and Danielli's monograph (1952) and need not be discussed in detail here. It should be emphasized, however, that in many of these classical methods the entry of the permeant was measured indirectly by observing the swelling of the cells (erythrocytes, marine eggs, and unicellular plants) as water followed the solute by osmosis. Cell volume changes were measured by direct observation using a microscope, by observing the light scattered by the cells—the amount of light scattered varying with the cell volume (see below)—or by measuring the time taken for the cells to swell to a certain critical volume at which they burst, the hemolysis point. Thus in these methods the permeability coefficient was measured under conditions of varying volume—which, as we have seen in Section 2.4, yields a parameter which combines both ω and σ. We shall consider here only the more recent studies in which it has proved possible to measure separately the three phenomenological coefficients. The results of such studies will be considered in detail in Chapter 3.

A. The Coefficient L_p

What is required here is a measure of the change with time of the volume of the cells (or organ) under the influence of a known hydrostatic pressure gradient. For living membranes, the use of pressure gradients generally involves serious experimental difficulties. We have seen, however, that if it is known or can be shown that a particular solute exhibits a value of $\sigma = 1$ for a particular membrane (that is, if the membrane is strictly semipermeable toward that solute), then the volume flow developed by an osmotic gradient will be identical with that developed by a hydrostatic pressure gradient, or $L_p = -L_{pD}$ from Eq. (2.11). Osmotic gradients are easily established by suitable manipulation of the solute concentrations so that most published measurements of L_p are indeed measures of L_{pD}. It is essential always to establish that $\sigma = 1$ for the solute used to develop the osmotic gradient.

Inasmuch as the volume flow acts to diminish the osmotic gradient, this gradient will change with time. It is necessary, therefore, to have available a mathematical analysis of the time rate of volume change as a function of the initial osmotic gradient, if the results of such measurements are to be interpreted. This analysis has been obtained by Jacobs, and an excellent summary of his important contributions is available in his review (Jacobs, 1952). More recent procedures are those of Sidel and Solomon (1957) and Hempling (1960). It is always assumed in these derivations that for solutes for which $\sigma = 1$ the cell behaves as an ideal osmometer. The correctness of this assumption has been demonstrated by LeFevre (1964) for the case of the red blood cell.

What is required finally is some method of measuring the volume of the cell or organ. This can be done in the case of large plant cells (Stadelmann, 1956, 1963, 1966) or sea urchin eggs (Dick, 1959) by direct microscopic observations of the cell. Measurements of the volume of a population of cells packed in a hematocrit tube by centrifugation (Hempling, 1960) or by direct weighing of an organ [for example, in the case of gall bladders (Diamond, 1964a)], can also be used. A piece of flat epithelial tissue can be used to isolate two fluid compartments, and the volume changes of these compartments can be measured as water flows across the tissue (Leaf, 1961). An ingenious device to translate the small volume changes occurring across single internodal cells of algae forming the boundary between two compartments into the linear move-
flat epithelial tissue can be used to isolate two fluid compartments, and
Ginzburg (1964a). The mechanism of this device is illustrated by Fig. 2.1.

An indirect method that has been much used for measuring cell volume changes is dependent on the observation (Ørskov, 1935) that the amount of light scattered by a solution of cells is a function of the cell volume. The amount of light emerging from a cell suspension will decrease as the light scattered increases, so the measurement of the optical density of a cell suspension can also be used as a measure of cell volume. A variety of experimental apparatus has been used for these measurements, that of Widdas (1953) being a good example. Commercially available spectrophotometers, recording optical density with time at a fixed wavelength, can be used effectively. Sidel and Solomon (1957) describe an excellent piece of equipment for measuring light scattering as a function of time. All the indirect methods of measuring volume have, of course, to be calibrated against cells of known volume, the volume here either being measured directly or calculated on the assumption that the cell behaves as an ideal osmometer.

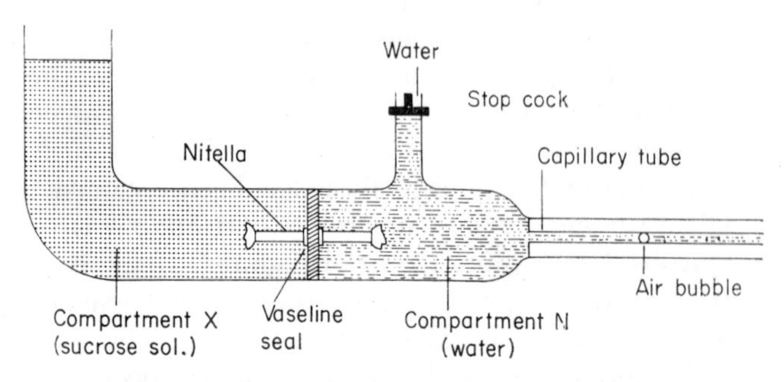

Fig. 2.1. Air bubble device for measuring transcellular osmosis. (Taken with kind permission from Dainty and Ginzburg, 1964a.)

For the very rapid entry of water (Chapter 3) into the red blood cell, measurements of cell volumes have to be made within much less than 1 sec of the establishment of the osmotic gradient. Sidel and Solomon (1957) used a rapid mixing device to ensure rapid equilibration of the cells with the added solute and a flow method to enable the volume changes to be recorded soon after mixing (Fig. 2.2). From a mixing chamber the cells and solute flow along a tube of variable length before entering the chamber in which the light scattering is measured. The tube

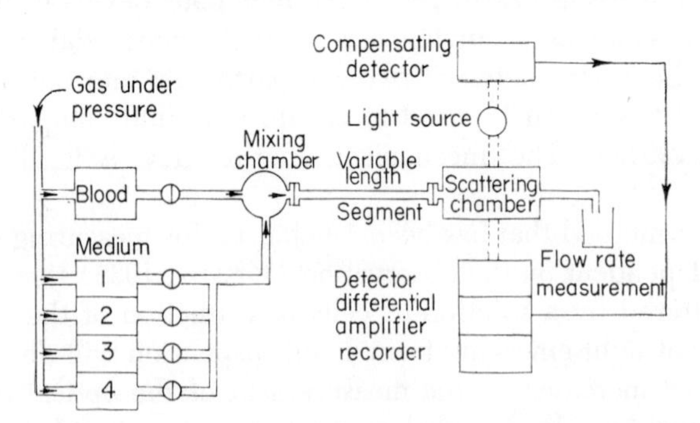

Fig. 2.2. Schematic diagram of rapid-mixing device for measuring osmotic permeability of erythrocytes. (Taken with kind permission from Sidel and Solomon, 1957.)

lengths were such that at the flow rates used readings of light scattering could be made at times of 50, 100, 155, and 215 msec after mixing of the cells and solute. Table 3.14 collects measurements of L_p for various cells and tissues. [Additional data: see Dick (1959) for a variety of membranes and Stadelmann (1963) for a comprehensive listing of the data on plant cell membranes.]

B. The Measurement of σ

Measurements of the parameter σ for biological membranes have been made only since the publication of Kedem and Katchalsky's theory (1958). (These authors show, however, how several of the older experiments described in the literature can be used to evaluate σ.) The essentials of any such method are to ensure that the volume flow across the cells is zero when the measurements are made, when from Eq. (2.6)

$$\sigma = \frac{\Delta p}{RT\,\Delta C_s} \tag{2.20}$$

The first experiments designed directly to measure σ for biological membranes were described in the important paper of Goldstein and Solomon (1960). The principle of their method was as follows: If red blood cells are suspended in a dilute solution of a penetrating solute, then as the solute enters the cell, water follows osmotically and the cell swells. If the solution is dilute enough, cell swelling will occur at all times. If, however, the osmotic effect of the solute is greater than that of the internal solutes of the cell, water will initially *leave* the cell and the cell will shrink, to be followed by a reswelling of the cell as the permeant enters. If cells are suspended in solutions of varying concentrations of the permeant, at one concentration the cells will neither shrink nor swell, at zero time. Then at this concentration of permeant C_s, the internal solute concentration C_i is exactly equivalent osmotically to the osmotic effect of the external concentration σC_s [Eq. (2.12)] so that

$$C_i = \sigma C_s \tag{2.21}$$

If it is known that for a particular solute $\sigma = 1$, that is, the membrane is strictly impermeable to this solute, then C_i can be found and thereafter σ determined for other solutes for which σ is unequal to 1. Goldstein and Solomon (1960) give a rigorous derivation of Eq. (2.21) based on the equations of irreversible thermodynamics. In their study, Goldstein and Solomon determined the cell volume changes by the light scattering apparatus of Fig. 2.2, as previously used by Sidel and Solomon. More recent studies from Solomon's laboratory have applied essentially the same method to obtain values of σ for numerous other cells and tissues (see Curran, 1963).

Dainty and Ginzburg (1964d) have used their capillary air-bubble technique (Fig. 2.1) to measure the volume changes taking place across the algal cells that they have studied and have similarly determined the solute concentrations that lead to zero initial rates of volume change. From these measurements the values of σ for the passage of various solutes across the plant cell membranes could be determined. Table 2.5 collects values of σ that have been determined for the passage of various solutes across a number of cell membranes.

C. Measurements of the Coefficient ω

Although many hundreds of solute permeabilities have been measured for a variety of cell membranes (Chapter 3), for the reasons discussed

TABLE 2.5

Values of the Reflection Coefficient σ for the Penetration of Solutes into Various Cells and Tissues (Together with the Derived Value of the Equivalent Pore Radius r Using the Renkin Equation, Section 3.6)

Cell or tissue	Solute and corresponding value of σ	Derived value of r in Å	References[a]
Chara australis (algal cell)	Urea 1.0, ethylene glycol 1.0, formamide 1.0, methanol 0.30, ethanol 0.27, *n*-propanol 0.22	—	(1)
Nitella translucens (algal cell)	Methanol 0.25, ethanol 0.29, *n*-propanol 0.16, isopropanol 0.27, *n*-butanol 0.11, methyl acetate 0.07, ethyl acetate 0.12, DHO 0.0061 at 4*M*, 0.0033 at 20*M*	—	(1)
Squid axon:			
Axolemma	Ethylene glycol 0.72, glycerol 0.96, urea 0.70, ethanol 0.63, formamide 0.44, methanol 0.25	4.25	(2)
Schwann cells	Ethylene glycol 0.05, glycerol 0.09	—	(2)
Necturus maculosus (an amphibian) kidney slices	Raffinose 1.0, sucrose 1.0, mannitol 1.0, erythritol 0.89, glycerol 0.77, urea 0.52	5.6	(3)
As above + 10 pressor units of "pitressin"	Sucrose 1.0, erythritol 0.77, glycerol 0.65, urea 0.52	6.5	(3)
Toad bladder	Thiourea 0.995, chloride ion 0.993, urea 0.79	—	(4)

TABLE 2.5—*Continued*

Cell or tissue	Solute and corresponding value of σ	Derived value of r in Å	References[a]
Frog, single muscle fibers	Mannitol 1.0, sucrose 1.0, glycerol 0.86, urea 0.82, formamide 0.65	4.0	(5)
Rabbit, corneal epithelium	Urea, NaCl, glucose, sucrose, raffinose, albumen—all 1	—	(11)
Rabbit, corneal endothelium	Glucose, sucrose, raffinose, albumen, all circa 1, urea, NaCl 0.6	—	(11)
Rat, luminal surface of intestinal mucosal cells	Mannitol 0.99, sucrose 0.99, erythritol 0.93, urea 0.81, ethylene glycol 0.27, formamide 0.22	4.0 (formamide and ethylene glycol too permeable)	(6)
Cat, capillaries of hind limb	Inulin 0.375, sucrose 0.058, glucose 0.04	—	(7)
Man, erythrocytes	Glycerol 0.88, propylene glycol 0.85, thiourea 0.85, malonamide 0.83, methylurea 0.80, propionamide 0.80, ethylene glycol 0.63, urea 0.62, acetamide 0.58	4.2 (acetamide too permeable; thiourea too impermeable)	(8)
Artificial membranes:			
Dialysis tubing	Sucrose 0.11–0.16, glucose 0.06–0.12, urea 0.006–0.013	—	(9)
Wet gel	Sucrose 0.036, glucose 0.019–0.024, urea 0.0016	—	(9)
Dialysis tubing[b]	Bovine serum albumen 1.02, inulin 0.76, raffinose 0.44, sucrose 0.37, glucose 0.20, urea 0.024, D_2O 0.002	23	(10)
Cellophane[b]	Bovine serum albumen 1.03, inulin 0.43, raffinose 0.089, sucrose 0.074, glucose 0.044, urea 0.006	41	(10)
Wet gel[b]	Bovine serum albumen 0.73, inulin 0.23, raffinose 0.035, sucrose 0.028, glucose 0.016, urea 0.004, D_2O 0.001	82	(10)

[a] References: (1) Dainty and Ginzburg (1964d). (2) Villegas and Barnola (1961). (3) Whittembury *et al.* (1960). (4) Leaf and Hays (1962). (5) Zadunaisky *et al.* (1963). (6) Lindemann and Solomon (1962). (7) Computed by Kedem and Katchalsky (1958) from data of Pappenheimer *et al.* (1951). (8) Goldstein and Solomon (1960). (9) Ginzburg and Katchalsky (1963). (10) Durbin (1960). (11) Mishima and Hedbys (1966).

[b] The values of r for these last three membranes were derived from a comparison of L_p and P_w for water permeability (see Table 3.14) but were shown to describe adequately the data on σ.

in Section 2.4, very few of those measurements can be used to derive an exact value for the solute diffusional coefficient ω. If ω is to be unambiguously measured (without reference to the measurements of σ and L_p), determinations have to be performed under conditions of zero volume change when Eq. (2.14) can be used without difficulty. This limits possible procedures to studies of the rate of exchange of isotopically labeled permeant, across cell membranes already in equilibrium with permeant—when there is no change in the concentration of permeant—or to determinations obtained by extrapolation to zero volume flow from conditions where volume flow is of necessity changing. This latter procedure has not been explored as yet. The methods using isotopically labeled permeant need no special consideration apart from the fact that, as the rates of exchange may be high, special rapid separation procedures may have to be used to remove the cells from the solution prior to the measurements of the intracellular radioactivity. [See Paganelli and Solomon (1957) for the exchange of tritiated water across the red cell membrane; see Dainty and Ginzburg (1964c) for measurements of the movement of ^{14}C-labeled alcohols across algal cell membranes.] Thus Paganelli and Solomon (1957) use a flow system similar in principle to that of Fig. 2.2 with the solution surrounding the cells being forced out of the flow system at various distances along the flow tube (and hence at various time intervals after mixing) across Millipore filters. Millipore filters inserted into a hypodermic syringe have been used effectively to withdraw cells from an incubation mixture (Mawe and Hempling, 1965), as has rapid filtration across Celite filter pads (Lacko and Burger, 1961). Rapid cooling can stop translocation if the temperature coefficient of the permeation process studied is high, the cells being separated thereafter by centrifugation. A procedure which could be applied more frequently is that of Keynes (1951) who mounted a nerve fiber directly over a Geiger-tube radioactivity counter and measured the rate of loss of previously accumulated isotope into a flowing stream of washing fluid. In the special case of the inhibitible specialized permeability systems to be discussed in Chapters 4, 5, and 6, the addition of an appropriate inhibitor will stop translocation and again the cells can be, at leisure, removed by centrifugation for subsequent analyses.

Methods for measuring P_s based on the chemical or isotopic analysis of the entering solute [such as the classic studies of Collander and Barlund (1933) on the nonelectrolyte permeability of *Chara* (Figs. 3.5 and 3.6)] will give, in general, an accurate measure of ωRT, even if conditions are such that the exchange of isotopically labeled permeant is not being measured—provided that minimal volume changes occur on the addition of solute. This will be the case if the concentration of the solute is low compared with the tonicity of the medium. Thus, in their studies on the penetration of alcohols across algal membranes, Dainty and Ginzburg (1964c) observed that there was no detectable difference in the rate of uptake of radioactively labeled solute as between the case (1) when the cells had been pre-equilibrated with unlabeled solute so that no volume flow was occurring, and the case (2) when the radioactive tracer was added directly to the cells.

In these experiments the concentration of the permeant was 0.1 molal, a concentration which might be expected to induce a substantial osmotic flow of water out the cell. These observations suggest that even with these permeants, where the reflection coefficients σ are of the order of 0.15 to 0.3 (Table 2.5), solvent-solute drag does not substantially modify the apparent value of ω. For less rapidly penetrating solutes, therefore, where values of σ not too far removed from unity are obtained, the measured values of the permeability coefficient P_s will be most unlikely to differ from ωRT by an order of magnitude and will in general be overestimated by only a fewfold. The relative ranking of permeants as judged by the coefficients P_s or ω will differ little if at all since, in general, P_s and ω will follow one another.

Available data on ω for water movement are collected in Table 3.14. Values of P_s are collected and exhaustively discussed in Chapter 3.

2.7 The Diffusion and Distribution of Ions

The charged nature of ions adds a due measure of complexity to the description of the movement of these particles across cell membranes. It is a necessary condition that the sum total of positive charges on one side of a membrane shall equal the sum total of negative charges on that same side. The reason for this is to be found in the relatively enormous amounts of energy that would be required to separate positive and negative charges across a distance corresponding to the cell membrane. If the total of positive and negative charges must balance, then if a positively charged ion (for example, the sodium cation) is to penetrate the membrane, it must do so either in exchange for another positively charged ion or it must penetrate accompanied by a negatively charged ion. Let us consider a simple system: the movement of sodium chloride across a cell membrane when only this salt is present. Now a *net* movement of sodium ions must be accompanied by a net movement of chloride since a charge-balancing movement of sodium outward in exchange for the movement of sodium inward would not show up in a *net* movement study. The net rate of sodium efflux will then depend in the first place on the local concentration of sodium ions $[Na_i]$—

Fick's law—and also on the local concentration of chloride ions $[Cl_i]$, for a simultaneous efflux of sodium and chloride is mandatory. The rate of efflux of sodium is therefore given by P_{Na} $[Na_i] \times [Cl_i]$ where P_{Na} is a permeability coefficient. The rate of influx of sodium is given by the corresponding product P_{Na} $[Na_e] \times [Cl_e]$, depending on the combined products of the external sodium and chloride concentrations. At the steady state we have the condition that efflux exactly balances influx so that the relationships in Eqs. (2.22) and (2.23) hold true.

$$[Na_i]\,[Cl_i] = [Na_e]\,[Cl_e] \tag{2.22}$$

$$\frac{[Na_i]}{[Na_e]} = \frac{[Cl_e]}{[Cl_i]} \tag{2.23}$$

There is, therefore, a reciprocal relationship between the anion and cation equilibrium distribution ratios. By Eq. (2.22) the net flux of any cation will depend on the concentration and also the mobility of the surrounding anions and vice versa.

Equation (2.23) can readily be generalized to more complex systems. To do so, it is more convenient to discuss the equilibria in terms of activities (Daniels and Alberty, 1961). The condition of equilibrium between two phases (for example, the two phases separated by the cell membrane) is that the activity of any component i is the same in both phases. Now the activity a_{MX} of a univalent salt MX is given by the product of the activities of the separate ions a_M and a_X, that is,

$$a_{MX} = (a_{M^+})(a_{X^-}) \tag{2.24}$$

Equation (2.23) follows directly from (2.24) if the activities in the two phases (internal and external) are equated, and the simplifying assumption is made of putting concentrations equal to activities. For more complex salts of the form $M_{\nu^+}X_{\nu^-}$ where ν^+ is the number of cations and ν^- the number of anions in the salt, the activity $a_{M_{\nu^+}X_{\nu^-}}$ of the salt is given by the expressions

$$a_{M_{\nu^+}X_{\nu^-}} = (a_M)^{\nu^+}(a_X)^{\nu^-} \tag{2.25}$$

so that the membrane equilibrium condition becomes

$$\left(\frac{a_{M_i}}{a_{M_e}}\right)^{\nu+} = \left(\frac{a_{X_e}}{a_{X_i}}\right)^{\nu-} \tag{2.26}$$

or

$$\left(\frac{[M_i]}{[M_e]}\right)^{+} = \left(\frac{[X_e]}{[X_i]}\right)^{-} \tag{2.27}$$

on replacing activities by concentrations. Equation (2.23) follows, of course, from Eq. (2.27) if the valency terms ν^+ and ν^- are set equal to unity.

Equation (2.27) and the simpler forms preceding it will refer only to the distribution of the mobile ions. If one of the ions cannot penetrate the membrane—and this is an exceedingly common situation in biological systems where charged macromolecules, impermeable to most membranes, abound—the electroneutrality condition requires that an equivalent number of opposite charges must be immobilized on the same side of the membrane as the fixed ion. In many cells fixed anions are commonly found. These require that an electrically equivalent amount of cation be maintained on the inside of the membrane and this raised concentration of cation ensures, by the operation of the still-valid laws of Eqs. (2.22) to (2.27), that the concentration of mobile anion on the inside of the membrane is lowered and that on the outside raised. If we have a concentration $[X_i]$ of fixed anion on the inside of the membrane and sodium chloride concentrations $[Na_i]$, $[Cl_i]$ and $[Na_e]$, $[Cl_e]$ on the inside and outside phases, respectively, then we have the following relations from the electroneutrality condition:

$$
\begin{aligned}
[Na_i] &= [Cl_i] + [X_i] \\
[Na_e] &= [Cl_e]
\end{aligned}
\qquad (2.28)
$$

while Eq. (2.22) still holds for the mobile ions. Solving these equations, we find that

$$
\mathbf{r} = \frac{[Na_e]}{[Na_i]} = \frac{[Cl_i]}{[Cl_e]} = \sqrt{\frac{[Cl_i]}{[Cl_i] + [X_i]}} \qquad (2.29)
$$

Thus the distribution ratio **r** at equilibrium depends on $[X_i]$, or rather on the relative concentrations of X_i and Cl_i. From Eq. (2.29), as $[Cl_i]$ (the salt concentration) increases with respect to $[X_i]$, the term under the square root sign in Eq. (2.29) tends to unity and the unequal ion distribution ratios disappear, that is, **r** becomes one. In those cases where the activities of the mobile ions on either side of the membrane are not equal, this activity difference will set up an electrical potential across the membrane, for an activity gradient of any such mobile charged species produces an electromotive force—we have, in effect, a "concentration cell" of the charged species (see the excellent discussion by Butler, 1951). If the sodium and chloride ions are the mobile species, the potential is given by Eq. (2.30):

$$
E_D = \frac{RT}{F} \log_{10} \frac{a_{Na_e}}{a_{Na_i}} = \frac{RT}{F} \log_{10} \frac{a_{Cl_i}}{a_{Cl_e}} \qquad (2.30)
$$

R is the gas constant, T the absolute temperature, and F the Faraday constant.

These phenomena involving fixed ions were first investigated by F. G. Donnan and the potential difference is commonly termed a "Donnan potential," while the finding of an activity ratio unequal to unity and determined by Eq. (2.30) is often referred to as a "Donnan distribution." We shall develop these considerations somewhat further in Chapter 6.

2.8 Active and Passive Transport

We have seen [Section 2.2, Eq. (2.2)] that if Fick's law of diffusion is obeyed, the unidirectional flux $J_{I \to II}$ of a penetrating species should be strictly proportional to the concentration C_I of permeant on the side I from which penetration is occurring. For a charged species (Section 2.7) the flux will, however, be determined both by the concentration and the electrical potential $E_I - E_m$ acting through the membrane where E_I is the potential on side I and E_m is the maximum potential within the membrane. These two effects can be combined if we substitute the words "electrochemical potential" for "concentration" in our statement of Fick's law—the flux $J_{I \to II}$ is proportional to the electrochemical potential A_I of the solute on the side I from which penetration is occurring. We have then

$$J_{I \to II} = PA_I = PC_I \exp\left[\frac{nF(E_I - E_m)}{RT}\right] \tag{2.31}$$

where P is a permeability coefficient, F the Faraday constant, n the number of positive charges on the solute, and R and T have their usual significance.

The flux in the opposite direction $J_{II \to I}$ is given by an expression of similar form on replacing the subscript I by II, so that

$$\frac{J_{I \to II}}{J_{II \to I}} = \frac{A_I}{A_{II}} = \frac{C_I}{C_{II}} \exp\left[\frac{nF(E_I - E_{II})}{RT}\right] \tag{2.32}$$

where now $E_I - E_{II}$ is the potential difference acting across the membrane. This result, first obtained by Ussing (1949a,b) is known as Ussing's flux ratio test. If the flux ratios accord with the predictions of Eq. (2.32), there is no proof that anything other than simple diffusion is operative for the solute under consideration. At equilibrium, the two oppositely directed fluxes are equal so that $A_I = A_{II}$ and there will be no electrochemical gradient across the membrane.

In many situations, however, it is found that the flux ratios are not those given by Eq. (2.32), and flux ratios higher or lower than those predicted are often found. We want here to consider particularly those cases where a flux ratio higher than A_I/A_{II} is found, that is,

$$(J_{I\to II}/J_{II\to I}) > (A_I/A_{II})$$

so that at the steady state, when the opposing fluxes across the membrane are equal, an enduring electrochemical gradient will be found across the membrane. There is an uphill transport of the solute from side II to side I. Such an electrochemical gradient is a source of energy—it can be used to drive a chemi-osmotic engine—so that if the second law of thermodynamics is not to be violated, the energy inherent in the electrochemical gradient must arise from the consumption of energy provided elsewhere in the cell. How this active transport of the solute can come about is clearly demonstrated by Kedem's analysis (Kedem, 1961) on the basis of irreversible thermodynamics.

The point is that Eq. (2.31) neglects any interaction phenomena that may be occurring between various penetrating solutes or between the penetrating solute and certain membrane components. The full equation for the flux of a solute species i is, rather, as follows:

$$J_{i(I\to II)} = R_{ii}A_{i,I} + R_{ij}J_{j(I\to II)} + R_{ir}J_{rI} \qquad (2.33)$$

where the phenomenological coefficient R_{ii} is the straight coefficient determining the permeability of i, while R_{ij} is a cross-coefficient describing the coupling of the flow of species i to the flow $J_{j(I\to II)}$ of some other species j, while R_{ir} determines the coupling between the flow of i and the rate J_r of some chemical reaction r in the membrane. It is clear that since a similar equation can be written for the oppositely directed flux $J_{i(II\to I)}$, the flux ratio will only be equal to the electrochemical ratio $A_{i,I}/A_{i,II}$ if the coefficients R_{ij} and R_{ir} are zero. If the flow of i is coupled in some way with the flow of a species j or with the occurrence of some chemical reaction r, an active transport of the solute i from side II to side I will be found. The energy for this active transport can be provided in two distinct ways as Eq. (2.33) makes clear: either by harnessing the free energy present in an electrochemical gradient of some second species j or by a direct coupling to an energy-yielding reaction r. We shall see in Chapters 5 and 6, respectively, that both these methods of providing free energy are indeed used in biological active transport systems.

If the measured flux ratio is lower than the ratio of electrochemical potentials, that is, $(J_{I\to II}/J_{II\to I}) < (A_I/A_{II})$ for A_I greater than A_{II}, the solute moves more slowly than is predicted by Fick's law. The rate of transit to the equilibrium (where $A_{\mathrm{I}} = A_{\mathrm{II}}$) will be delayed but this equilibrium situation will be reached eventually. The finding that

$$\frac{J_{\mathrm{I}\rightarrow\mathrm{II}}}{J_{\mathrm{II}\rightarrow\mathrm{I}}} < \frac{A_{\mathrm{I}}}{A_{\mathrm{II}}} \qquad (A_{\mathrm{I}} > A_{\mathrm{II}}) \tag{2.34}$$

is evidence for the operation of some saturable system operative in transport, that is, one of the facilitated diffusion systems, as shall be shown in Chapter 4.

CHAPTER 3

The Molecular Basis of Diffusion across Cell Membranes

3.1 Introduction

In the subsequent chapters of this book, we shall consider the properties and mode of action of the many specialized systems that move metabolites across cell membranes. Before doing so, however, we must clearly understand the manner in which the cell membrane itself acts as a permeability barrier. In this chapter, therefore, we shall consider the model for the cell membrane that we were led to by our analysis of Chapter 1—the membrane as the lipid-protein ensemble of Fig. 1.8—and explore the probable consequences of this model for diffusion. We shall consider first and at some length the problem of diffusion within liquids and then apply this knowledge to diffusion across the membrane, following closely the arguments of Zwolinski *et al.* (1949). We shall consider some of the quantitative data on cell permeabilities to test whether or not our view of Fig. 1.8 as a model for the cell membrane is correct. We shall then take another view of cell permeability—that this takes place through water-filled channels in the cell membrane—and test this view by the quantitative data. Finally, we shall attempt to synthesize these two major current hypotheses of membrane diffusion, to establish a secure basis for the arguments of the succeeding chapters.

The historical background of the development of the rival theories of membrane permeability and a comprehensive account of these theories is given in the excellent (but rather neglected) review of Wartiovaara and Collander (1960), who have themselves contributed many of the data which the theories set out to explain. We shall consider here only two of the numerous viewpoints that Collander and Wartiovaara list but shall do so in a fairly detailed fashion and, as far as possible, on the basis of a detailed molecular model of diffusion and of the cell membrane.

3.2 Diffusion in Liquids as Movement within a Lattice

Less is known about the physics of liquids than about the gaseous or solid states, yet the brilliant work of Eyring and his school, and of Frenkel, has led to a fairly adequate theory of diffusion in liquids (Glasstone *et al.*, 1941; Frenkel, 1946; Jost, 1960). This theory views the molecules in a liquid to be, to a first approximation, in a quasi-crystalline array and applies the results and viewpoins of both solid state theory and transition state theory to the analysis of liquids. The ordered crystal lattice of a solid is considered to be only partially disrupted on melting, so that local regions of more or less strict crystallinity remain. It seems, then, that in liquids as well as in solids, a relative movement between molecules—such as occurs in diffusion or in viscous flow—can occur only when a molecule of the crystalline or quasi-crystalline lattice possesses sufficient thermal energy to shake free of the bonds anchoring it in its position in the lattice and, in addition, finds an adjacent hole in the lattice into which it is now free to move. Diffusion is thus a process in which a molecule moves from hole to hole in the crystal lattice (Fig. 3.1). On the basis of this lattice model,

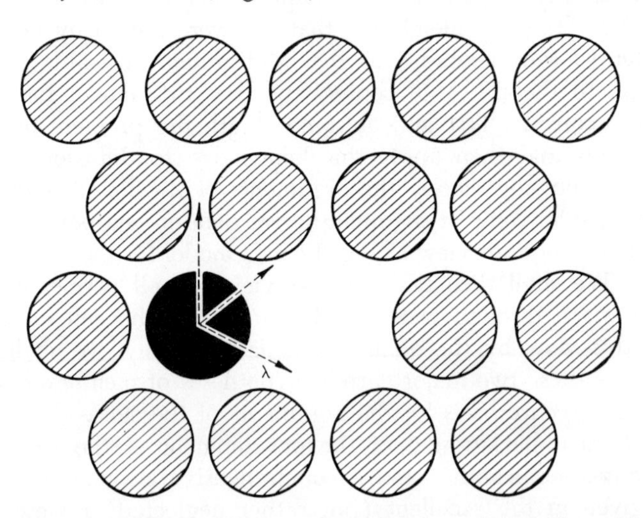

Fig. 3.1. Lattice model for diffusion. The molecule in heavy fill is able to move into the adjacent vacant position in the lattice. Three orthogonal lattice parameters λ_1, λ_2, λ_3 are defined by the center-to-center distance of the molecules forming the lattice. These parameters are, of course, statistical averages of rapidly fluctuating variable distances.

Glasstone *et al.* (1941) show that the self-diffusion coefficient D (which describes the diffusion of an isotopically labeled molecule in a pure liquid) is given by

$$D = \frac{\lambda_1 kT}{\lambda_2 \lambda_3 \eta} \quad (3.1)$$

where λ_1, λ_2, and λ_3 are the coordinates of the crystal lattice of Fig. 3.1, η is the viscosity of the liquid, while k is Boltzmann's constant, and T is the absolute temperature. This is of the same form as the Stokes-Einstein relation

$$D = \frac{kT}{6\pi r \eta} \quad (3.2)$$

which applies to the diffusion of large particles of radius r in a liquid medium. The diffusion coefficient D is here inversely proportional to the radius of the diffusing particle and hence to the *cube root* of the molecular weight, as in the following equation:

$$DM^{1/3} = \text{constant} \quad (3.3)$$

Analysis of the lattice model of Fig. 3.1 shows that while Eq. (3.3) may be expected to hold if the radius of the molecule is large compared with the lattice constant λ for an iso-dimensional lattice, at comparable values of λ and r, the relation (3.4) should apply:

$$DM^{1/2} = \text{constant} \quad (3.4)$$

Figure 3.2 shows a set of experimental data for diffusion of molecules of varying molecular weights in water. Clearly, up to a molecular weight of some 100, the relation of Eq. (3.4) is followed; thereafter, Eq. (3.3) holds. This transition is accounted for on the lattice model as follows (Stein, 1962a):

When the radius r of the diffusing molecule is considerably greater than λ, the solute will diffuse largely by the diffusion of the small solvent molecules in the opposite direction. From Fig. 3.3 it will be apparent that the rate of movement of such a large particle is inversely proportional to its surface area or to r^2. Since, by random walk theory (Einstein, 1905), the diffusion coefficient is proportional to the square root of the rate of flow, D is inversely proportional to the radius r, if r is great compared with λ. However, for a molecule of lesser dimension, comparable with the lattice spacings (Fig. 3.4), the assumption that the number of empty "holes" necessary for movement is related to the surface area of the molecule no longer holds. For example, if molecule A in Fig. 3.4 is to move one space to the right, two lattice spaces must be (serially)

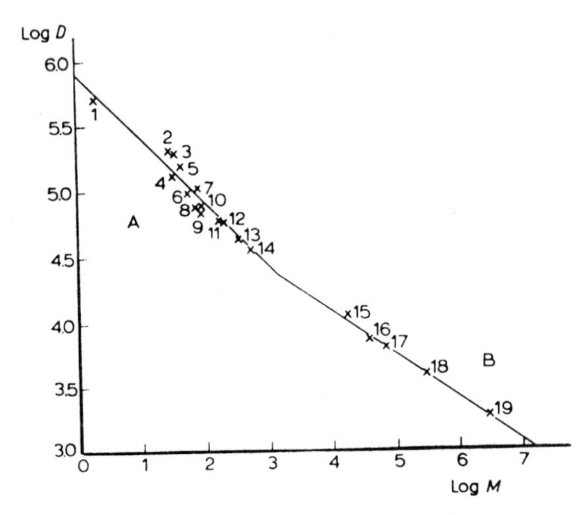

Fig. 3.2. The diffusion coefficient D as a function of molecular weight M. Diffusing molecules as follows: 1, hydrogen; 2, nitrogen; 3, oxygen; 4, methanol; 5, carbon dioxide; 6, acetamide; 7, urea; 8, *n*-butanol; 9, *n*-amyl alcohol; 10, glycerol; 11, chloral hydrate; 12, glucose; 13, lactose; 14, raffinose; 15, myoglobin; 16, lactoglobulin; 17, hemoglobin; 18, edestin; 19, erythrocruorin. All at or near 20°C. A is the relation $DM^{1/2}$ constant, B the relation $DM^{1/3}$ constant. (Taken from Stein, 1962a.)

available. The rate of movement of A is thus approximately one-half that of a molecule the size of the lattice hole. Now, the molecular weight of A is twice that of a molecule the size of a lattice hole, thus the velocity of movement of A is inversely proportional to the molecular weight and its rate of diffusion therefore inversely proportional to the square root of the molecular weight as in Eq. (3.4). Similar arguments apply to mole-

Fig. 3.3. Diffusion of a large molecule on the lattice model. The molecule can move one lattice space to the left since an entire half-shell of solvent molecules has diffused away from the left-hand side. (Taken from Stein, 1962a.)

cules B and C of Fig. 3.4 (Stein, 1962a). For r larger than 2.5λ, the inverse one-third power law becomes the better approximation, the inverse one-half power holding below these values of r, so that for diffusion in water, a molecular weight of some 250 appears on these rough theoretical grounds to be the transition point consistent with Fig. 3.2.

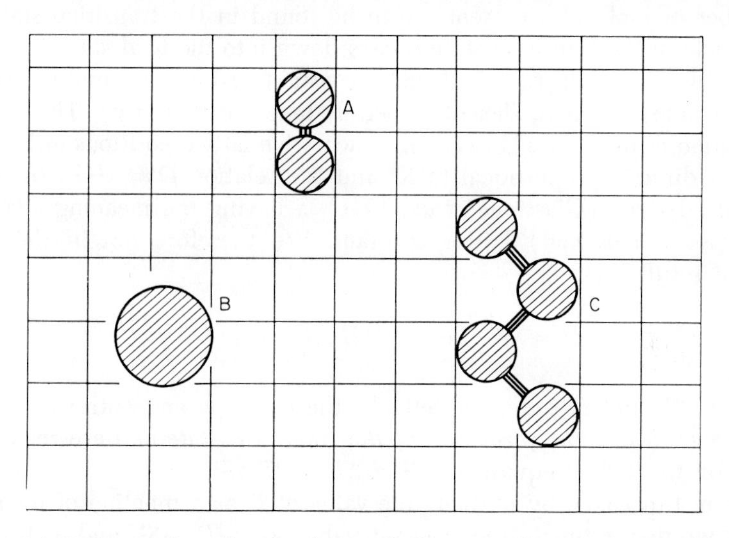

Fig. 3.4. Diffusion of small molecules on the lattice model. The lattice is depicted as a checkerboard; A, B, and C are molecules whose dimensions are such that for diffusion they require more vacant lattice spaces than would be expected from their volume alone.

For one particular molecule diffusing in different liquids we therefore expect by Eq. (3.1) that the product $D\eta$ will be constant for these liquids, while for different molecules diffusing in a single liquid the products $DM^{1/2}$ (for low M) or $DM^{1/3}$ (for high M) will be constant.

A. The Temperature Coefficient of Diffusion in Liquids

To account for the effect of increasing temperature on the rate of diffusion we use the transition state theory for rate processes (Glasstone *et al.*, 1941), the rate process considered here being the translation of the diffusing molecule from one position in the lattice to another. Between the initial and final states of such a system is a transition state, characterized by a higher free energy. Following Eyring, the transformation of the initial state into the transition state is considered as being a typical equilibrium process characterized by an equilibrium constant $K^{\ddagger}$. The free energy $\Delta F^{\ddagger}$ characterizing this transformation is thus related to $K^{\ddagger}$ by the well-known relation

$$\Delta F^{\ddagger} = -RT \ln K^{\ddagger} \tag{3.5}$$

Being an equilibrium constant, $K^{\ddagger}$ determines what fraction of the total number of molecules present are to be found in the transition state. A molecule in the transition state breaks down into the final state at a rate given by $\mathsf{k}T/h$ (where h is Planck's constant and k is a number known as the transmission coefficient, generally taken to be unity). The rate of transition of molecules between the successive lattice positions in Fig. 3.1 is thus directly proportional to $K^{\ddagger}$ and the relation $D = \lambda^2(\mathsf{k}T/h)K^{\ddagger}$ is readily derived (Glasstone *et al.*, 1941), λ having the meaning again of the (iso-dimensional) lattice constant. We therefore obtain the very useful result that

$$D = \lambda^2 \frac{\mathsf{k}T}{h} \exp(-\Delta H^{\ddagger}/RT) \exp(\Delta S^{\ddagger}/R) \tag{3.6}$$

where $\Delta H^{\ddagger}$ and $\Delta S^{\ddagger}$ are, respectively, the enthalpy and entropy changes per mole *during the formation of the transition state* and are related to $\Delta F^{\ddagger}$ by the Gibbs equation $\Delta F^{\ddagger} = \Delta H^{\ddagger} - T\,\Delta S^{\ddagger}$.

From Eq. (3.6), by studying the value of D as a function of temperature, we may estimate the relevant values of $\Delta H^{\ddagger}$, $\Delta S^{\ddagger}$, and λ. Clearly, $\Delta H^{\ddagger}$ determines the exponential term in the classical Arrhenius expression for the relation between diffusion and temperature, viz.,

$$D = D_0 \quad e^{-A/RT} \tag{3.7}$$

where A is the activation energy. (Actually, A found experimentally differs by a small factor from the true value of $\Delta H^{\ddagger}$ defined as "the enthalpy change for the formation of the transition state at the absolute zero of temperature," the following expression being valid (Glasstone *et al.*, 1941):

$$A = RT + \Delta H^{\ddagger} \tag{3.8}$$

It is clear from Eq. (3.6) that the *rate* of diffusion (as opposed to its temperature coefficient) is determined by the *free energy* change $\Delta F^{\ddagger}$ for the formation of the transition state and not merely by the *heat of activation* (A or $\Delta H^{\ddagger}$) for this change. The values of $\Delta H^{\ddagger}$ and $\Delta F^{\ddagger}$ can be very different from one another depending on the contribution of any *entropy* change $\Delta S^{\ddagger}$ to the stability of the transition state. We will need to bear in mind this distinction in our discussion of the transition state for entering the cell membrane. Simple substitution yields the result that the temperature coefficient Q_{10} at 27°C is given by $Q_{10} \simeq e^{0.055A}$, where A is the activation energy in kilocalories per mole.

B. Diffusion and Viscosity in Associated Liquids

If values of A (that is, $RT + \Delta H^{\ddagger}$) are computed in this way for a variety of liquids, both from diffusion and from viscosity data, a number of noteworthy points emerge (see Table 3.1). The polar liquids have high values for A, values for the nonpolar liquids being in general far lower. The commonly accepted interpretation of these facts (Butler, 1951) is that the molecules in the polar liquids are bound to one another more

TABLE 3.1

Values of the Activation Energy A Determining the Temperature Dependence of Viscosity (or of Diffusion)[a]

Gas or Liquid	Activation energy (kcal mole^{-1})	Reference[b]	Liquid	Activation energy (kcal mole^{-1})	Reference[b]
Oxygen	0.40	(1)	Water:		
Carbon monoxide	0.47	(1)	0°C	10.2	(1)
			50°C	9.6	(1)
Nitrogen	0.45	(1)	100°C	8.98	(1)
Argon	0.52	(1)	150°C	8.28	(1)
Methane	0.72	(1)	Acetic acid	2.6	(2)
Ethylene	0.79	(1)	Aniline	5.3	(2)
			Methanol	2.8	(2)
Carbon disulfide	1.28	(1)	Ethanol	3.4	(2)
Ether	1.61	(1)	Formic acid	3.4	(2)
Chloroform	1.76	(1)	Glycerol	15	(2)
Acetone	1.66	(1)	Mannitol (diffusion) in water:		
Benzene	2.54	(1)			
Cyclohexane	2.89	(1)			
Pentane	1.58	(1)	0–10°C	6.6	(3)
Heptane	1.91	(1)	10–20°C	5.5	(3)
Decane	2.60	(1)	20–30°C	4.46	(3)
Hexadecane	4.01	(1)	Copper sulfate in dilute sulfuric acid (diffusion)	5.14	(3)

[a] Nonassociated fluids given in left-hand columns and associated fluids in right-hand columns.

[b] *References:* (1) Ewell and Eyring (1937). (2) Author's calculations on data collated by Partington (1951, p. 109). (3) Taylor (1938).

firmly than are the molecules in the nonpolar liquids. The polar liquids are *associated*. The energy of activation for the polar liquids is that to be expected for the breaking of single *hydrogen* bonds (4–5 kcal/mole) (Pauling, 1948). It can be shown (Glasstone *et al.*, 1941) that the major contribution to $\Delta H^{\ddagger}$ for nonassociated liquids is required for *hole making* rather than to enable the molecules to move into prepared holes, while the reverse situation is true for the associated liquids. For the polyhydric alcohols the value of A, while high, is not much higher than for water, suggesting that only single hydrogen bonds have to be broken when a molecule of, for example, mannitol diffuses; that is, it creeps in the manner of a caterpillar from hole to hole rather than diffusing as a perfectly free molecule.

This view—that the molecules in the polar liquids are associated with one another—is supported also by the absolute values of the viscosities and by the data on the boiling points of the polar liquids, both sets of data being in general anomalously high for molecules of such molecular weight. Each molecule in a polar liquid is embedded, therefore, in the quasi-crystalline lattice and anchored there by effective hydrogen bonds. These factors dominate the properties of such liquids, and it is from this viewpoint then that we must consider the molecular basis of transfer across the cell membrane.

3.3 A Lattice Model for Diffusion into and across the Cell Membrane

Let us first consider, on the basis of the above discussion, how far the model of the cell membrane as a bimolecular lipid leaflet (Fig. 1.8) enables us to systematize the available data on membrane permeability. Accepting the view put forward in Section 3.2 as to the probable state of the solute and solvent molecules in a polar liquid, we have now to face the problem posed by the presence of the phase discontinuity at the cell membrane-water interface (and at the water-membrane interface for the emerging molecule). We cannot tell from studies on the isolated systems whether, for instance, a molecule such as glycerol, which can move step by step breaking only single hydrogen bonds in water, can so creep across the phase boundary at the lipid-water interface, or whether it must break simultaneously all three hydrogen bonds that its hydroxyl groups make with the surrounding water molecules. We cannot tell—even accepting that the structure of the membrane is given by some such model as Fig. 1.8—whether diffusion through such a comparatively strictly ordered system will obey laws similar to those operating in less ordered liquids. But our previous analysis poses for us these problems and shows us how to go about answering them. Thus, taking a given cell type, we can measure Q_{10}, that is, A or $RT + \Delta H^{\ddagger}$, for the permeability of a number of different substances entering such cells, and can then ask whether or not $\Delta H^{\ddagger}$ increases with the number of hydrogen bond accepting or donating sites in the molecules studied. We can also study the variation of the permeability constant P as the structure of the penetrating substance is varied and hence study, using Eq. (3.6), the dependence on structure of the term $\Delta F^{\ddagger}$ for the formation of the transition state. The variation of the structural contribution to $\Delta F^{\ddagger}$ with temperature will give us both $\Delta H^{\ddagger}$ and $\Delta S^{\ddagger}$ for the contribution of this structural modification to the transition state, and the magnitude of these parameters can then be compared with the number of hydrogen bonds that the permeant can make with water.

Finally, we may hope to obtain the diffusion constant D for the penetration within the membrane when the processes at the phase discontinuity have been accounted for. This value of D (and its variations with temperature) may tell us something of the nature and structure of the diffusion barrier.

Analysis of the Data on Simple Diffusion into Plant and Animal Cells

Probably the most comprehensive set of accurate measurements assessing the variation of the permeability constant with the chemical structure of the penetrating species are the often-quoted results of Collander and Barlund (1933) on the plant cell *Chara ceratophylla*. These and many other earlier permeability determinations are collected in Volume 19 of *Tabulae Biologicae* (Handovsky, 1941). Figure 3.5 is a plot of these data reproduced from Collander (1949) where $PM^{1/2}$ (P is the permeability constant in centimeter hour^{-1} and M the molecular weight of the penetrating substance) is plotted on a logarithmic scale against the partition coefficient for the distribution of the permeant between olive oil and water. The relatively good correlation so observed has led Collander to view penetration as being governed by the lipid solubility of the penetrating substance, suggesting the lipidic nature of the membrane. Since it would appear (Fig. 3.5), however, that small molecules (methanol, water, formamide, and ethylene glycol) penetrate faster than would be predicted from their oil-water coefficients, the membrane—in Collander's view—must in addition possess a sievelike character.

We shall see that it is possible with the same data, but interpreting the results in terms of the lattice model of Section 3.2, to understand at a molecular level the basis both for the correlation of $PM^{1/2}$ and the partition coefficient and also the sieving effect on small molecules, and

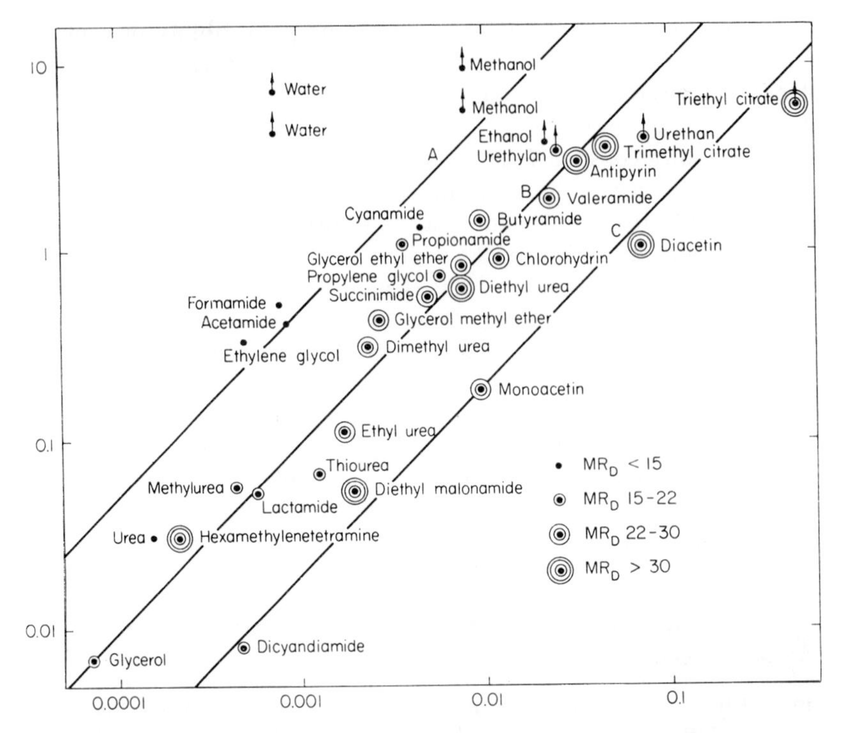

Fig. 3.5. The permeability of cells of *C. ceratophylla* to organic nonelectrolytes of different oil solubility and different molecular size. Ordinate: $PM^{1/2}$ (P in cm hr^{-1}); abscissa: olive oil-water partition coefficients. MR_D is the molar refraction of the molecules depicted, a parameter proportional to the molecular volume. (Taken with kind permission from Collander, 1949.)

to obtain various quantitative parameters relating to the energetics and mechanics of membrane diffusion.

It is clearly correct to study the variation with structure of $PM^{1/2}$ rather than P since, whether the barrier to diffusion is an aqueous channel or is within the lipid of the membrane, the same law—$DM^{1/2}$ is constant—will be obeyed for molecules of relatively low molecular weight (cf. Fig. 3.2). Perhaps, however, for very large molecules the molecular weight should be present as the cube root rather than as the square root.

The argument at the beginning of this section suggested that if we want to determine whether all anchoring hydrogen bonds need be broken during the passage of a permeant across the phase barrier between water and lipid we must plot some function of $PM^{1/2}$ against the number (N) of hydrogen bonds that the permeant makes in water. Then, if all anchoring bonds need be broken, a (negative) correlation between permeability and this number N will be obtained. Alternatively, if, as for diffusion within the aqueous phase, only a single hydrogen bond needs to be broken, $PM^{1/2}$ will be independent of N. What function of $PM^{1/2}$ needs to be plotted against N can be seen by the following argument: Each hydrogen bond that must be broken must raise the free energy of the transition state by Q cal/mole. We have then for N hydrogen bonds, a total increment NQ cal/mole in the free energy. This will decrease the equilibrium constant $K^{\ddagger}$ according to $K^{\ddagger} = \exp[(-\Delta F^{\ddagger}/RT)(-NQ/RT)]$. If $K^{\ddagger}$ is decreased, this again decreases the diffusion constant D in strict proportion according to Eqs. (3.5) and (3.6). Thus we have

$$D = \frac{kT}{h} \exp[(-\Delta F^{\ddagger}/RT)(-NQ/RT)] \tag{3.9}$$

or

$$\ln D = \ln \frac{kT}{h} - \frac{\Delta F^{\ddagger}}{RT} - \frac{NQ}{RT} \tag{3.10}$$

Thus if we plot the logarithm of our diffusion constant (here as $PM^{1/2}$) against N, the number of hydrogen bonds broken, a straight line should be obtained with slope determined by Q, the strength of each hydrogen bond.

Assigning a correct number N to each of the permeants of Fig. 3.5 is a difficult task. As a preliminary step, the assignments in Table 3.2 have been made. These assignments require further comment. They assume that an —OH group can be both a donor—through the hydrogen atom—and an acceptor—at the oxygen atom—of hydrogen bonds. For instance, a water molecule in ice is held in the crystal lattice by four hydrogen bonds (Pauling, 1948). We assume, therefore, that a molecule such as methanol does not form only a single hydrogen bond with water but that indeed two such bonds (an acceptor and a donor) are formed. A number of other difficult points remain. If water makes four hydrogen bonds with its neighbors, the acceptor oxygen providing sites for two of these bonds, then perhaps in alcohols the oxygen of the hydroxyl function might be expected similarly to be bivalent. We have taken the view here that the bulk of the methyl group will, for steric or other reasons, forbid the formation of a hydrogen bridge on the methyl side of the molecule. Similarly, the oxygen atom in the carbonyl function of the amides listed is considered as being able to accept only a single hydrogen bond. (We return to this point later in this section.) The ether oxygen is considered to form no hydrogen bonds, while in esters the relatively nonpolarized carbonyl function is given a weight of one-half of a hydrogen bond.

Figure 3.6 is a plot of log $PM^{1/2}$ (from the data of Collander and Barlund, 1933) against N, the number of hydrogen bonds assigned in the above fashion. There is clearly a good negative correlation between log $PM^{1/2}$ and N, a marked exception being the value for water which is anomalously high. (Note that the value used here by Collander for the permeability of water is only an estimated value, the true rate being probably even higher and, indeed, too fast to measure accurately.) At

TABLE 3.2
ASSIGNMENT OF N VALUES TO HYDROGEN-BONDING FUNCTIONS

Function	Group in which present	Assigned value of N
—OH	Alcohols, sugars, glycols	2
	Carboxylic acids	2
H—O—H	(Water)	4
$—NH_2$	Primary amines	2
	Primary amides	2
—N(R)H	Secondary amines	1
	Secondary amides	1
—CO—	Carboxylic acids	1
	Amides	1
	Aldehydes	1
	Esters	½
—O—	Ethers	0
—C≡N	Nitriles	1
	Dicyandiamide	1

any value of N, the range of values of $PM^{1/2}$ embraces not more than 1.5 log units or a range of 30-fold, a reasonably good fit for this preliminary analysis. The demonstration of this inverse proportionality between $PM^{1/2}$ and N suggests that all (or a constant fraction) of the "anchoring hydrogen bonds" have to be broken for a permeating molecule to move into the transition state during penetration across the membrane. The slope of the curve of log $PM^{1/2}$ against N gives the free energy increment for the formation of the transition state for every hydrogen bond that is to be broken when this state is formed. This value—from the best straight line of Fig. 3.6—is

$$\Delta F^{\ddagger} = 0.95 \text{ kcal/mole}$$

and corresponds to a fivefold decrease in $PM^{1/2}$ for each hydrogen bond that needs to be broken. The scatter of points is such that this value cannot be considered as reliable. We do not yet at this stage of the discussion know the more readily interpretable $\Delta H^{\ddagger}$ and $\Delta S^{\ddagger}$ values.

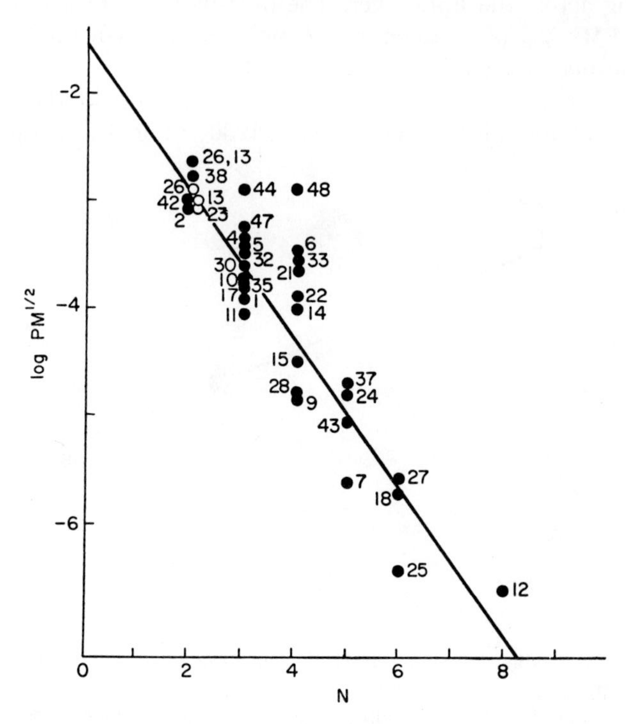

Fig. 3.6. The permeability data of Fig. 3.5 (●) replotted to show the variation of log $PM^{1/2}$ (P in cm sec^{-1}) on the ordinate with the number N of hydrogen-bond-forming groups in the permeant on the abscissa, assigned according to Table 3.2. ○, Data of Dainty and Ginzburg (1964c) on the related alga *C. australis*. The straight line is drawn by least squares in this and subsequent figures of similar nature. Permeants have been assigned the following numbers here and in subsequent figures: 1, acetamide; 2, antipyrin; 3, butanol; 4, butyramide; 5, cyanamide; 6, diacetin; 7, dicyandiamide; 8, diethylene glycol; 9, diethyl malonamide; 10, diethyl urea; 11, dimethyl urea; 12, erythritol; 13, ethanol; 14, ethylene glycol; 15, ethyl urea; 16, ethyl urethane; 17, formamide; 18, glycerol; 19, glycerol + CO_2; 20, glycerol + Cu^{2+}; 21, glycerol ethyl ether; 22, glycerol methyl ether; 23, isopropanol; 24, lactamide; 25, malonamide; 26, methanol; 27, methylol urea; 28, methyl urea; 29, monoacetin; 30, monochlorohydrin; 31, propanol; 32, propionamide; 33, (α,β)-propylene glycol; 34, (α,γ)-propylene glycol; 35, succinimide; 36, tetraethylene glycol; 37, thiourea; 38, triethyl citrate; 39, triethylene glycol; 40, trihydroxybutane; 41, trihydroxybutane + Cu^{2+}; 42, trimethyl citrate; 43, urea; 44, urethane; 45, urethylan; 46, urotropin; 47, valeramide; 48, water; 49, 2,3-butylene glycol.

The intercept with the vertical axis of the best straight line through the experimental points of Fig. 3.6 gives the value expected for log $PM^{1/2}$ when there are no hydrogen bonds anchoring the permeant in the aqueous phase, and is thus the maximum value of log $PM^{1/2}$ for *any* permeant penetrating across the lipid layer. The data of Fig. 3.6 give a value for this term $PM^{1/2}{}_{max}$ of 0.026 cm sec^{-1} mole$^{1/2}$. We shall further consider the significance of such data in Section 3.4.

The data on penetration into the unfertilized eggs of the sea urchin *Arbacia punctulata* (Jacobs and Stewart, 1936) (Fig. 3.7), on penetration

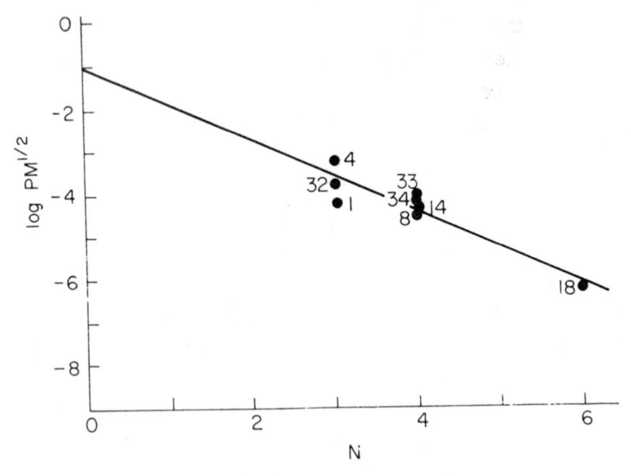

Fig. 3.7. Penetration of nonelectrolytes into *Arbacia* eggs as a function of the number of hydrogen-bonding groups. 22°C. Ordinate: log $PM^{1/2}$ (P in cm sec^{-1}); abscissa: number of hydrogen-bonding groups in permeant, assigned as in Table 3.2. Numbering of permeants as in Fig. 3.6. (Data from Jacobs and Stewart, 1936.)

into the internodal cells of the alga *Tollypellopsis stelligera* (Wartiovaara, 1942) (Fig. 3.9b), on penetration into bovine erythrocytes, recalculated from Jacobs *et al.* (1935) (Fig. 3.8), and on penetration into the "erythrocytes" of *Phascolosoma* (W. E. Love, 1953) (Fig. 3.9a) are presented below. In all these cases it is clear that a strict inverse proportionality between $PM^{1/2}$ and N applies, with only two exceptions: (1) water generally, and (2) urea for the erythrocytes, and we must assume that to account for these exceptions a more complex model of the membrane will be required.

Table 3.3 collects the relevant data for the increase in free energy $\Delta F^{\ddagger}$ in the transition state per potential hydrogen bond and the limiting value $PM^{1/2}{}_{max}$ where no hydrogen bonds are formed. From the work of Jacobs *et al.* (1935) we can set up a plot such as that of Fig. 3.9 for data

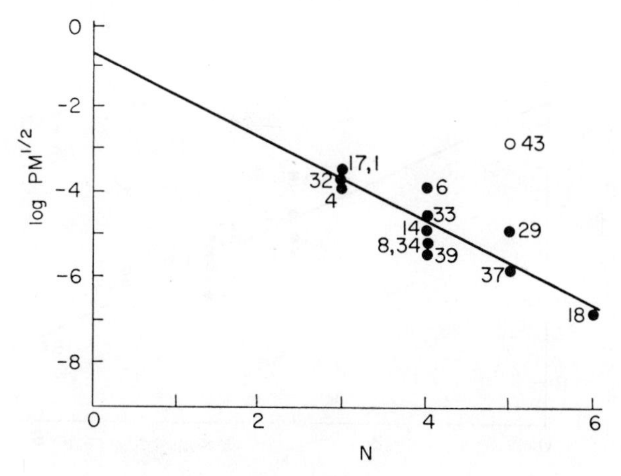

Fig. 3.8. As Fig. 3.7 but penetration into bovine erythrocytes at 20°C. (Data recalculated from Jacobs *et al.*, 1935.)

at four different temperatures 0°, 10°, 20°, and 30°C, and hence obtain $\Delta F^{\ddagger}$ and $PM^{1/2}{}_{\max}$ at these temperatures. Table 3.3 includes the parameters so found. From the values of $\Delta F^{\ddagger}/T$ per potential hydrogen bond, plotted against the reciprocal of the absolute temperature T, $\Delta H^{\ddagger}$ and

TABLE 3.3

PARAMETERS FOR THE TRANSITION STATE IN MEMBRANE DIFFUSION[a]

Type of cell	Temp. (°C)	Reduction in P for each pair of H bonds	$\Delta F^{\ddagger}$ (kcal mole^{-1})	$PM^{1/2}{}_{\max}$ (cm sec^{-1} mole$^{1/2}$)	$P_{\max}$ for mol. wt. of 100 (cm sec^{-1})
Chara	20	24	1.9	0.026	2.6×10^{-3}
Tollypellopsis	20	90	2.7	0.49	4.9×10^{-2}
Arbacia	20	48	2.3	0.098	9.8×10^{-3}
Bovine erythrocyte	20	100	2.8	0.25	2.5×10^{-2}
Human erythrocyte	25	38	2.2	0.080	8.0×10^{-3}
Phascolosoma	25	73	2.6	0.040	4.0×10^{-3}
Bovine erythrocyte	0	130	2.6	0.020	2.0×10^{-3}
(Glycerol, gly-	10	73	2.4	0.028	2.8×10^{-3}
cols, thio-	20	60	2.4	0.078	7.8×10^{-3}
urea only)	30	46	2.3	0.10	1.0×10^{-2}

[a] Computed from Figs. 3.6 to 3.9.

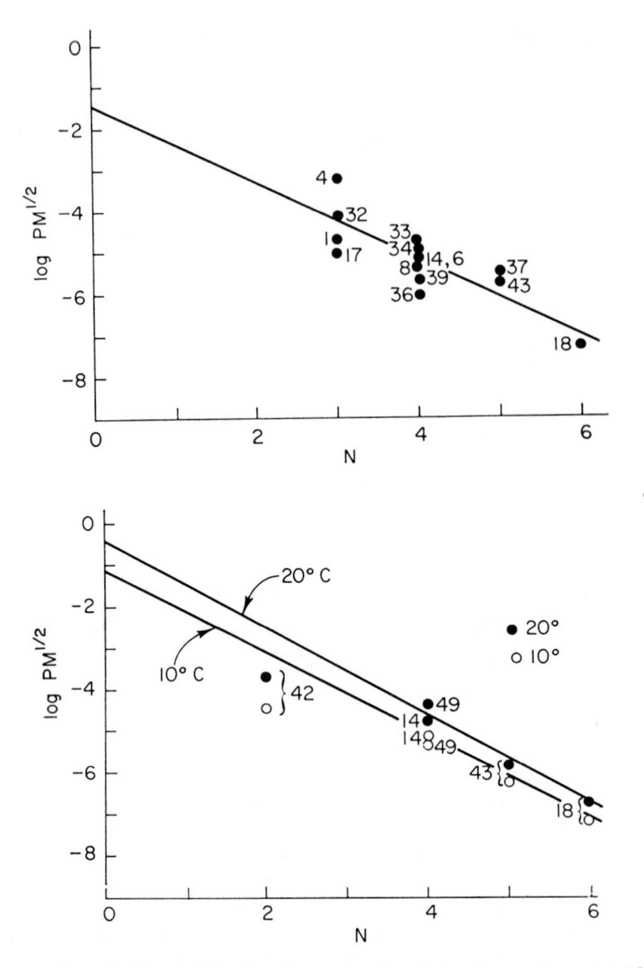

Fig. 3.9. As Fig. 3.7 but: (a) (top) penetration into the erythrocyte-like cells of *Phascolosoma* at 25°C. (Data from W. E. Love, 1953.) (b) (bottom) penetration into cells of the alga *Tollypellopsis* at both 10° and 20°C. (Data from Wartiovaara, 1942.)

hence $\Delta S^{\ddagger}$ can be found. The derived values at the absolute zero of temperature are as follows:

$$\Delta S^{\ddagger} \text{ per H bond} = 11.0 \text{ eu/mole}$$

$$\Delta H^{\ddagger} \text{ per H bond} = 5.6 \text{ kcal/mole}$$

Further comment on the variation of the parameter $PM^{1/2}{}_{max}$ with temperature will be reserved until Section 3.4, when we shall also discuss the significance of the $\Delta S^{\ddagger}$ and $\Delta H^{\ddagger}$ terms.

TABLE 3.4

$\Delta H^\ddagger$ DERIVED FROM THE VALUE OF Q_{10}

Permeant	Q_{10}	A [Eq. (3.7)] (kcal mole^{-1})	$\Delta H^\ddagger$ [Eq. (3.8)] (kcal mole^{-1})	No. of H bonds (Table 3.2)	$\Delta H^\ddagger$ per H bond (kcal mole^{-1})
(a) Penetration into Bovine Erythrocytes[a]					
Glycerol	3.9	24.5	24	6	4
Ethylene glycol	2.8	19	18½	4	4½
Diethylene glycol	2.8	19	18½	4	4½
Triethylene glycol	3.2	21	20½	4	5
1,2-Propandiol	3.0	20	19½	4	5
1,3-Propandiol	2.9	19.5	19	4	5
Propanol	1.3	5	4½	2	2
Thiourea	2.2	14	13½	4	3½
Urea	1.4	6.5	6	5	1
(b) Penetration into Eggs of *Arbacia*[b]					
Ethylene glycol			23.6	4	6
Propionamide			21.6	3 (or 4?)	7 (or 5½?)
Butyramide			22.8	3 (or 4?)	7½ (or 5½?)
(c) Penetration into Ascites Tumor Cells[c]					
Urea			16.5	5	3½
Ethylene glycol			15.0	4	4
Diethylene glycol			15.6	4	4
Triethylene glycol			18.5	4	4½

[a] Data taken from Jacobs *et al.*, 1935.
[b] Data taken from Stewart and Jacobs, 1936.
[c] Data taken from Hempling, 1959.

Data on the variation of P with temperature (that is, Q_{10} data for the entry of various substances into ox erythrocytes) are also available from the study of Jacobs *et al.* (1935). These are recorded in Table 3.4 together with the computed values of A, the activation energy applying Eq. (3.7), and for $\Delta H^\ddagger$ applying the small correction suggested by Eq. (3.8). Using the values recorded in Table 3.2 for the number of potential hydrogen bonds, we can obtain once again the contribution to the value of $\Delta H^\ddagger$ for the formation of the transition state for each hydrogen bond broken—*if it is assumed that this requirement is the major determinant of the total* $\Delta H^\ddagger$. Apart from the value for urea, the values obtained for $\Delta H^\ddagger$ per hydrogen bond range from 2.3 to 5.1 kcal/mole, with most of the more accurate values lying between 4 and 5 kcal/mole in excellent agreement with the value from Table 3.3.

We can make one other useful interpretation of these data. If we consider a homologous series of substances with the same number and type of functional group (for example, the series of aliphatic amides or alcohols), we can investigate what contribution to $\Delta F^{\ddagger}$ of the transition state will be made by increments in the number of —CH_2— residues in the aliphatic side chains. Figure 3.10 presents the required plot of log $PM^{1/2}$ against the number of —CH_2— groups in the side chain for the penetration of a series of amides into *Chara* (Collander and Barlund, 1933), a series of alcohols at 0° and 20°C into *Nitella* (Wartiovaara, 1949), of

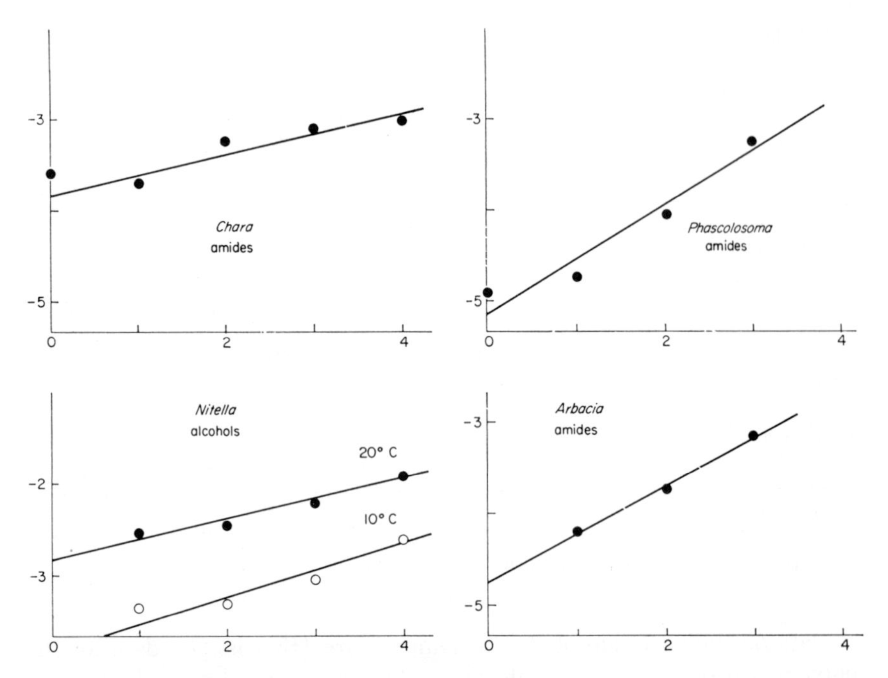

Fig. 3.10. Influence of length of hydrocarbon chain on the rate of penetration of some amides and alcohols. Abscissa: number –CH_2– residues in permeant. Ordinate: log $PM^{1/2}$ (P in cm sec^{-1}). [Data taken from Collander and Barlund (1933) for *Chara*, W. E. Love (1953) for *Phascolosoma*, Wartiovaara (1949) for *Nitella*, and Jacobs and Stewart (1936) for *Arbacia*.]

amides into *Arbacia* (Jacobs and Stewart, 1936), and the "erythrocytes" of *Phascolosoma* (W. E. Love, 1953). Relatively good straight lines are obtained for these plots, although in most cases it would appear that the straight line relation begins only with the second member of the series. Table 3.5 collects the values of the relevant parameter $F^{\ddagger}$ per —CH_2— residue, the data for propanol and butanol being used for the alcohol series. Since $\Delta F^{\ddagger}$ is negative, the presence of additional —CH_2— residues

TABLE 3.5

INCREMENT IN PERMEABILITY COEFFICIENT PER UNSHIELDED —CH_2— GROUP[a]

Cell	Substrate	Temp. (°C)	Increase in $PM^{1/2}$ per —CH_2—	$\Delta F^{\ddagger}$ per —CH_2— (cal. mole^{-1})
Chara	Amides	20	1.67	320
Arbacia	Amides	20	3.2	690
Nitella	Alcohols	20	1.8	360
Nitella	Alcohols	0	2.0	430
Phascolosoma	Amides	25	3.9	810

[a] From the slope of log $PM^{1/2}$ plotted against the number of —CH_2— groups in Fig 3.10.

stabilizes the transition state. The data for alcohols were obtained at two different temperatures, so that we can separate the contributions of enthalpy and entropy to the total free energy and obtain values of

$$\Delta H^{\ddagger} = -1.5 \text{ kcal/mole —CH}_2\text{— residue}$$
$$\Delta S^{\ddagger} = -3.7 \text{ eu/mole —CH}_2\text{— residue}$$

Generally, the effect on $PM^{1/2}$ of the introduction of each —CH_2— residue is approximately to double this term.

In a similar fashion we can compare values of $PM^{1/2}$ chosen for *pairs* of permeants, differing only in the introduction of a single —CH_2— residue. Some of these data are collected in Table 3.6. Here again the introduction of an extra —CH_2— residue approximately doubles the value of $PM^{1/2}$. (We note, however, that only a twofold increase is found in cases such as the pair triethyl citrate/trimethyl citrate and the pair diethyl urea/dimethyl urea where it would appear that three or two *groups* of additional —CH_2— residues were being added rather than a

TABLE 3.6

INCREASE IN PERMEABILITY COEFFICIENT PER UNSHIELDED —CH_2— GROUP, FROM A CONSIDERATION OF HOMOLOGOUS PAIRS OF PERMEANTS[a]

Homologous pair	Ratio of $PM^{1/2}$
Glycerol ethyl ether/glycerol methyl ether	2.0
Diethyl urea/dimethyl urea	2.1
Ethyl urea/methyl urea	2.3
1,2-Propylene glycol/ethylene glycol	2.5
Triethyl citrate/trimethyl citrate	1.7

[a] Data for the cell *Chara*, taken from Collander and Barlund (1933).

single residue.) Taking this rough average figure of a twofold increase in $PM^{1/2}$, this corresponds to an extra stability—a lowering of the potential energy—of the transition state by 700 cal/mole for each $—CH_2—$ residue added. It may be significant to compare this value with the figure of 800 cal/mole for the energy of adsorption per mole of $—CH_2—$ residue at the hydrocarbon-water interface (Haydon and Taylor, 1960), although such a comparison may be hazardous if the determining $\Delta H^{\ddagger}$ and $\Delta S^{\ddagger}$ values are unknown. One other interesting feature of these data is that to be effective the added $—CH_2—$ residue must be at the end of the molecular chain. Thus the values of $PM^{1/2}$ for ethylene glycol and for 1,3-dihydroxypropane penetrating both *Arbacia* and the erythrocyte are very close to one another and always some 2- to 3-fold lower than the value for 1,2-dihydroxypropane. The additional $—CH_2—$ residue in the 1,3-isomer is seemingly not able to contribute to the stability of the transition state. This is understandable if, when it is in the transition state, the penetrating molecule is partially inserted between the lipid chains of the membrane. Only those $—CH_2—$ residues which are not screened by hydroxyl groups can contribute to the stability of the transition state. Some such explanation may be valid also for the data on the diethyl versus dimethyl pairs of Table 3.6.

A simple plot such as that of Fig. 3.6 does not allow for the effect of structural modification in the permeant other than the introduction of hydrogen-bond-forming groups. We can amend this by correcting the true $PM^{1/2}$ values by a term to compensate for the now known effect of the introduction of $—CH_2—$ groups. Figure 3.11 is a plot of the data of Fig. 3.6 corrected for the effect of added $—CH_2—$ groups, the value of $PM^{1/2}$ used being the measured value divided by our empirical factor of 2 for each bare $—CH_2—$ residue. The inverse correlation between $PM^{1/2}$ and the number of hydrogen bonds is now very good. No point (the value for water being excepted) is more than 0.7 log units, that is, a factor of 5, from the straight line. (Note that all the values of $PM^{1/2}$ for the amides, urea, and the substituted ureas fall a little below the line for the alcohols and polyalcohols, suggesting that the assignment of 1 for the hydrogen-bonding potential of the carbonyl function is too low.)

For the data on *Chara*, therefore, values of

(1) $PM^{1/2}{}_{max} = 8.9 \times 10^{-3}$ cm sec^{-1} mole$^{1/2}$,
(2) $\Delta F^{\ddagger} = 1.6$ kcal/mole, for the reduction in stability of the transition state per pair of potential hydrogen bond forming groups, and
(3) $\Delta F^{\ddagger} = -700$ cal/mole, for the increase in stability of the transition state per bare $—CH_2—$ residue,

suffice to predict the value for practically any permeability coefficient (except water) within a factor of five, when the permeability constants themselves vary over a range of 3×10^4.

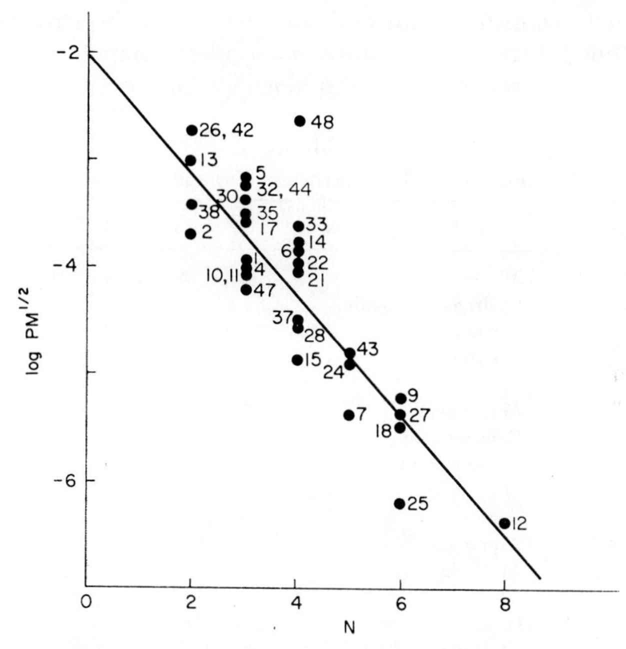

Fig. 3.11. Permeability data for *Chara* adjusted to take account of the enhancing effect of $-CH_2-$ groups. For each bare $-CH_2-$ group in the permeant, the permeability constant (data of Collander and Barlund, 1933) is divided by a factor of two. Scales and numbering as in Fig. 3.6.

3.4 Molecular Significance of the Parameters $\Delta F^‡$, $\Delta H^‡$, $\Delta S^‡$, and $PM^{1/2}{}_{max}$

The advantage of the treatment presented here over some earlier interpretations is that a molecular basis can be given to the numerical parameters determined from the permeability data. We can hope, in particular, to interpret the parameters $\Delta F^‡$, $\Delta H^‡$, $\Delta S^‡$, and $PM^{1/2}{}_{max}$ and the contributions to $\Delta H^‡$ and $\Delta S^‡$ brought about by various modifications in the structure of the permeant.

We have seen that the introduction of each pair of hydrogen bonds assigned as in Table 3.2 raises the free energy of the transition state complex by a value of $\Delta F^‡ = 2–2.7$ kcal/mole, and the data on the variation of this value of $\Delta F^‡$ with temperature lead us to values (for the ox erythrocyte) of $\Delta H^\ddagger = 5.6$ kcal/mole, $\Delta S^\ddagger = 11$ eu/mole per hydrogen bond. The activation energy determinations are also consistent with this value of some 5 kcal/mole for $\Delta H^\ddagger$ per hydrogen bond formed.

Now, this value of $\Delta H^\ddagger = 5$ kcal/mole per hydrogen bond broken is itself perfectly consistent with the value found for the strength of the hydrogen bond between two water molecules, computed by Pauling (1960) from data on the heat of sublimation of ice (Table 3.7). The

TABLE 3.7

THERMODYNAMIC PARAMETERS OF HYDROGEN BONDS[a]

Bond type	Molecular species	$-\Delta H$[c]	$-\Delta S$[c]
—O—H···O[b]	Water	5 (1), 5.2 (1), 5.0 (2)	15 (2)
	Hydrogen peroxide	5 (1)	
	Methanol	6.05 (1), 4.5 (2)	16 (2)
	Ethanol	5.0 (1)	10 (2)
	Formic acid	7.1 (1), 7.0 (2)	25 (2)
	Acetic acid	7.6 (1), 7.0 (2)	28 (2)
	Benzoic acid	4.2 (1), 4.3 (2)	10 (2)
	o-Toluic acid	4.7 (1)	16 (2)
—N—H···O	Amide	3.9 (2)	
—N—H···N	Ammonia	1.3 (1), 4.4 (2)	15 (2)
	Methylamine	3.4 (2)	11 (2)
—F—H···F	Hydrogen fluoride	6.7 (1), 6.8 (2)	14 (2)
—F—H···F	Hydrogen difluoride ion	58 (1)	
—C—H···N	Hydrogen cyanide	4.6 (1), 3.3 (2)	10 (2)

[a] All ΔH values are given in kilocalories per mole of H bond; ΔS is in eu per mole when the dissociation constant is in moles per liter.

[b] For O—H···O bonds (other than acids), Pimentel and McClennan (1960) find that the relation $-\Delta H = 0.26(-\Delta S) - 0.36$ is reasonably well obeyed.

[c] *References:* (1) Pauling (1960). (2) Pimentel and McClennan (1960).

hydrogen bond between pairs of alcohol molecules in solid methyl and ethyl alcohol is considered to be somewhat stronger ($\Delta H^\ddagger = 5$–6 kcal/mole, again on Pauling's computations). We should expect, therefore, that the enthalpy of formation of hydrogen bonds between water and hydroxyl groups should be between 4.5 and 6.1 kcal/mole, as we find for the permeability data. (The hydrogen bond between ammonia molecules in crystalline ammonia, the —N—H···N— bond, is, however, of lower enthalpy—some 1.3 kcal/mole, according to Pauling.)

The value for the entropy change can be considered as follows: Breaking a hydrogen bond between two molecules leads to the formation of two independent units in the system rather than the single unit present previously. There is now an increase in disorder in the system, and the entropy contribution resulting from this increased disorder can be computed. For the formation of two kinetic units from a single such unit the increase in entropy is some 8 eu/mole (Lumry, 1959). This would lead to a value of $\Delta S^{\ddagger} = 8$ eu/mole in the present situation in relatively good agreement with the experimental value of 11 eu/mole. This analysis may be an oversimplification, but the absolute value found for $\Delta S^{\ddagger}$ is consistent with experimental determinations for $\Delta S^{\ddagger}$ for the breaking of a hydrogen bond (Table 3.7). Thus both enthalpy and entropy changes are as predicted for the compulsory breaking of all hydrogen bonds during the formation of the transition state. Absolute reaction rate theory applied to our model for membrane transfer will allow us now to compute, for any permeant, both the absolute value of the permeability constant (given by the $\Delta F^{\ddagger}$ term) and its temperature coefficient (given by the $\Delta H^{\ddagger}$ term), provided we are able to compute or to estimate the $PM^{1/2}{}_{\max}$ term. We must now consider this latter problem.

The term $PM^{1/2}{}_{\max}$ gives the maximum permeability that the (simple) membrane will allow for any molecule, since this value is for a molecule free of anchoring hydrogen bonds. It would be of much interest to determine $PM^{1/2}{}_{\max}$ experimentally for the diffusion of, for example, hydrocarbons into the red blood cell but no such data are in fact available.

If it can be converted from a permeability coefficient into a diffusion coefficient, $PM^{1/2}{}_{\max}$ can be given a molecular significance. We can do this if we make the assumption—very broadly supported by the structural studies on cell membranes, but nevertheless still an assumption—that the resistance to diffusion occurs across a distance of some 50 Å, the width of the bimolecular lipid leaflet (Chapter 1). Then we can convert the measured concentration differences into concentration gradients and in this way derive values for $DM^{1/2}{}_{\max}$ ranging from 2.5×10^{-7} cm^2 sec^{-1} mole$^{1/2}$ for *Tollypellopsis* to 1.5×10^{-8} cm^2 sec^{-1} mole$^{1/2}$ for *Chara*. Thus, $DM^{1/2}{}_{\max}$ is of the order of 1×10^{-7} cm^2 sec^{-1} mole$^{1/2}$. For water or methanol $M^{1/2}$ is of the order of 5 so that here $D_{\max}$ will be of the order of 2×10^{-8} cm^2 sec^{-1}. We can now compare this computed diffusion coefficient with some examples of values found in other systems. These values are collated in Table 3.8, the data being taken from Glasstone *et al.* (1941) and Jost (1960) and chosen to show the range of values found. It is clear that $D_{\max}$ for the cell membrane is some 100- to 1000-fold less than would be expected for diffusion across an aqueous barrier or for diffusion across a barrier composed of small nonpolar molecules—but is of the order found for diffusion through natural and synthetic rubber membranes—highly polymerized, long-chain organic "phases."

If the general relation between the diffusion coefficient and the viscosity coefficient [Eqs. (3.1) and (3.2)] can be expected to hold within the membrane, the 100- to 1000-fold decrease in D can be expected to be paralleled by a 100- to 1000-fold increase in viscosity. The viscosity of water being 1 cP at 20°C, the components of the cell membrane might, on this argument, have a viscosity of some 100–1000 cP. This value is in the range expected for oils and similar viscous fluids. Thus at 20°C, representative viscosities are (in centipoises): water 1, methanol 0.6, amyl alcohol 3.5, phenol 10, cyclohexanol 70, olive oil 80, rape oil 160, castor oil 990, and glycerol 1500 (Hodgman, 1954).

TABLE 3.8

DIFFUSION COEFFICIENTS D FOR NONELECTROLYTES

Diffusant	Solvent	Temp. (°C)	$10^5 \times D$ ($cm^2\ sec^{-1}$)	References[a]
Bromoform	Ether	20	3.4	(1)
Bromoform	Benzene	20	1.7	(1)
Bromoform	Amyl alcohol	20	0.5	(1)
Methanol	Water	18	1.4	(1)
Glycerol	Water	18	0.8	(1)
Glucose	Water	18	0.6	(1)
Hydrogen	Neoprene (vulcanized)	17	0.1	(2)
Nitrogen	Neoprene (vulcanized)	27	0.02	(2)
Nitrogen	Butadiene-acrylonitrile interpolymer	17	0.007	(2)
Phenol	Solvent-free rubber	20	0.001	(2)
D_{max} for cell membrane (ox erythrocyte)		20	0.002	(3)

[a] *References:* (1) Glasstone *et al.* (1941). (2) Jost, 1960. (3) Present study.

We could, perhaps, reverse this argument and consider that if the barrier to diffusion through the cell membrane is indeed the bimolecular lipid layer then, since its viscosity—as an oil—may be some 100- to 1000-fold higher than that of water, diffusion through this layer should be at a rate 100- to 1000-fold less than through an equivalent layer of water; and thus we could compute a value for D_{max}. This is, in principle, the approach used by Danielli in computing the permeability coefficients for a lipid membrane (Davson and Danielli, 1952). In any event, the value found for D_{max} is consistent with that to be expected for a bimolecular lipid leaflet.

We can, however, obtain a little more information as to the probable structure of the diffusion barrier within the membrane. The data in Table 3.3 record values for $PM^{1/2}{}_{max}$ obtained at four different temperatures for the ox erythrocyte. If we assume a constant thickness for the diffusion barrier over this temperature range, we obtain the variation of D_{max} with temperature. From this we can find from Eq. (3.7) the value of A, the activation energy of diffusion within the membrane—a value of 10 kcal/mole being found—together with the term D_0 of Eq. (3.7). Equation (3.6) shows that D_0 gives a value for the product of the lattice distance λ within the membrane, with the term $[\exp(\Delta S^{\ddagger}/R)]^{1/2}$ since kT/h is known. These values, together with those found for some other systems studied (data from Glasstone *et al.*, 1941; Jost, 1960), are collected in Table 3.9. Apparently the value found for the activation energy is similar

TABLE 3.9

LATTICE PARAMETERS COMPUTED FROM THE DIFFUSION COEFFICIENT D ACCORDING TO EQS. (3.6) AND (3.7)[a,b]

Diffusion system	D_0(cm² sec⁻¹)	A(cal mole⁻¹)	$\lambda(e^{\Delta S^{\ddagger}/R})^{1/2}$
H_2O in H_2O	2×10^{-1}	5,300	11.0
Mannitol in H_2O	1.2×10^{-2}	4,450	2.8
Phenol in H_2O	3.4×10^{-3}	3,150	1.4
$C_2H_2Br_4$ in $C_2H_2Cl_4$	1.7×10^{-3}	3,365	1.0
Br_2 in CS_2	0.43×10^{-3}	1,540	0.4
H_2 in butadiene-acrylonitrile copolymer	56	8,700	182
N_2 in butadiene-acrylonitrile copolymer	28	11,500	130
N_2 in neoprene	78	11,900	215
H_2 in neoprene	9	9,250	74
N_2 in butadiene-polystyrene interpolymer	9×10^{-1}	8,900	24
D_{max} for ox erythrocyte membrane	1×10^{-1}	10,000	7.7

[a] Symbols as defined by Eqs. (3.1), (3.6), and (3.7).
[b] Data from Jost (1960) and from present study (ox erythrocyte only).

to those values found for diffusion in highly polymerized networks, rather than the values for diffusion through a phase of small molecular weight substances. The value found here for $D_{0_{max}}$ and, therefore, $\lambda[\exp(\Delta S^{\ddagger}/R)]^{1/2}$ is low for typical members of the highly polymerized class and is closer to that found for diffusion in water, itself atypical. Eyring points out that high values for $\lambda[\exp(\Delta S^{\ddagger}/R)]^{1/2}$ suggest that there is an appreciable entropy contribution for the formation of the transition state during diffusion, since a lattice distance of 110 Å in water is quite unexpected, as are the very high values in column 4 of Table 3.9 for the rubber membranes.

Unless for the case of diffusion within the membrane the $\Delta S^{\ddagger}$ term is negative, a finding which would be quite unlike that found for diffusion in other systems that have been studied, the figure of 7.7 Å for $\lambda[\exp(\Delta S^{\ddagger}/R)]^{1/2}$ sets an upper limit of 7.7 Å for the lattice constant λ for diffusion within the membrane. Until the term $\Delta S^{\ddagger}$ can be computed the relation of this value to the "pore size" of 4 Å found by Solomon and others (see Curran, 1963) for diffusion through the cell membrane can only be surmised. As an *upper limit* this value of 7.7 Å is by no means unreasonable and in general it would appear, therefore, that the bimolecular lipid leaflet model gives an accurate molecular picture of the unspecialized permeability properties of the cell membrane. For such permeability we have the following:

(1) Diffusion within the membrane is a process with kinetic parameters most similar to diffusion in highly polymerized rubber membranes and results in a 100- to 1000-fold reduction of transfer rate in comparison with an equivalent thickness (50 Å) of water.

(2) To enter the membrane, each hydrogen-bonding acceptor or donor group that the permeant molecule makes with the water molecules of the aqueous phase has to be broken, a step which lowers the transfer rate by a further 6- to 10-fold.

(3) Finally, in the transition state the permeant molecule is perhaps oriented at the aqueous-lipid interface and each bare —CH_2— group in the permeant will increase the transfer rate by some twofold.

The many "exceptions" to these rules, systems in which some specialized component of the cell membrane is apparently involved in transfer, remain to be discussed.

3.5 The Movement of Ions

A. Ion Transfer through a Simple Lipid Lattice

We might expect that the membrane model we have been considering would have to be modified somewhat to account for the movement of ionic particles across cell membranes. We have considered in Section 2.7 how the *equilibrium distribution* of ions will be determined by the electrical potential and in particular the Donnan potential across the membrane, but we must now take up the problem of the *rates* of movement of ions.

To apply our lipid lattice model to the data on ion permeabilities, we shall have to affix to each ion a value of N, the number of hydrogen-bonding groups. This is an extremely difficult task. It is well known (Harned and Owen, 1958) that ions in aqueous solution are hydrated, that is, they are surrounded by a sheath of more or less firmly bound water molecules, the number **n** of such bound water molecules depending in a characteristic fashion on the size and charge of the ion. A high charge and a small ion size—both factors leading to a high charge density at the surface of the ion—are associated with an increase in the degree of hydration. Now, these bound water molecules are much more firmly bound than the hydrogen-bonded water molecules that we have been considering for the nonelectrolytes. The free energies of hydration of ions have been computed (Butler, 1951) to be between 46 (for the iodide ion) and 1062 (for the trivalent chromium ion) kcal/mole of ion. We shall see below that each such ion binds between 1 and perhaps 20 water molecules so that the free energy of interaction between an ion and a single water molecule is some 40 kcal/mole, as opposed to a value of some 2 kcal/mole for a hydrogen-bonded water molecule. It is because interaction between an ion and a water molecule is a charge-dipole interaction, that it is far stronger than the dipole-dipole interactions which are the basis of hydrogen bond formation. It is most unlikely, therefore, that any substantial quantity of the naked, unhydrated ions will exist in aqueous solution, and it is rather the penetration of hydrated ions that we must consider. Each penetrating ion, then, will be surrounded by a firmly bound hydration shell and it is these bound water molecules that will, by hydrogen bonding, be anchored in the aqueous phase and will determine the parameter N.

Values for the number **n** of water molecules of hydration have been determined, generally rather indirectly, and Table 3.10 collects some of

TABLE 3.10

ESTIMATES OF ION HYDRATION NUMBERS **n**[a]

Solute	**n** from activity measurements	**n** from diffusion data
RbCl	1.2	—
NH_4Cl	1.6	0.2–0.5
KCl	1.9	0.6–0.8
NaCl	3.5	1.1–3.5
LiCl	7.1	2.9–6.3
LiBr	7.6	2.9–5.6
LiI	9.0	—
$MgCl_2$	13.7	—
$Zn(ClO_4)_2$	20.0	—

[a] Data collected by Robinson and Stokes (1959).

these data. The number **n** varies with the method used to characterize **n** (Bockris, 1949). Understandably, if an equilibrium method (such as the study of the dependence of activity coefficients on concentration) is used, a larger number is obtained for **n** than if a dynamic method (for example, the study of ionic diffusion) is used, since in a dynamic study, the outermost water molecules will be left behind the traveling ion. Presumably, the correct value for membrane permeability is closer to those obtained from dynamic determinations. In Table 3.10 the total hydration number for a *salt* is recorded, since the relevant numbers for the individual ionic species cannot be obtained separately. On theoretical grounds, the degree of hydration of potassium and of fluoride or chloride ions may be expected to be roughly equivalent (Bockris, 1949). We might therefore draw the following conclusions from Table 3.10: The chloride and potassium ions will each bind between one-half and two molecules of water; bromide and iodide will bind perhaps a little less water, while sodium will carry between 1½ and 3 molecules, lithium between 3 and 6 molecules, and magnesium up to 10 molecules of water. Now, each firmly bound water molecule will be anchored to perhaps 2 more water molecules by hydrogen bonding. We might expect then the following values for N: K^+, 1–2; Cl^-, 1–2; Br^-, 1–2; I^-, 1–2; Na^+, 3–6; Li^+, 6–12; Mg^{++}, 20. If each unit increase in N decreases the permeability by some eightfold—as the data on nonelectrolyte permeabilities would suggest—we might expect the following to be the relative orders of magnitude of the permeabilities of the ions (if we take in each case the lower value of N in the range quoted above): $K^+ : Cl^- : Br^- : I^- : Na^+ : Li^+ : Mg^{2+} :: 1 : 1 : 1 : 1 : 10^{-3} : 10^{-4} : 10^{-17}$. The assumptions made above in deriving the values of N are extremely gross, but it is perhaps worth comparing these predictions with the rather limited quantitative data available for cell membranes.

B. Passive Ion Transfer Across Nerve and Muscle Cell Membranes

We take first the data which best support the predictions we have made—the studies on nerve and muscle membranes. The *relative* ionic permeabilities of these tissues have been determined by studying the variation of the electrical potential across the membrane as the concentrations of the bathing ions are varied. The assumption is made that the potential across the membrane is given by the sum of the potentials determined by the various ions, each ion contributing to the potential according to its flux across the membrane. [An extended discussion of the assumptions made in these studies is given in Harris' monograph (E. J. Harris, 1956).] For the squid axon the data are best fitted by the assumption that the relative permeabilities of potassium, chloride, and sodium ions are as 1 : 0.45 : 0.04, or at high potassium concentrations, as 1 : 0.3 : 0.025 (Hodgkin and Katz, 1949). In frog muscle a similar study (Jenerick, 1953) gave values of $P_K : P_{Cl} : P_{Na}$ of 1 : 0.23 : 0.027, and similar results were found for the South American frog, *Leptodactyla* (E. J. Harris and Martins-Ferreira, 1955). Adrian (1956) finds that a ratio of 1 to 30 for the potassium ion to sodium ion permeabilities is applicable to frog muscle. Calcium and magnesium ions penetrate far more slowly than do the univalent cations. Table 3.11 records some data obtained

TABLE 3.11

RELATIVE NET ENTRANCE RATES OF IONS INTO FROG MUSCLE AND INTO KIDNEY SLICES[a,b]

	Relative permeability	
Series	Frog muscle	Kidney slices
Cation		
KCl	100	100
RbCl	38	25
CsCl	0	5
NaCl	0	3
LiCl	0	16
$CaCl_2$	0	—
$MgCl_2$	0	—
Anion		
KCl	100	—
KBr	63	—
KNO_3	17	—
K phosphate	4	—
KOOC CH_3	3	—
$KHCO_3$	1	—
K_2SO_4	0	—

[a] Value for KCl set equal to 100.

[b] Data taken from Conway, 1954; Whittembury *et al.*, 1960.

for frog muscle by Conway (1954). Again Na^+ and also Li^+ penetrate far more slowly than potassium (and rubidium). The highly charged (and hence heavily hydrated) amines also penetrate slowly. These data then, as far as they go, fit well with the predictions made on our naive analysis of the lipid lattice model for ions.

We might ask, however, whether our model which correctly accounts for the relative order of ion permeabilities can account also for the absolute magnitudes of these permeabilities. We need to know the term $PM^{1/2}{}_{max}$. Unfortunately, we have to approach this by collecting data from various sources rather than from a single study. Nonelectrolyte permeabilities for the squid axon have been determined by Villegas and Villegas (1960) and by Villegas *et al.* (1962). Values of P_s at 22° to 24°C for water, ethylene glycol, and glycerol can be calculated from their data to be 14×10^{-5}, 2.7×10^{-5}, and 0.26×10^{-5} cm sec^{-1}, respectively. [Compared with determinations of P_s for *Chara,* for example, the value for glycerol is tenfold higher than, while the values for water and for ethylene glycol are of the same order of magnitude as for, the algal cell. There is some evidence (Villegas and Villegas, 1962) that a specialized system for glycerol transfer is present in the Schwann cells surrounding the axon and this fact might complicate the interpretation of glycerol permeabilities.] The influx and efflux of potassium from resting squid axons have been measured by Hodgkin and Keynes (1955a,b). At an external concentration of potassium of 10.5 m*M* and an internal concentration (measured at the end of each experiment) of 267 mmole/liter axoplasm, the average influx was 21 mμmole cm^{-2} sec^{-1} while the average efflux was 28 mμmole cm^{-2} sec^{-1}. In squid axons poisoned with 0.2 m*M* dinitrophenol to eliminate the active transport component of the fluxes, the efflux was 3.0 mμmole cm^{-2} sec^{-1}. To convert these fluxes to permeability coefficients we will have to take into account (Section 2.8) both the ambient concentration of potassium and the prevailing electrical potential. In resting nerve at the above concentrations of internal and external potassium a potential difference of some 60 to 70 mV (inside negative) can be expected. This potential will enhance the rate of influx of the potassium ions some 12-fold but will decrease the rate of efflux by a like factor. If we take the product of influx and efflux, the effect of the potential difference will be balanced out. We have, therefore, efflux $\times$ influx $= P_s^2 C_e C_i$ from which we obtain a value for the permeability coefficient of the passive movement of potassium ion through the membrane as 17×10^{-5} cm sec^{-1}, which we can now compare with the values obtained by Villegas *et al.* for nonelectrolytes (Villegas *et al.*, 1962). Clearly, the permeability coefficient for potassium ion is of the same order of magnitude as that of water, in good agreement with our suggestion that between a half and one molecule of water is firmly bound to each potassium ion.

We must immediately point out, however, that any agreement between the experimental results and our predictions may be entirely fortuitous. It is quite clear that the potassium permeability and the sodium permeability of the nerve axon membrane and of the muscle cell membrane are physiological variables rather than constants. The brilliant studies of Hodgkin, Huxley, Keynes, and others (Hodgkin, 1958) have shown that during the passage of the nerve impulse there is a brief period of vastly increased sodium permeability (such that the rate of penetration of sodium exceeds that of potassium) followed by a period of increased potassium permeability. Both sodium and potassium permeability coefficients increase markedly as the membrane is depolarized, whether this occurs either by the application of an external electrical potential or by increasing the potassium concentration in the external medium. Considering some representative data taken from Hodgkin and Keynes (1955b), at a membrane potential of 88 mV (inside negative), the efflux of potassium is some 2 mμmole cm^{-2} sec^{-1} while at 30 mV (inside negative), the efflux is some 500 mμmole cm^{-2} sec^{-1}. A change of 58 mV in applied potential should increase the potassium flux by tenfold, if the only effect were the increasing of electrochemical potential of potassium according to Eq. (2.31). The flux, however, increases 250-fold, thus the permeability coefficient itself has increased 25-fold, following this depolarization of the membrane. The simple lipid lattice model provides no grounds for predicting these changes in permeability coefficient. It could still be argued, however, that the permeability of the resting nerve membrane is dominated by the properties of a simple lipid lattice but that a parallel specialized system comes into action when the membrane is depolarized. This is a plausible point of view but a major problem still remains: the flux ratios for potassium ions even in the resting membrane do not accord with the predictions of the Ussing flux ratio criterion (Section 2.8). This phenomenon has been carefully studied by Hodgkin and Keynes (1955b). Taking the data we have already cited for the potassium fluxes in dinitrophenol-poisoned nerve axons we have

$$\frac{\text{Potassium efflux}}{\text{Potassium influx}} = \frac{27}{3.0} = 9$$

while the corresponding term required for the Ussing flux ratio criterion

$$\frac{\text{External concentration of } K^+}{\text{Internal concentration of } K^+} \exp\left(\frac{E_{\text{II}} - E_{\text{I}}}{RT}\right)F = \frac{267}{10.4} \times \frac{1}{12} = 2.2$$

If the movement of potassium were occurring by the simple diffusion of independent ions, these two ratios would be equal as we have seen in Section 2.8. (When the membrane was depolarized, the increased fluxes occurring also did not obey the Ussing flux ratio criterion for the independence of fluxes. Deviation from the Ussing prediction occurred at all values of the membrane potential differences, and the deviations were everywhere of much the same order of magnitude.) A high external potassium ion concentration at constant membrane potential decreased the rate of efflux of potassium.

Since these data were obtained with poisoned axons we are probably entitled to assume that any transport-linked vectorial chemical reactions have been abolished, and our irreversible thermodynamic analysis (Section 2.8) leads us then to conclude that an interaction between the potassium flux and some other fluxes must be occurring [that is, some coefficient R_{ij} of Eq. (2.33) must be unequal to zero]. The explanation proposed by Hodgkin and Keynes (1955b) is that the potassium effluxes and influxes are interfering with one another. One mechanism by which this could occur is if the transfer of potassium ion occurred through narrow tubes or channels in the membrane, each channel being only a little wider than the hydrated potassium ion, but yet sufficiently long so that several ions could be expected to be in the channel at any moment of time. This is the "long pore" model. Movement of potassium ions occurring in one direction, say, efflux, would tend to sweep along a column of water, making it easier for other potassium ions to move in the same direction but hindering movement in the opposite direction (here, influx).

We can arrive at a more general mechanism by introducing the frictional coefficients of the irreversible thermodynamic analysis previously discussed in Section 2.5. In addition to the coefficients defined in Section 2.5, describing the solute-membrane, solvent-membrane, and solute-solvent drags, namely, f_{sm}, f_{wm}, and f_{sw}, we clearly have to consider here the mutual drag of the penetrating solute molecules, that is, a solute-solute coefficient f_{ss}. Then the long pore model is one particular example of a phenomenon which can be expressed more generally as the finding of a nonzero coefficient f_{ss}. Another model for this phenomenon, one which retains the postulate of the lipid bilayer, is as follows: If we accept that the solute species, here the potassium ions, are hydrated in water and *retain their innermost hydration shell* within the membrane, then these hydrated ions will exert a significant drag upon one another by hydrogen bonding. The outermost water molecules will have a strong tendency to associate into complexes. This hydrogen bonding will be far more intense within the hydrophobic environment of the lipid membrane interior than in the external water, where competition by surrounding water molecules effectively weakens the hydrogen bond. Then a high rate of diffusion of hydrated potassium ions from the outside of the membrane to the inside will tend to depress a contrary efflux of potassium ions—as the hydrated ions interact within the membrane.

We should note, however, that although the lipid lattice model accounts adequately for the *relative* permeabilities of the ions, this evidence provides no proof that such a model is in fact valid. Any model which requires that the degree of hydration of the ions should dominate their rate of passage across the membrane would give much the same sort of prediction. In particular, the long pore model of Hodgkin and Keynes (1955b), requiring as it does that the diameter of the pores be only a little bigger than that of the hydrated potassium ions, would predict that the rate of potassium and chloride ion movements would be comparable but that both sodium and the divalent ions should enter the pores with difficulty, if at all, and should thus penetrate slowly.

[In this connection we refer the reader to the very interesting treatment of membrane permeability that Blank and his associates (Blank and Britten, 1965; Blank, 1965) are currently developing on the basis of a theoretical study of lipid monolayers. Blank's view, concordant with that developed in the present section, is that "free spaces in the monolayer become available for permeation from the natural free area in a lattice, from the equilibrium fluctuation in monolayer density (local expansions) at a gas molecule-monolayer collision site, and from the work of expansion that the permeant molecule can perform against the monolayer forces." Applying this model in particular to ion fluxes, quite reasonable values for these fluxes and for the change in fluxes on membrane depolarization can be derived.]

C. Passive Ion Movements across Erythrocyte Membranes

If we come now to the data on ion movements across the red cell membrane, we shall see that the simple lipid lattice gives a less satisfactory explanation of the experimental findings. For these cells, chloride ion movement is very rapid, times of less than one second for the exchange of chloride and bicarbonate ion having been reported in the earlier literature, reviewed in Davson and Danielli's monograph (1952). A more recent study by Tosteson (1959) using a rapid reaction technique (compare Section 2.6) leads to a value of 3.1 sec^{-1} for the rate constant for chloride ion permeability. These values are only two orders of magnitude below the rate of permeability of water across the red cell membrane (Section 3.3). In contrast, the rate of potassium ion movement is slow. We can again derive values for the permeability coefficient for potassium ion by combining results of efflux and influx measurements, thereby eliminating the effect of the membrane potential. Tosteson and Hoffman (1960) have used radioactive tracer methods to measure the influx and efflux of both potassium and sodium ions for the erythrocytes from two strains of sheep, the high potassium (HK) strains and the low potassium (LK) strains. [We shall consider the problem of osmotic balance in these erythrocytes, on the basis of Tosteson and Hoffman's study, in some detail in Chapter 7. Here, we consider only data on the passive component of ion fluxes; i.e., we restrict our analysis to data on erythrocytes treated with strophanthidin, a drug which inhibits the active transport of these cations (see Chapter 8)]. Table 3.12 records these influx and

TABLE 3.12

PASSIVE CATION MOVEMENTS ACROSS SHEEP ERYTHROCYTE MEMBRANES (STROPHANTHIDIN-POISONED)[a]

Parameter	Unit	Value for High potassium (HK) strain	Value for Low potassium (LK) strain
	Sodium Ions		
External conc.	mM	165	165
Internal conc.	mM	37	137
Passive efflux	mmole(liter red cells)$^{-1}$ hr^{-1}	[2.65	5.0
Passive influx	mmole(liter red cells)$^{-1}$ hr^{-1}	2.8	3.9
Flux ratio (experimental)[b]	—	0.95	1.28
Flux ratio (computed)[c]	—	0.15	0.56
Mean permeability coefficient[d,e]	cm sec^{-1}	4.4×10^{-10}	3.7×10^{-10}
	Potassium Ions		
External conc.	mM	5.0	5.0
Internal conc.	mM	121	17.4
Passive efflux	mmole(liter red cells)$^{-1}$ hr^{-1}	0.67	1.53
Passive influx	mmole(liter red cells)$^{-1}$ hr^{-1}	0.04	0.12
Flux ratio (experimental)[b]	—	16.8	12.7
Flux ratio (computed)[c]	—	16.3	2.4
Mean permeability coefficient[d]	cm sec^{-1}	0.84×10^{-10}	5.7×10^{-10}

[a] Data taken, or computed, from Tosteson and Hoffman, 1960.

[b] Experimental flux ratio calculated from the ratio of passive efflux to passive influx.

[c] Computed flux ratio calculated from $\frac{2}{3} \times$ internal cation conc./external cation conc., the factor $\frac{2}{3}$ being the concentration ratio of internal chloride to external chloride, this ion being assumed to be distributed according to the prevailing Donnan potential.

[d] The mean permeability coefficient is given by the square root of the product of the passive efflux and influx, the volume of the red cell being taken as 30 μ^3 and its area as 67 μ^2 (Jacobs, 1952).

[e] The flux values for sodium ion, as reported here, are augmented to an unknown extent by the presence of considerable exchange diffusion. In Table 7.2 the values of the ratio of the leak rate constants for sodium and potassium, measured in a situation which largely eliminates exchange diffusion, are recorded.

efflux measurements on strophanthidin-poisoned erythrocytes, and records the prevailing ionic concentrations, the derived values of the (mean) permeability coefficients, and both the flux ratio and the predicted flux ratio from the Ussing relation (Section 2.8). We see that in all cases the permeabilities are exceedingly low in comparison with the value of 530×10^{-5} cm sec^{-1} found for the penetration of water across the (human) erythrocyte membrane (Section 3.5), that there is little selectivity for potassium ion over sodium ion—indeed for the HK strain, the permeability coefficient for potassium ion is so reduced that it may now be only one-fifth of that for the sodium ion—and that the Ussing flux ratio predictions are not borne out. We might suggest again that the solute molecules interact in some way during membrane translocation. The fluxes here are of the order of 10^{-14} mole cm^2 sec^{-1}, while for nerve axon the passive entry fluxes are, as we have seen, some 10^{-9} mole cm^2 sec^{-1}. E. J. Harris (1954) has computed figures for the passive permeability coefficients of Na^+ and K^+ transfer in human erythrocytes. The coefficient for K^+ transfer is only 0.62 times that for sodium transfer. Finally, Maizels (1954) finds that the rate constant for the (passive) influx of lithium ion is again of the same order of magnitude as those for Na^+ and K^+, yet the lithium ion binds considerably more water molecules than do the other cations.

There is, indeed, good evidence (marshaled by Passow, 1964) that the potassium impermeability of the erythrocyte is a physiological specialization of this cell, and that this functional impermeability is under metabolic control. In the presence of metabolic inhibitors such as fluoride ions and iodoacetate, and provided sufficient calcium ions are present, the permeability of potassium ions rises some twentyfold, the sodium ion permeability being largely unaffected. The potassium-sodium selectivity thus reverts to the level typical of nerve and muscle cell membranes. (At high calcium ion concentrations, however, both sodium and potassium ions can permeate rapidly and the erythrocytes lyse.)

It is clear that the simple lipid lattice model cannot account for these ion permeability data. A simple long pore model will, however, likewise not be able to differentiate in this way between these cations, nor between cations and anions. To understand this latter specificity, we must now consider the possibility that the interior of the membrane is itself charged.

If the interior of the membrane contains regions of high positive charge, these will act as a potential energy barrier for the movement of cations but will allow, and in fact aid, the penetration of negatively charged ions. Such an explanation for the cation impermeability of cell membranes has been proposed and formally worked out by Meyer and Sievers (1936a,b) and by Teorell (1935), these and subsequent studies being adequately reviewed in the monographs of Davson and Danielli (1952) and E. J. Harris (1956). Figure 3.12 is an example of the type of structure that could account for these findings. A lattice of fixed positive charges exists in the interior of the membrane. These charges effectively repel the mobile cations but attract any mobile negatively charged ions. The concentration of cations within the membrane (and

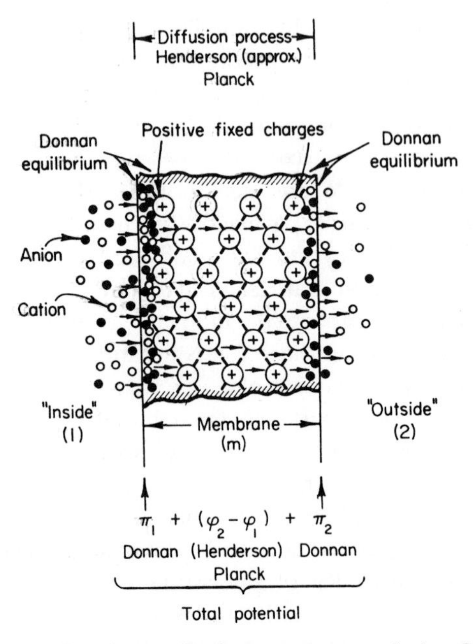

Fig. 3.12. Diagram illustrating the basic concepts of the fixed charge theory. (Taken with kind permission from Teorell, 1956.)

hence the fluxes) are reduced, while those of the anions are both increased. We shall find it necessary to consider, however, whether such fixed charges can occur only as groups lining porous channels through the membrane or whether a lipid lattice could accommodate a set of fixed cationic groups.

Let us first consider further permeability experiments which support a fixed charge model. Passow (1964), for instance, has described experiments in which the flux of sulfate ions (measured using $^{35}SO_4^{2-}$) across erythrocyte membranes was measured as a function of the pH and chloride ion concentration of the suspension medium. The relationship between the sulfate ion flux and the pH followed an S-shaped titration curve with an inflection point at about pH 7.2, the flux increasing as the hydrogen ion concentration increased. Clearly, increasing the hydrogen ion concentration here increased the charge on some proton-acceptor (base) within the membrane, resulting in an increase in the rate of penetration of anion in accordance with the fixed charge theory. An increase in the prevailing chloride concentration, in contrast, decreased the rate of sulfate flux. This is also to be expected on the fixed charge theory. The mobile chloride ions will compete against sulfate ions in binding to the positively charged groups within the membrane. Passow could account for his data on the assumption that the membrane contained an effective concentration of some 3 moles/liter of the basic groups, these groups having a pK of about 9, consistent with their being amino groups. Further, studies on the rates of transfer of a series of dicarboxylic anions (Giebel and Passow, 1960) were consistent with penetration being limited by the size of the anions, a result to be expected if passage occurred through narrow channels. These and other studies on ion and water transfer across erythrocyte membranes have been reviewed by Passow (1964).

E. J. Harris (1954) has shown how the effect of changes in the ambient pH on the rates of (passive) sodium influx in human erythrocytes can be analyzed to yield the pH dependence of the sodium ion permeability coefficients. The problem here is that the potential difference across the membrane is itself determined by pH, so that the passive cation influx will in the first place vary as a result of this potential difference change. This effect can be corrected for, however, by the simultaneous analysis of the chloride ion distribution ratio which responds to, and hence measures, the membrane potential. Harris' computations show that a drop of about 0.5 pH unit decreases the rate constant for sodium influx by 15%. This accords in general with the predictions of the fixed charge theory, the change in sodium influx here being of course opposite in direction to the change in the rate of sulfate ion transfer found by Passow. Again, if the simultaneous entry of two cations is measured, the presence of one cation depresses the rate of penetration of the other (Solomon, 1952), a result difficult to explain if penetration is indeed occurring through a simple lattice—but consistent with a fixed charge model.

D. The Ionic Permeability of a Synthetic Lipid Membrane

Some very interesting and relevant experiments with synthetic model cell membranes have been reported by Bangham *et al.* (1965). When shaken with water and salt solutions, dry lecithin forms a liquid-crystalline suspension which, on electron microscopic examination, appears to consist of microspheres of diameters varying up to 2000 Å, these spheres consisting of concentric shells of bimolecular lipid leaflets. Optical birefringence studies confirmed this picture. (We have discussed in Section 1.5 similar phospholipid suspensions prepared and studied by Green and his associates.) When the lecithin is swollen in salt solution the microspheres occlude salt into their interior, and exhaustive dialysis against distilled water leads to the loss of only a small part of this salt. The remainder is extremely resistant to loss by dialysis. If, however, these well dialyzed suspensions are now dialyzed against salt solutions, the cations and anions present within the microspheres will exchange with the ions of the dialysate. Bangham and his associates showed, by incorporating into the phospholipid-water emulsion defined quantities of positively or negatively charged detergents, that the charge on the lipid bilayers could be altered and that the effect on ion transport of charging up the lipid layers could therefore be studied. In each case the prevailing electrical potential at the surface of the microspheres was measured by an electrophoretic method. Figure 3.13 shows how the efflux of potassium ion depended strictly on this surface charge. For pure lecithin, where the microspheres have no net charge, a small but definite efflux of potassium occurs, but the addition of as little as 5 mole-% of a long-chain cationic detergent completely prevented potassium loss. The addition of an anionic detergent greatly increased this loss of potassium. These results

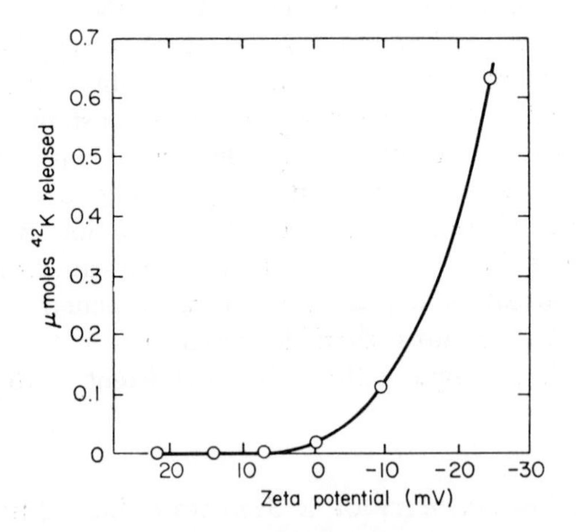

Fig. 3.13. Release of $^{42}K^+$ from lecithin/long-chain ion mixtures as a function of the measured zeta potential, on dialysis against 0.145*M* KCl (unlabeled). Abscissa: zeta potential in mV. Ordinate: release of $^{42}K^+$ in μmoles, following 30 min dialysis at 37°C. (Taken with kind permission from Bangham *et al.*, 1965.)

are a beautiful demonstration of the importance of the membrane charge in determining ion transfer rates.

When different cations (Li^+, Na^+, K^+, Rb^+, and choline) were studied, little difference in the rates of transfer was found. For the anions, the rate of transfer was in general far more rapid than that of the cations, the rate decreasing with the size of the hydrated ion in the order

$$Cl^- = I^- > F^- > NO_3^- = SO_4^{2-} > HPO_4^{3-}$$

and being independent of the surface charge in the microspheres. (This latter statement is made but without supporting data, in the paper of Bangham *et al.*) Water traveled extremely rapidly across these synthetic membranes. The properties of these synthetic membranes are thus remarkably like those of erythrocytes. For our present purposes we might perhaps emphasize the following three important features of these results:

(1) These studies give strong support to the fixed charge model.

(2) Unless the synthetic membranes indeed contain more or less permanent pore structures—a most unlikely situation—these studies suggest that the charge at the surface of the membrane (as opposed to charge within the membrane) can control the ionic permeability of the bilayer. Thus the phenomenon of ion selectivity does not require for its explanation the assumption of a porous membrane.

(3) The data give us experimental support for the attempt to reconcile the behavior of the nerve membrane and the erythrocyte membrane—a difference in composition of these membranes, such that the red cell membrane contained fixed positive charges, could explain the high anion selectivity of this cell.

It is clear that the details of the composition of various cell membranes (Section 1.2) and in particular the presence or absence of groups possessing a net charge at physiological pH will determine the ion permeability of the membranes.

E. The Temperature Coefficient of Cation Movements

Before we attempt to draw a final conclusion from all these data, we might consider briefly one further set of relevant experimental findings—the data on the temperature dependence of ion transfer rates. In Table 3.13 we collect some of the derived values for the activation energies for passive movements of potassium and sodium ions across a number of cell membranes. For comparison, the activation energies for potassium efflux in the model membrane system studied by Bangham and his associates

TABLE 3.13

TEMPERATURE COEFFICIENT OF THE PASSIVE CATION PERMEABILITY OF VARIOUS CELL MEMBRANES

Cell studied	Flux	Q_{10}	Activation energy (kcal mole^{-1})	References
Human erythrocyte	Linear component of K^+ influx	2.3	15	Glynn (1956)
Human erythrocyte	K^+ efflux	2.5	15.8	Sheppard and Martin (1950)
Human erythrocyte	K^+ efflux	2.0	12.4	Solomon (1952)
Human erythrocyte	Na^+ influx	3.0	20.2	Solomon (1952)
Human erythrocyte	Na^+ influx	2.3	15	Clarkson and Maizels (1956)
Ascites tumor cells (mouse)	Na^+ influx	2.5 to 4.0	17 to 25	Maizels *et al.* (1958)
Ascites tumor cells (mouse)	Na^+ influx	1.2	3.3	Hempling (1958)
Ascites tumor cells (mouse)	K^+ efflux	4.1	26	Hempling (1958)
Nerve cell (squid axon)	Na^+ influx	1.2 to 1.6	3.3 to 9.7	Hodgkin and Keynes (1955a)
Nerve cell (squid axon)	K^+ efflux	1.1	1.6	Hodgkin and Keynes (1955a)
Nerve cell (squid axon)	Passive component of K^+ influx	1.0	0	Hodgkin and Keynes (1955a)
Synthetic membrane phospholipid micelles	K^+ efflux	2.3	15	Bangham *et al.* (1965)
Free diffusion in water	K^+	1.26	4.2	Longsworth (1955)
Free diffusion in water	Na^+	1.3	4.7	Longsworth (1955)
Diffusion in ion exchange resins	K^+	1.4	6.5	Boyd and Soldano (1954)
Diffusion in ion exchange resins	Na^+	1.3	5.22	Boyd and Soldano (1954)
Diffusion in ion exchange resins (highly cross-linked)	Na^+	1.6	8.62	Boyd and Soldano (1954)

and the activation energies for the free aqueous diffusion of these ions are also recorded. The data on membrane transfer, for erythrocyte and synthetic membranes, are consistent; a value of the order of 15 kcal/mole is found—some threefold the relevant value for free diffusion. [The temperature dependence for the penetration of Na^+ into a highly cross-linked cation exchange resin is, however, also high, the data being consistent with an activation energy of 8½ kcal/mole for this process (Soldano, 1953).] The data on other cells cannot be so concisely summarized but, in general, the activation energies clearly suggest that membrane penetration is not occurring by a process of free diffusion through wide and water-filled channels. If penetration is indeed occurring through a simple lipid lattice then some three hydrogen bonds (each 5 kcal/mole energy) need to be broken to form the transition state for membrane transfer (consistent with an average value of one and one-half molecules of water being firmly bound to each ion) while, if a narrow pore system is being traversed, again an outer loosely bound shell of one to two water molecules has to be shed before the pore can be entered. Alternatively, these activation energies could reflect a loosening of the bonds holding the pores together as the temperature is raised, with a consequent widening of the pores. This alternative explanation would predict a decrease in the size selectivity of the pores as the temperature is raised, and it might predict the disappearance of the long pore effect at high temperatures.

F. Conclusions

We should note first that the predictions of the simple lattice model are not borne out by the data on the synthetic lipid membranes. These membranes distinguish too strongly between cations and anions and do not distinguish sufficiently strongly within the (univalent) cation series for our predictions to hold. In fact, the lack of a dependence of transfer rate on the hydrated ion size is difficult to account for on any model other than that of a system of very wide channels. The data do not, however, allow us to assert that penetration here necessarily occurs through aqueous channels. Transfer can as easily be accounted for by accepting that the charge at the membrane surface will dominate the energy relations of the transition state for entry into the lipid. We cannot state that the fixed charges that we have seen to be necessary to explain the anion selectivity of these membranes are groups lining the surfaces of aqueous channels. The presence of charged head groups of the phospholipid cations at the water-oil interface would itself be sufficient to determine such selectivity. It is clear that much more attention will have to be paid to these aspects of the problem if we are to understand the ion permeability of the natural membranes.

We must conclude, therefore, that the studies on the synthetic lipid membranes offer a good model for the erythrocyte data, but that we do not fully understand the molecular basis of the behavior of this model.

For the data on nerve and muscle cell membranes, the absolute and relative rates of ion transfer accord with our simple lipid lattice model, but they do not accord with the behavior of the synthetic membrane systems. The flux ratio data, however, uniformly suggest that there is appreciable interaction between penetrating solute molecules. A model in which penetration occurs through longish narrow pores would account for such interaction, and this model could account also for the data on the rates of ion transfer—provided the necessary assumptions were made as to the size and number of these pores. But the flux ratio data do not exclude the possibility that a lipid bilayer model is valid since, as we have seen, solute-solute interaction within the bilayer would be expected.

It is obvious that the data on ion transfer across cell membranes do not agree sufficiently among themselves to allow broad statements to be made as to the behavior of natural membranes in general, and it is also clear that the model systems are not yet sufficiently understood for any very strong statements on their molecular basis to be valid. But it has been argued that a simple lipid lattice does not account for the ion transport data, and we must now proceed to consider other evidence which is perhaps in discord with this model.

3.6 An Alternative Model–Pores in the Cell Membrane

We have seen that a model of the membrane as a simple bimolecular layer of lipid accounts adequately for the assembled data on the permeability of a number of cell types, but we have been already forced to take note of three sets of exceptions to this general finding. First, the rate of water penetration is generally higher than would be expected from a molecule that can make four hydrogen bonds with (other) water molecules. Second, the rate of penetration of urea across the animal cell membranes is anomalously high while, third, the penetration of ions is clearly a complex problem with no solution in terms of a simple model. If now we attempt to apply the analysis of Section 3.3 to the available data for human erythrocytes, a further set of anomalies is found. Figure 3.14 presents these data where (1) the results of Höber and Ørskov (1933) obtained by a method which involves the measurement of the rate of hemolysis of red cells in solutions of the permeant in 0.12% saline; (2) the results of Jacobs *et al.* (1935) on measurements of the rate of hemolysis of cells in isotonic solutions of the permeant alone, and (3) the results of Stein (1956, 1962b) from measurements of the rate of swelling of cells in solutions of the permeant in 0.17% saline are recalculated to give values of the permeability coefficient in a common unit (cm sec^{-1}). It should be noted that agreement between these sets of data is reasonably good.

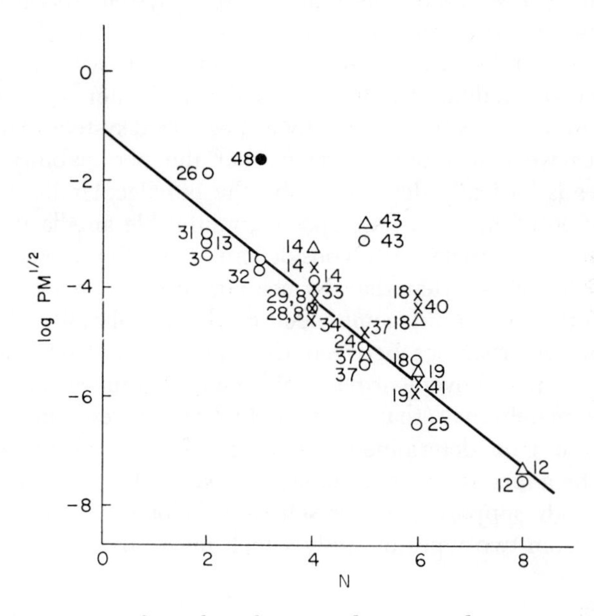

Fig. 3.14. Penetration of nonelectrolytes into human erythrocytes, as a function of the number of hydrogen-bonding groups. Scales and numbering as in Fig. 3.6. ○, Data of Höber and Ørskov (1933); ●, Paganelli and Solomon (1957); △, Jacobs *et al.* (1935); ×, Stein (1956, 1962b). All at 20° to 25°C.

It is clear that the human red cell is as permeable (for the values of $PM^{1/2}{}_{max}$ see Table 3.3) as the ox erythrocyte. But it is also clear that, although a general parallelism between log $PM^{1/2}$ and the number of putative hydrogen bonds remains, there is a less strict adherence of the data to this rule. In particular, the data for water (48 in Fig. 3.14) and urea (43) remain anomalously high but, in addition, glycerol (18) and 1,2,4-butantriol (40) are clearly high, and ethylene glycol (14) and 1,3-propandiol (34) are rather high. If these are "exceptions" to the rule that we are testing—that permeability is a simple function of N—then these exceptions now form an appreciable fraction of the available data (6 exceptions of 19 individual substances). That there may be some justification for the view that the behavior of these substances is indeed exceptional can be seen from a consideration of the pairs 18 and 19, 18 and 20, 40 and 41, where in each case the addition of acid or of copper ions has the effect of reducing the permeability of glycerol (or its homologue 1,2,4-butantriol) to a value consistent with that expected for a lipid membrane. There is now indeed a considerable body of evidence (LeFevre, 1948; Stein and Danielli, 1956; Stein, 1962c) which confirms that a specialized membrane transport system operating on glycerol or 1,2,4-butantriol and active also toward ethylene glycol and propylene glycol, is present in human erythrocytes, this system being one of the well-known facilitated diffusion systems (Chapter 4). There is evidence that urea, too, is transported by a specialized system (Hunter *et al.*, 1965). Thus we can adopt the view that the permeability of human erythrocytes is basically determined by the bimolecular lipid membrane but superimposed upon this are specific, inhibitible ancillary systems.

But we must be prepared to consider an alternative model. Let us for the moment reject the view that permeation occurs by entrance into, and diffusion across, the lipid component of the membrane (that is, in a hydrophobic environment between the lipid hydrocarbon chains) and suggest rather that movement takes place within aqueous channels penetrating the membrane (that is, in a hydrophilic environment). Here diffusion would be determined largely by the size of the penetrating molecule, the aqueous channels acting as a size-selective sieve. This view is most strongly supported by the school of Solomon (1960) and derives its support from two types of study which must now be considered.

A. The Relation between Hydrodynamic, Osmotic, and Diffusion Flows

The first such study arises from the suggestion of Koefoed-Johnsen and Ussing (1953) that a comparison of the rate of movement of water when measured as a bulk flow under the influence of an osmotic gradient (symbolized by L_p) and the rate of movement by diffusion when tracer D_2O or T_2O movement is followed (symbolized by P_w) can give an estimate of the size of the pores in a membrane (see also Pappenheimer *et al.*, 1951). A consideration of Fig. 3.15 will make the argument clear. A membrane M separates the two phases A and B. We can measure P_w, the rate of diffusion across M, by adding tracer D_2O to A and analyzing compartment B. We can measure L_p, the bulk flow of water, by applying a hydrostatic pressure to A and measuring the rate of increase of volume of compartment B. Then, in the one limiting case where M is strictly impermeable to water, the rates of transfer by diffusion and by bulk flow are equal to each other, both being zero. At the other limiting case, where the membrane is suddenly withdrawn, that is, rendered infinitely permeable, bulk flow will be very fast—in fact, a shock wave will result—but flow by diffusion will occur moderately slowly at a rate governed by the diffusion coefficient of water in water and the surface area of M. Clearly at any intermediate permeability of M, the ratio of L_p to P_w will give a measure of the permeability of the membrane. If movement occurs through aqueous channels of radius r, the area available for diffusion is $n\pi r^2$ where n is the number of channels and the rate of diffusion can be calculated from the diffusion coefficient D, the available area $n\pi r^2$, and the concentration gradient. In contrast, movement along such

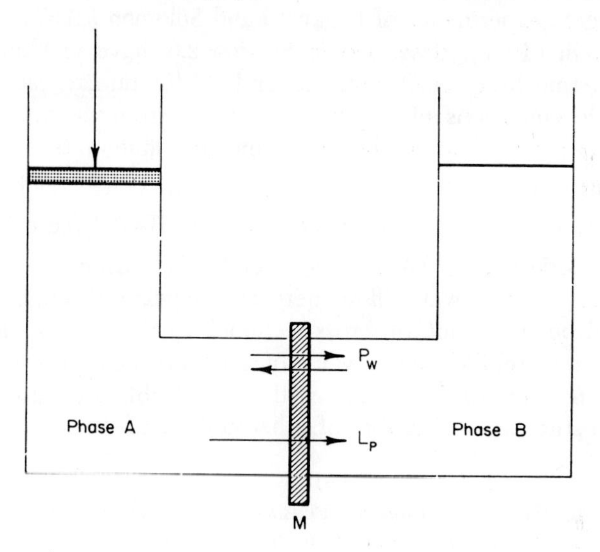

Fig. 3.15. A schematic diagram of an apparatus for comparing the two water permeability coefficients P_w and L_p (see text, Section 3.6).

channels by bulk flow can be expected to be governed by Poiseuille's law, when the flow is given by $(n\pi r^4)/(8\eta\bar{V})$ times pressure gradient, where η is the viscosity of water in the channels and $\bar{V}$ is the molar volume of water. From these two formulas a value of r, the radius of the pores, can be found as the single unknown if D and η are available.

For work on cell membranes, a hydrostatic pressure gradient across the membrane often cannot be set up but a gradient of osmotic pressure can be imposed. That flow by an osmotic pressure gradient is not identical with flow by hydrostatic pressure has been proposed by Chinard (1952) and by E. J. Harris (1956) but, as we have seen in Chapter 2, these arguments are rendered untenable by Kedem and Katchalsky's (1958) analysis based on irreversible thermodynamics. Molecularly, we can follow a suggestion of Onsager, mentioned in Mauro (1957), that an osmotic gradient gives rise to a pressure gradient because the solvent molecules in the pore are shielded by the walls of the aperture of the pore from bombardment by the solute molecules. The molecules within the pore on the solute side are thus at a lower hydrostatic pressure than molecules within the pore on the side opening into the pure solvent—where all the momentum of the solvent molecules (and hence the entire available *pressure*) can be transmitted by bombardment to the molecules within the pore. A differential *hydrostatic* pressure is thus set up within the pore as a result of the osmotic pressure gradient.

The elegant experiments of Paganelli and Solomon (1957) and of Sidel and Solomon (1957), described in Section 2.6, have yielded values for the two permeability coefficients L_p and P_w for human red cells under comparable conditions of temperature, ionic strength, and pH. These values are as follows when converted into the same units:

$$L_p = 1.5 \times 10^{-14} \text{ ml } H_2O \quad \text{sec}^{-1}(\text{cm } H_2O)^{-1}(\text{red cell})^{-1}$$

$$P_w = 0.62 \times 10^{-14} \text{ ml } H_2O \quad \text{sec}^{-1}(\text{cm } H_2O)^{-1}(\text{red cell})^{-1}$$

Hence, osmotic flow is 2.4 times greater than diffusion flow. On the assumptions (1) that water flow here is occurring through water-filled cylindrical pores of uniform cross-sectional area, and (2) that the bulk flow follows Poiseuille's law, a value of 3.5 Å for the radius of these pores has been obtained for the human erythrocyte. Table 3.14 records the data from such studies on a number of other cell membranes.

TABLE 3.14

VALUES OF THE HYDRAULIC (OSMOTIC) PERMEABILITY COEFFICIENT L_p AND THE DIFFUSIONAL COEFFICIENT P_w, TOGETHER WITH THE DERIVED VALUES FOR THE EQUIVALENT PORE RADIUS r, FOR THE PENETRATION OF WATER ACROSS THE MEMBRANES OF A NUMBER OF CELLS AND TISSUES[a]

Cell or tissue	$10^4 \times L_p$ (cm sec^{-1})	$10^4 \times P_w$ (cm sec^{-1})	L_p/P_w	Derived value of r (Å)	Reference[b]
Single cells:					
Amoeba	0.37	0.23	1.61	(2.1)	(1)
Frog, ovarian egg	89.1	1.28	70	(30)	(1)
Frog, body cavity egg	1.30	0.75	1.74	(2.8)	(1)
Xenopus, body cavity egg	1.59	0.90	1.77	(2.8)	(1)
Zebra fish, ovarian egg	29.3	0.68	43	(23)	(1)
Zebra fish, shed egg	0.45	0.36	1.25	(1.3)	(1)
Dog, erythrocyte	—	—	6.3	(7.5)	(2)
Beef erythrocyte	—	—	3.0	(4.3)	(2)
Human (adult) erythrocyte	127	53	2.4	3.5	(3, 4)
Human (adult) erythrocyte	116	41	2.9	4.1	(5, 6)

TABLE 3.14—*Continued*

Cell or tissue	$10^4 \times L_p$ (cm sec^{-1})	$10^4 \times P_w$ (cm sec^{-1})	L_p/P_w	Derived value of r (Å)	Reference[b]
Human (fetal) erythrocyte	61	23	2.7	3.9	(5, 6)
Squid, axon (axolemma)	—	—	7.8	(8.5)	(7)
Aplysia, neurone	1.9	—	—	—	(18)
Frog, single muscle fiber	130	—	—	—	(8)
Tissues:					
Frog, gastric mucosa	—	—	20	(15)	(9)
Toad, bladder:					
No vasopressin	4.1	0.95	4.3	8.5	(10)
With vasopressin	188	1.6	118	40	(10)
Rat, luminal surface of intestinal mucosal cells	83	—	—	—	(11)
Rat, kidney					
Proximal tubule	2400 (unchanged in diuresis)				(12)
Distal tubule	1100 (1/10 this value in diuresis)				(12)
Goat, ventricular walls (cerebrospinal fluid-brain barrier)	270	2.8	96	(36)	(13)
Rabbit, choroidal epenchyma	38	—	—	—	(14)
Synthetic membranes:					
Bimolecular lipid membrane	8.3–14.4	2.3	3.6–6.3	(5.0)–(7.5)	(15)
Dialysis tubing	230	—	—	23	(16)
Cellophane	870	—	—	41	(16)
Wet gel	3400	—	—	82	(16)
Dialysis tubing	380	10.9	35	23	(17)
Wet gel	1200	19.2	62.5	31	(17)

[a] Values are converted to the common unit cm sec^{-1}, using Tables 2.1 and 2.2. Values of r given in parentheses have been computed for the present book according to the relation $r^2 = (L_p/P_w - 1)(8\eta\ D\bar{V})/RT$ (see Section 3.6). Other values of r have been computed by the authors cited, using the same formula. Where r is of the same order of magnitude as w, the radius of a water molecule, a correction must be made for the restricted diffusional area. The right-hand side of the above equation is divided by the term $2 - [1 - (w/r)]^2$ (Paganelli and Solomon, 1957). If w is taken to be 1.5 Å, $\eta = 9.36 \times 10^{-3}$ poise and $D = 2.59 \times 10^{-5}$ cm^2 sec^{-1}, we obtain the simple relation

$$r = -1.5 + \sqrt{4.5 + [(L_p/P_w) - 1] \times 14.5}, \quad \text{giving r in Å}$$

[b] *References:* (1) Prescott and Zeuthen (1953). (2) Villegas *et al.* (1958). (3) Sidel and Solomon (1957). (4) Paganelli and Solomon (1957). (5) Sjölin (1954). (6) Barton and Brown (1964). (7) Villegas and Villegas (1960). (8) Zadunaisky *et al.* (1963). (9) Durbin *et al.* (1956). (10) Hays and Leaf (1962a). (11) Lindemann and Solomon (1962). (12) Ullrich *et al.* (1964). (13) Heisey *et al.* (1962). (14) Welch *et al.* (1966). (15) Hanai *et al.* (1965). (16) Durbin (1960). (17) Ginzburg and Katchalsky (1963). (18) Austin *et al.* (1966).

We can immediately proceed to test whether this model of the cell membrane can accommodate the accumulated data on cell permeability as successfully as could our previous analysis in Section 3.3. If the 3.5 Å radius pores exist and select between permeants according only to diameter, that is, if the pores have no special "affinity" for water molecules, then it will be possible to predict for any molecule of known dimensions its permeability. The theoretical basis for such predictions is given by Renkin (1954) who has obtained from simple geometrical considerations the following relation: The ratio of area available for diffusion A_s for a solute of effective radius s, to the geometrical area of the pore A where the radius of the pore is r, is given by

$$A_s/A = (1 - \mathrm{s/r})^2[1 - 2.104(\mathrm{s/r}) + 2.09(\mathrm{s/r})^3 - 0.95(\mathrm{s/r})^5] \quad (3.11)$$

If, therefore, we compare the permeability P_s for a solute s with the value P_w for water and assume that the pores distinguish between permeants *only* on the basis of size according to Eq. (3.11), then substituting in Eq. (3.11) we obtain Eq. (3.12), where w is the effective radius of the water molecule:

$$\frac{P_s}{P_w} = \frac{(1 - \mathrm{s/r})^2[1 - 2.104(\mathrm{s/r}) + 2.09(\mathrm{s/r})^3 - 0.95(\mathrm{s/r})^5]}{(1 - \mathrm{w/r})^2[1 - 2.104(\mathrm{w/r}) + 2.09(\mathrm{w/r})^3 - 0.95(\mathrm{w/r})^5]} \quad (3.12)$$

The value of P_w is available to us from the study of Paganelli and Solomon (1957) referred to above while Fig. 3.14 records available data on P_s. For the effective radii s and w we can use the values obtained by Goldstein and Solomon (1960) from measurements made on scale models. The values used are listed in Table 3.15. In Fig. 3.16 we plot the

TABLE 3.15

MOLECULAR RADII OF NONELECTROLYTE PERMEANTS IN Å

Permeant	From Catalin molec. models[a]	From viscosity measurements
Glycerol	2.74	3.1[b]
Propylene glycol	2.61	
Thiourea	2.18	
Malonamide	2.57	
Methyl urea	2.37	
Propionamide	2.31	
Ethylene glycol	2.24	
Urea	2.03	1.8[c]
Acetamide	2.27	
Erythritol	3.06	3.4 to 3.7[d]

[a] From Goldstein and Solomon, 1960. [b] From Miner and Dalton, 1953. [c] From Jones and Talley, 1933. [d] From Schultz and Solomon, 1961.

data for these substances as log P_s/P_w against r. The solid line in the figure is the theoretical relation in Eq. (3.12) for a value of r = 3.5 Å while the dashed line is for r = 2.8 Å. Clearly, a pore size of 3.5 Å completely fails to account for the available data. In particular, such a pore *overestimates* the permeability for slower permeants such as erythritol and malonamide by a factor of more than 10,000. Were the model to underestimate the permeability, such lack of accord could be explained by postulating some additional specialized system of penetration in parallel with penetration by pores, but overestimating the permeability can only be accounted for if the pores are *selective* for factors other than size—or if the size of the pore is grossly misjudged. The 2.8 Å pore gives apparently a better fit to the data but still overestimates the permeability for malonamide by one-hundredfold and now requires the additional assumption that a selective parallel system is required for erythritol.

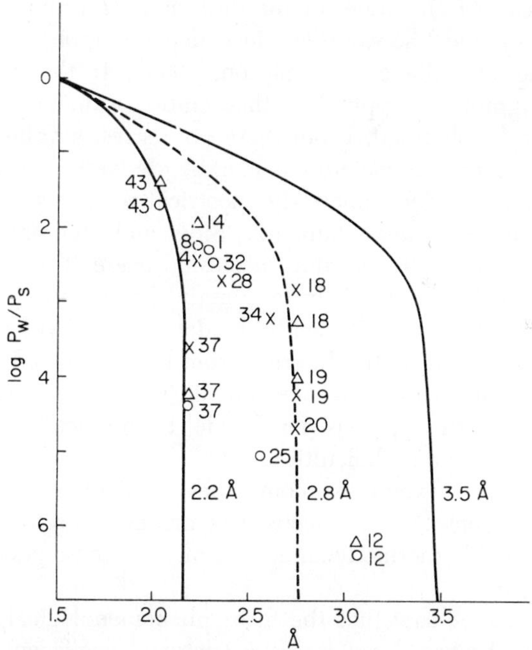

Fig. 3.16. Permeability of nonelectrolytes into human erythrocytes as a function of the molecular radius of the permeant. Ordinate: logarithm_{10} of the ratio of the permeability of water P_w to the permeability of the solute P_s, in the same units (data of Fig. 3.14). Abscissa: radius of permeant in Å, values as assigned by Goldstein and Solomon (1960). The lines drawn are the theoretical predictions of the Renkin (1954) equation in the form of Eq. (3.12), where the radius of the equivalent right circular pore is taken to be 2.2, 2.8, or 3.5 Å, as indicated. (Numbering as in Fig. 3.6.)

The derived value of r will be in error if either (1) the assumption that Poiseuille's law holds at low values of r is invalid, or (2) the Renkin equation does not hold at low r, or (3) the bulk values of the diffusion coefficient D and the viscosity coefficient η cannot be used for transport within such narrow pores or, finally, (4) the molecular dimensions used in Fig. 3.16 are incorrectly assigned. But these hesitations about the actual value to be assigned to r are subsidiary to the main doubt, whether the finding that L_p/P_w is greater than one, makes it indeed obligatory to postulate the existence of pores.

There are a number of situations where, in model systems (presumably possessing a homogeneous, nonporous structure), ratios of L_p/P_w greater than one have been found. Thus Thompson (1964) has prepared synthetic bimolecular lipid leaflets which separate two aqueous phases, and has measured the movement of water across these membranes both by radioactive tracers (P_w) and by observing a net flow caused by an osmotic gradient (L_p). Values of the ratio of L_p/P_w ranging from 4.4 to 104 have been found, the value obtained depending on the nature of the lipids forming the bilayer (Thompson, 1966). If the analysis of the previous paragraphs is applied to these ratios, values of the operative pore sizes can be obtained. From these pore sizes, a value for the electrical resistance of the membrane can be derived, but these derived values very much underestimate the electrical resistance that Thompson found for such membranes. Similarly, Sidel and Hoffman (1962) have measured values of L_p across films of mesityl oxide. The fairly thick film of mesityl oxide is very unlikely to contain the more or less permanent aqueous channels that the Poiseuille model requires, yet here again the value of L_p was significantly different from that predicted for free diffusion. In both biological membranes and in synthetic systems, therefore, the application of the aqueous pore model to the measurement of water transfer leads to serious difficulties.

We can to some extent overcome these difficulties if we avoid the explicit formulations of the aqueous pore model and return rather to the rigorous irreversible thermodynamic treatment of these phenomenological coefficients.

In Section 2.5 we saw that the three phenomenological coefficients σ, ω, and L_p can be translated into the frictional coefficients f_{sw}, f_{sm}, and f_{wm}, describing the solute-solvent drag and the friction between solute and membrane and solvent and membrane, respectively. In particular, Eq. (2.19) of that section, giving the ratio between the permeability coefficient ω for a solute and L_p, is of utmost relevance to the present discussion. In the case of water transfer, ω is correctly given by the permeability coefficient P_w, determined using isotopically labeled water, while the frictional coefficients f_{sw} and f_{sm} can with little error be equated with f_{ww} and f_{wm}. We assume that the isotopically labeled water behaves much as $^1H_2{}^{16}O$ insofar as its frictional drag with the membrane and with other water molecules is considered. [A small isotope effect does, however, exist (Wang *et al.*, 1953). Thus the rate of diffusion in water of THO is 14% smaller than that of $H_2{}^{18}O$.] Equation (2.19) then transforms into

$$\frac{\omega \bar{V}_w}{L_p} = \frac{f_{wm}}{f_{ww} + f_{wm}} \tag{3.13}$$

or

$$\frac{L_p}{\omega \bar{V}_w} = \frac{f_{ww} + f_{wm}}{f_{wm}} = 1 + \frac{f_{ww}}{f_{wm}} \tag{3.14}$$

The left-hand side of this equation is the quantity we are considering the ratio of L_p to P_w for water, and Eq. (3.14) shows, therefore, that the difference between this ratio and unity expresses the ratio of the water-water frictional coefficient to the water-membrane frictional coefficient. If, as for the human red blood cell, the ratio of the water transport rates is 2.4, we obtain the result that

$$f_{ww}/f_{wm} = 1.4$$

This result has a fairly rigorous thermodynamic basis. It is in the interpretation of the result and its application to model systems that grounds for controversy can arise.

The finding that the ratio f_{ww}/f_{wm} is not zero requires that the water molecules interact with one another during their movement across the membrane. It is established, therefore, that they do not penetrate as isolated molecules. Yet this association behavior is precisely what our analysis of the structure of liquids would lead us to expect (Section 3.2). Water molecules will form hydrogen bonds with other water molecules, both in the aqueous phase and within the membrane, and we should not be surprised at the finding that groups of water molecules travel together during bulk flow across the membrane. Since the hydrogen bond interaction between water molecules is likely to be more vigorous than the van der Waal's interaction between water molecules and the lipid side chains, we might well predict that f_{ww} is greater than f_{wm}. We should note, however, that these frictional coefficients measure in effect an overall value of the drag between the interacting molecules. This drag is made up of two parts, interactions taking place within the membrane and, in addition, those at the membrane-water interface.

Thus, the finding that L_p is greater than P_w implies that water molecules within the membrane are associated with one another. One interpretation of this result would be that the association occurs within more or less permanent aqueous channels extending through the membrane. We have seen that this model leads to difficulties if the probable permeability of such channels to other solutes is taken into consideration. Another interpretation of the permeability coefficient ratio, however, would be simply that, within the membrane, clusters of water molecules forming, breaking, and re-forming at random are to be found. The drag between water molecules within such a cluster is likely to be substantial and, in particular, to be greater than that between an isolated water molecule and the lipid side chains. Under conditions of bulk flow, these clusters of water molecules move through the membrane, whereas in exchange flow the clusters form pools with which the tracer exchanges during the passage through the membrane.

It should perhaps be emphasized that the relative validity of these two models—the assumption of permanent channels or the assumption of shifting aqueous pools within the membrane—is not indicated by the data on the water transport coefficients. It is only when the movement of nonelectrolytes is considered as well, that the aqueous channel model becomes less satisfactory. [We might note that a third explanation for the difference between the osmotic and the diffusional coefficients has been put forward by Dick (1959, 1964). Dick is of the opinion that diffusion within the bulk of the cytoplasm, if correctly taken into account, could explain these differences.]

B. The Reflection Coefficient σ

Additional evidence for the presence of pores in the human erythrocyte membrane comes from the studies of Goldstein and Solomon (1960) who have measured the reflection coefficient σ for a number of the permeants listed in Table 3.15. This reflection coefficient is, as we have seen in Section 2.3, a measure of the selectivity that the membrane demonstrates between a solute and solvent. If the assumption is made that permeation occurs through aqueous channels, these values of σ can be compared with theoretical plots of σ against r, the pore radius, derived on the basis of the Renkin (1954) equation. Figure 3.17, taken from Goldstein and Solomon, depicts the data plotted in this fashion and shows the accord with a pore size of 4.2 Å.

In spite of the good accord of these data with the 4.2 Å pore, it is clear from Fig. 3.16 that the analyses of the studies on σ and P_s are inconsistent—a pore size of 4.2 Å will overestimate several of the values for P_s by 10^4- to 10^5-fold. Conversely, a 2.8 Å pore underestimates σ for most species. The problem, then, is to reconcile these studies.

The two sets of data appear to be inconsistent when we test the assumption that penetration of all these permeants occurs through pores. We might, therefore, expect that a consistent picture would be obtained if the penetration of at least some of these permeants occurred rather by a parallel route, for example, by dissolution in the membrane. We can test this view by using Kedem and Kalchalsky's (1958) analysis (Section 2.4) of the interrelationships between the coefficients ω, L_p, and σ. It

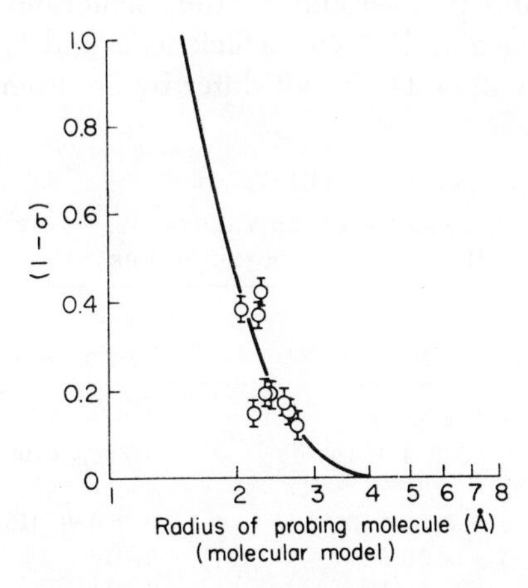

Fig. 3.17. Experimental values of $(1 - \sigma)$ for nine permeant molecules as a function of the radius of the permeant. The theoretical curve is drawn for a pore of radius 4.2 Å. (Taken with kind permission from Goldstein and Solomon, 1960.)

will be remembered that for a membrane in which both solute and solvent travel together through a capillary system,

$$\sigma = 1 - \frac{\omega \overline{V}_s}{L_p} - \frac{A_s}{A_w} \tag{3.15}$$

where $\overline{V}_s$ is the partial molar volume of the solute, L_p is the coefficient for the penetration of water under an osmotic gradient, while σ is the permeability coefficient of the solute measured under conditions where there is no volume flow. For a membrane where solute and solvent enter by separate pathways, for example, the solute penetrates by dissolution in the lipid, while the solvent water moves through pores,

$$\sigma = 1 - \frac{\omega \overline{V}_s}{L_p}$$

or (3.16)

$$1 - \sigma = \frac{\omega \overline{V}_s}{L_p}$$

In Table 3.16 we have listed values of $1 - \sigma$, taken from Goldstein and Solomon (1960) and computed corresponding values of $\omega\bar{V}_s/L_p$ from values of these parameters, taken from the literature. [Note that the values of ω that should be used in Eqs. (3.15) and (3.16) are, by the criteria of irreversible thermodynamics, the permeability coefficient P_s measured in the absence of volume flow, whereas in all the determinations recorded in Fig. 3.14 a volume flow occurred. We have seen, however, that if the water permeability of the membrane is high compared with that of the permeant, the two coefficients ω and P_s—when expressed in the same units—will certainly not differ by an order of magnitude.]

TABLE 3.16
CALCULATED AND OBSERVED VALUES OF $1 - \sigma$ FOR THE HUMAN ERYTHROCYTE MEMBRANE[a,b]

Permeant	P_s/P_w	$1 - \sigma$ calc. by Eq. (3.16)	$1 - \sigma$ observed
Urea	$1.7\text{–}4 \times 10^{-2}$	0.05–0.11	0.38
Thiourea	$0.89\text{–}2.4 \times 10^{-4}$	0.0003–0.008	0.15
Ethylene glycol	$3.0\text{–}5.3 \times 10^{-3}$	0.009–0.016	0.37
Acetamide	5.4×10^{-3}	0.016	0.42
Propionamide	4.0×10^{-3}	0.014	0.20
Methyl urea	2.1×10^{-3}	0.007	0.20
Malonamide	4.9×10^{-6}	0.000019	0.17
Glycerol	$0.084\text{–}1.3 \times 10^{-3}$	0.0003–0.005	0.12

[a] Observed values taken from Goldstein and Solomon (1960); calculated values from Eq. (3.16), using values of P_s as recorded in Fig. 3.14.
[b] *Note:* Equation (3.16) requires that values of ω be used to calculate $1 - \sigma$. Here we use instead the values of P_s, as values of ω are not available.

In all cases except that of urea the experimental value of $1 - \sigma$ is at least a hundredfold the value predicted from the experimental values of P_s and in some cases (for example, malonamide) 10^4-fold this value. For urea the accord is somewhat better, the experimental value being only twentyfold that predicted by Eq. (3.16). In fact, in this instance, there is a suggestion that the measured value of P_s reported in Fig. 3.14 is too low. Savitz, reported by Dainty and Ginzburg (1964b), has obtained a value of 10^{-3} cm sec^{-1} for the permeability coefficient of urea for human red blood cells and from this value a value for $1 - \sigma$ of 0.2 would be calculated, compared with the value of 0.38 found experimentally.

The available data for σ and P_s are thus not fitted by Eq. (3.16) and presumably Eq. (3.15) applies. Both solvent and solute, therefore, penetrate by the same route. If the assumption is made that this penetration takes place through water-filled pores, the values of $1-\sigma$ will be determined largely by the ratio of the filtration areas A_s/A_w (Section 2.5). This is, of course, the assumption made by Goldstein and Solomon (1960) in the analysis depicted in Fig. 3.17. But the view that permeation occurs through water-filled pores of 4.2 Å radius leads to the contradiction posed by Fig. 3.16—a pore of this size cannot account for the permeability coefficients P_s. It is clear then that either our theoretical analysis is too restrictive or else the *data* on σ and P_s are irreconcilable.

We can remove the restrictions that the pore model imposes by returning once again to the formalism of irreversible thermodynamics. If we combine Eqs. (2.15) and (2.16) of Section 2.5 we obtain a relation for the difference between $1-\sigma$ and the term $\omega\overline{V}_s/L_p$ as

$$1-\sigma-\frac{\omega\overline{V}_s}{L_p}=\frac{Kf_{sw}}{\phi_w(f_{sw}+f_{sm})} \qquad \text{[from Eq. (2.15)]}$$

$$=\frac{\Delta x}{(\phi_w)^2}\cdot K\cdot\omega\cdot f_{sw} \qquad (3.17)$$

or from Eqs. (2.15) and (2.17),

$$1-\sigma-\frac{\omega\overline{V}_s}{L_p}=\frac{\overline{V}_w}{\phi_w L_p f_{wm}}\cdot K\cdot\omega\cdot f_{sw} \qquad (3.18)$$

In both formulations the difference between $1-\sigma$ and $\omega\overline{V}_s/L_p$ is given by the product of four parameters, the first of which is a constant for the membrane, while the last three terms are characteristic of the solute. Table 3.16 shows that $\omega\overline{V}_s/L_p$ is small in comparison with $1-\sigma$, so that the left-hand sides of Eqs. (3.17) and (3.18) are effectively equal to $1-\sigma$. (Again, note that we have been forced to use values of P_s rather than ω in Table 3.16.) Now, for the solutes listed in Table 3.16, $1-\sigma$ increases as P_s increases, as Eqs. (3.17) and (3.18) predict. The increase in $1-\sigma$ is, however, only fourfold over the range from the slowest to the most rapidly permeating substance while the increase in P_s is over a five-thousandfold range. Furthermore, the data of Fig. 3.5 show that P_s, the permeability coefficient, and K, the partition coefficient, are proportional to one another, so that the product $K\,.\,\omega$ may well vary over a range of some 10^6. Thus the frictional coefficients for the solutes listed in Table 3.16 must decrease very sharply as ω increases. Now, the fastest permeants are, as we have seen in Section 3.3, those bearing the smallest number of water-binding groups, and hence those for which f_{sw} can be expected to be smallest. The prediction that ω and f_{sw} should be negatively correlated may well be borne out, and to this extent Eqs. (3.17) and (3.18) can be considered to fit the data, and the data on P_s and on σ may be reconcilable. We do not know the relevant values of f_{sw}, but it would be most unexpected if, for these solutes, we were to find the millionfold range that Eqs. (3.17) and (3.18) would appear to demand. We can make no firm decision on this point at this stage, yet it seems most unlikely that the range of values of $1 - \sigma$ found for these permeants is only 4-fold. This argument leads one to suspect that it is possibly the method of obtaining σ that is not wholly reliable.

The data on P_s are a collection of those from three different laboratories obtained by three somewhat different methods and are consistent among themselves. We have pointed out (Section 2.4) that these methods do not measure the true term ω for Eqs. (3.15) and (3.16), but there is little doubt that for the slower permeants the terms P_s and ω do not differ greatly. (It is clearly essential, however, to check this by measurements of the true ω for a wide variety of permeants.) The data on σ have been obtained as yet by a single method (albeit by a most ingenious one) and in a single laboratory, and it would be as well to attempt to measure σ here by some quite independent means. In the event that on further study the values of σ and of ω remain as they are at present reported, a further analysis of the theoretical formulations of Eqs. (3.15) and (3.16) will clearly become necessary. Until then, there is no reason to accept that the values derived for σ for the red cell membrane *demonstrate* the existence of pores in this cell membrane. Taken alone, however, the values of σ are indeed consistent with this view.

3.7 The Permeability of Water

Two considerations suggest that we should discuss more fully the available data on the permeability of water. First, it is the one substance that, almost uniformly, permeates at least an order of magnitude faster than a simple lipid structure appears to allow. Second, it is mainly the studies on water movement that have led to the proposal of the aqueous channels model of the cell membrane.

The data of Figs. 3.6–3.9 and 3.11, suggest that water penetrates some fifty times faster than would be expected from the number of hydrogen bonds by which it is presumed to be held within the solvent (aqueous) lattice. This lack of accord may be due to one or more of a number of reasons:

(a) We may here have incorrectly assigned N, the number of hydrogen bonds.

For the alcohols we have assumed that only a single hydrogen bond is accepted by each oxygen atom, while for water we assume that two such bonds are accepted. Perhaps then, it may be more correct to assign an N value of 3 to water. The permeability is now underestimated by a factor of 6 or so. Accord with the straight line of Fig. 3.11 is achieved only if an N of 2 is assigned to water which certainly does not seem a reasonable figure.

(b) We may have incorrectly designated the penetrating species.

Water exists at a very high molar concentration—55M—in most solutions. In such circumstances interaction between water molecules to form dimers and other more highly associated species must occur to a great extent.

The importance of such dimer formation is apparent from the analysis of some rather striking effects found when one studies the penetration of mixtures of nonelectrolyte solutes rather than the pure solutes alone. Consider, for example, Fig. 3.18 (taken from Stein, 1962b) which shows

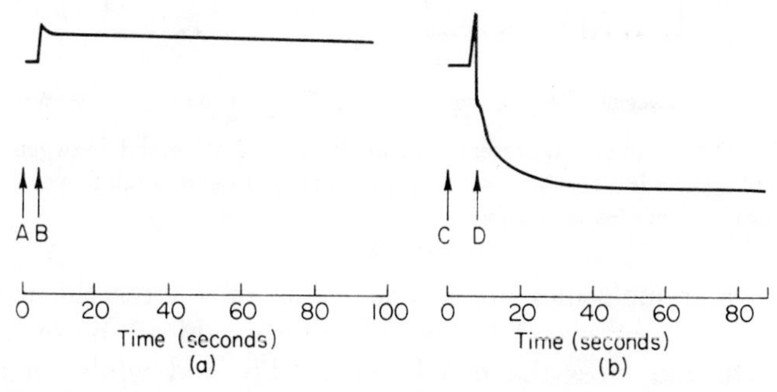

Fig. 3.18. The effect of dihydroxypropane on the penetration of glycerol into bovine erythrocytes. (a) Glycerol alone added to a cell suspension at point B. (b) Glycerol added at point D, to a second suspension of cells, pre-equilibrated with a molar solution of 1,3-dihydroxypropane. The tracings are records of the time course of the absorbancy change of the cell suspensions, a downward deflection indicating entrance of the permeant. (Taken from Stein, 1962b.)

a photometric record of the time course of swelling of bovine erythrocytes in two circumstances. In Fig. 3.18a, the washed cells are placed in 0.1M glycerol and the tracing obtained is consistent with a very slow entry of glycerol into the cell. In Fig. 3.18b, cells already equilibrated in 1M 1,3-dihydroxypropane are placed in a mixture of 0.1M glycerol and 1M 1,3-dihydroxypropane. We observe a dramatic rise in the rate of entry of glycerol. (There is no movement of 1,3-dihydroxypropane as this is always in equilibrium across the cell membrane.) Similar effects are found when glycerol entry (or exit) is studied in the presence of other glycols. The author has interpreted these results as indicating that an interaction occurs (by hydrogen bonding) between solute molecules in the bulk aqueous phase, when these are present in high concentrations. An appreciable formation of internally bonded species (such as are depicted in Fig. 3.19) will be found. If the resulting complex has fewer free hydroxyl groups than the parent monomers an apparent increase in permeability results. Studies with molecular models show that internally bonded complexes can be readily formed between glycerol and 1,3-dihydroxypropane. Since the anchoring hydrogen-bond-forming groups of the nonelectrolytes determine the rate of permeation of these substances across the cell membrane, if these anchoring groups are suppressed, an increase in permeability results. We have seen (Section 3.6) that measurements of the coefficients ω and L_p for water permeability indicate

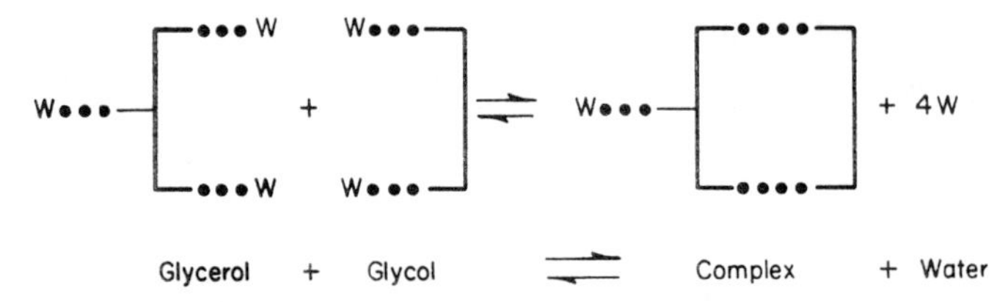

Fig. 3.19. Diagrammatic representation of the complex formed between a glycerol and a glycol molecule. The symbol W represents hydrogen-bonded water molecules displaced during complex formation.

that the water molecules are associated as they cross the membrane. Dimers, trimers, and even tetramers of water would still be small enough to penetrate and cross the membrane readily and while, in such circumstances, the number N of hydrogen bonds broken per penetrating n-mer is still probably 3 or 4, the average number broken per monomer—and it is this that determines P_w—is reduced.

(c) The interior of the membrane may present less of a barrier to water diffusion than the $DM^{1/2}$ law would predict.

The $DM^{1/2}$ law rests on the assumption (Section 3.2) that the size of the penetrating molecule is of the same order of magnitude as a "hole" in the lattice. A water molecule may, however, be somewhat smaller than the average "hole" size in the membrane lattice. If, in addition, a range of "hole" sizes is available, and especially if the distribution of sizes about the mean is skew, an appreciable number of "holes" may be available only for the smallest molecules, whose movement would be correspondingly faster than expected. Since the "holes" in the lattice are formed by the spaces between the lipid chains, it is quite likely that a large number of these spaces are such that water and other small molecules could slip through—but not molecules of much greater size. No estimate seems to be available of the distribution of "hole" sizes in a lattice formed by hydrocarbon chains, but these computations would be of much interest.

(d) Penetration may occur through narrow water-filled channels of a more or less permanent nature.

We have already discussed this very plausible explanation in Section 3.6.

Some evidence bearing on these alternative possibilities comes from an analysis of the available data on the temperature coefficient of water permeability. Thus an estimate of the enthalpy change $\Delta H^{\ddagger}$ for the transformation into the transition state can be arrived at by using Eqs. (3.7) and (3.8). (The available data are collected in Table 3.17.) We

TABLE 3.17

ACTIVATION ENERGIES FOR MOVEMENT OF WATER ACROSS CELLS AND TISSUES

Cell or tissue	Coefficient measured	Activation energy (kcal mole^{-1})	$\Delta H^{\ddagger}$	References
Arbacia punctulata (sea urchin egg)				
Unfertilized	L_p	14.7–16.6	14.1–16.0	McCutcheon and Lucké (1932)
Fertilized	L_p	20.3	19.7	McCutcheon and Lucké (1932)
Ascites tumor cell	L_p	9.6	9.0	Hempling (1960)
Bovine erythrocyte	P_w	5.4	4.8	Paganelli (1962)
Nitella translucens (algal cell)	L_p	8.5	7.9	Dainty and Ginzburg (1964a)
Turtle bladder				
No vasopressin	P_w	9.8	9.2	Hays and Leaf (1962b)
With vasopressin	P_w	4.1	3.5	Hays and Leaf (1962b)
Water (self-diffusion of DHO)	P_w	4.6	4.0	Wang (1951)

have seen that a value of 4–5 kcal/mole would be expected for the breaking of a single bond between a pair of water molecules (Section 3.2). The more recent estimates of the enthalpy change in Table 3.17 suggest therefore that two hydrogen bonds anchor each water molecule in the aqueous lattice, while the *a priori* analysis of Section 3.3 assigned an N of 3 or 4 to water. The experimental data do not allow an unequivocal assignment of N for water, but the value of 4 assigned from first principles is perhaps high. We can argue, therefore, that part of the reason for the apparently too rapid rate of water penetration may be due to our incomplete knowledge of the structure of the water lattice. The postulate that a range of pore sizes exists within the lattice of the membrane might well account for the remaining part of the apparently enhanced rate. It will be necessary to do the required computations before it can be stated with certainty that a homogeneous lipid membrane model cannot account for high rates of water permeability.

Meanwhile, if water does indeed penetrate through narrow pores, the data of Table 3.17 have some bearing on the processes involved. For the diffusion of water in water, the enthalpy change is 4–5 kcal/mole (Table 3.1), consistent with the single bond breakage model for diffusion. Since an enthalpy of twice this value is found for membrane penetration, apparently water diffusion in the narrow pores of the membrane is not identical with water diffusion in a bulk aqueous phase. Too much should not be made of this point, however, for a part of the explanation of the high enthalpy change might arise from some temperature-dependent change in membrane structure. Thus Dainty and Ginzburg (1964a) suggest that a narrowing of the pores in the membrane with increased temperature might lead to high values for the activation energy of water permeability, while Hempling (1960) similarly stresses that a part of this enthalpy change may be due to a structural alteration in pore geometry. The derived values for the enthalpy change, then, while low if water penetrates across a homogeneous lipid membrane, are high if water penetrates through narrow pores—but the data cannot be said to exclude either view, and the molecular basis of water movement remains obscure.

3.8 Conclusions

It is clear that no simple monistic view of the cell membrane is tenable. The model of a bimolecular lipid leaflet with permeation occurring by dissolution in the lipid layer leads to verifiable predictions as to the transition state for penetration into the lipid, to the characterization of cell membranes in terms of a maximum permeability $PM^{1/2}{}_{max}$, and correlates a large amount of the available data on permeability coefficients. But a number of additional *ad hoc* assumptions must be made to account for the high permeability of water and of ions in most membranes, of urea in animal cells and of various metabolites. Specific membrane transport systems must be postulated for some of these permeants. For the metabolites—and also for urea—additional evidence that such specific systems occur comes indeed from studies on the selective inhibition of these systems (Chapter 4). For water, the statement that water permeability is higher than would be expected for a lipid membrane might as much reflect our ignorance of the structure of liquid water as our ignorance of the structure of the membrane.

Equally, the view that the membrane is composed of water-filled pores, selecting permeants by virtue of size alone, is untenable. The presence of the specific inhibitible transport systems must imply that the pores are able to be blocked selectively against these permeants. No single value of the radius of the pore can adequately account for the data on the nonspecific systems and, as we have seen, a more profound analysis of hydrodynamic flow across a homogeneous lipid layer indicates that a porous membrane is not necessary to account for the difference between osmotic and diffusional flow.

It is clear, however, that an extended study of the permeability coefficients ω and σ for a well-chosen set of permeants, chosen so that (1) their size, (2) their ability to form hydrogen bonds, and (3) their ability to form hydrophobic bonds, could be independently varied, would enable a clear test to be made between these opposing views on the fundamentals of membrane structure—setting aside for the present the specific systems. The theoretical basis for such a definitive test between the two available models has been laid by the irreversible thermodynamic analysis. It remains to provide and to interpret the experimental data.

In the meantime, the model that the cell membrane is basically composed of hydrocarbon chains—with penetration occurring by diffusion through a lattice formed by the spaces between these chains—is on the whole tenable. The proviso must be made that the lattice spaces need not be rigidly fixed at some particular constant value. Rather, a range of "hole" sizes will be available, with rather more small "holes" than large "holes" being present. In succeeding chapters we shall have to modify still further this model in order to account for the mechanism of action of the specific metabolite transport systems.

CHAPTER 4

Facilitated Diffusion—the Kinetic Analysis

4.1 Introduction

We saw in Chapter 3 (Fig. 3.14) that certain molecules of physiological importance may penetrate cell membranes at a rate faster than would be predicted from a naive consideration of membrane structure. For instance, glucose, if present at a concentration of 1 mM at 25°C, crosses the erythrocyte membrane of man some 10^4 times faster than would be predicted from $PM^{1/2}_{max}$ for this membrane, with the value $N = 5$ for the number of hydrogen-bond forming groups present in glucose. Glycerol penetrates this cell some 10^2 times faster than would be predicted for the relevant value $N = 3$. When Davson and Danielli, in 1943, published their classic monograph, "The Permeability of Natural Membranes," (Davson and Danielli, 1943) a number of cases of such accelerated transport had already been clearly recognized, their action being ascribed to the presence of "active patches" in the cell membrane. Now, some twenty-five years later, we have a clearer picture of the criteria for the identification of such specialized membrane transport systems. This type of transport has been given a name, "facilitated diffusion," and a fair number of such systems have been adequately characterized in a number of cell types. We are still, however, only a little further advanced in our knowledge of the molecular mechanism by which these systems act. Their physiological role is plain. The cell membrane (we presume a lipid barrier) can, as we have seen, distinguish between permeants only by virtue of their size ($M^{1/2}$), the number of their hydrogen-bonding groups (N) and the number of their bare —CH_2— groups. To allow easy access for specific metabolites from the environment (such as energy sources, building materials, and vitamins) and to facilitate easy removal of waste products, the cell must have a greater flexibility of response than a simple lipid membrane can give. The specific permeability systems provide such a flexibility and their function is to control selectively the movement (egress and ingress) of specific components of the environment. We shall discuss in Chapters 5 and 6 more complex systems which allow a control of the concentration of specific metabolites inside and outside the cell. In this chapter we discuss—perhaps arbitrarily—only those systems which control the *rate* at which selected permeants enter the cell down a preexisting concentration gradient and it is these systems which have been termed by Danielli (1954) the "facilitated diffusion" systems.

4.2 Criteria for Identifying Facilitated Diffusion Systems

The experience of a number of investigators (see the reviews of Bowyer, 1957; Wilbrandt and Rosenberg, 1961; Stein, 1964a) leads to the following list of properties shared by many facilitated diffusion systems which may be considered as the defining characteristics of this class:

(a) Such systems operate on an existing electrochemical gradient of the permeant and lead to the disappearance of this gradient. They require no other input of free energy, except—in the long term view—for that required for the maintenance of the *structure* of the cell membrane. This criterion excludes active transport systems (see Section 2.8).

(b) The rate of penetration of a permeant is greater than could reasonably be expected from the number of hydrogen-bonding groups present in the permeant, $PM^{1/2}_{max}$ for the membrane being known (see Section 3.3). Likewise, the temperature coefficient of permeation is *less* than would be expected from the number of hydrogen-bonding groups in the permeant. If the permeant is optically active, its optical enantiomorph is likely to have a very different rate of penetration if this occurs by facilitated diffusion—by simple diffusion these rates should not differ.

(c) The rate of penetration may be expected not to be directly proportional to concentration but to reach a limiting (saturation) value as the concentration is increased. Fick's law in the simple form of Eq. (2.2) of Chapter 2 is thus not obeyed. We shall discuss this point in detail in this chapter.

(d) The rate of penetration may be markedly reduced by the presence of molecules structurally analogous to the permeant considered. This is the phenomenon of competition, as we shall discuss later in this chapter. We will see that criterion (c) is a special case of (d) where the competition is between identical, as opposed to similar, molecules.

(e) The rate of penetration may also be markedly and specifically reduced by the presence of substances differing chemically from the permeant, these inhibitors being often chemical reagents also active as enzyme poisons. The amount of such inhibitor present will often be insufficient to cover more than a small portion of the surface of the cell so treated. The significance of such observations will be discussed in Chapter 8.

(f) Certain clear differences between the rate of penetration of permeant measured on the one hand as net transfer of permeant or, on the other hand, as the unidirectional flux (cf. Section 2.1) of isotopically labeled permeant, may present themselves (see Section 4.4,C).

(g) It may be possible in certain cases to link the facilitated movement of permeant down its electrochemical gradient with the movement in the opposite direction of a structurally analogous molecule. This second molecule is thus driven up its electrochemical gradient. This is the phenomenon of "counter-transport" which we shall further consider below (Section 4.4,E).

Of these criteria, some may be considered as "weak." Thus criterion (b) is somewhat arbitrary—we do not include a statement of the amount by which a system must accelerate the rate of movement of permeant for it to be considered as a true facilitated diffusion system. Nor would we wish to include such a statement, inasmuch as this might exclude a slowly operating system or one which operates sluggishly on the particular permeant studied, when mechanistically a true distinction from simple diffusion arises. Criterion (c) is relatively strong, although at high permeant concentrations a lack of accord with Fick's law can be found if association occurs between molecules of the permeant in the aqueous phase. On the other hand, the absence of criterion (c) will not exclude categorization as a facilitated diffusion system, since the system may (as many do) saturate only at inaccessibly high permeant concentrations.

Criteria (d) and (e) are strong, provided that the demonstration that the penetration rate of the permeant under investigation is reduced by the inhibitor is accompanied by the control demonstration that a number of other permeants are not likewise affected, that is, that the lipid membrane barrier is not modified to any extent. Criteria (f) and (g) are very strong—it will be difficult not to conclude from the demonstration of one of these criteria that a facilitated diffusion system is present.

4.3 Distribution of the Facilitated Diffusion Systems

In Table 4.1 we record those systems of cells and substrates for which good evidence exists that a facilitated diffusion system operates. We record also which of the criteria (a) through (g) have been used to make this identification.

Facilitated diffusion obviously occurs in a wide range of cell types and for a smaller range of metabolites. Sugars, amino acids, and ions are the common substrates, as might have been expected. Intermediary metabolites, such as the phosphorylated sugars, intermediates on the pathways of carbohydrate breakdown and synthesis, amino acid anabolic and catabolic intermediates, and the "high energy phosphate" compounds are notable absentees from this list. There are few records of vitamins being transferred by facilitated diffusion. It is not clear whether these substances are absent from Table 4.1 because their permeabilities have not been studied or because there are indeed no specialized systems for their transfer. In many cases there are good physiological reasons for believing that the latter view will hold, but evidence bearing on this problem would perhaps be worth accumulating.

4.4 A Preliminary Kinetic Analysis of Facilitated Diffusion

A. The Unidirectional Michaelis–Menten Equation

It has been found experimentally (for example, Fig. 4.1 and see references in Table 4.1) that, for many systems classified by criteria (a) through (g) as facilitated diffusion systems, the unidirectional flux J of permeant across the membrane is given by an expression of the form

$$J = \frac{SV_{max}}{K_m + S} \tag{4.1}$$

where S is the concentration of permeant at the face of the membrane from which flow is occurring, and V_{max} and K_m are constants considered further below. Fick's law would give rather the simple form

$$J = PAS \tag{4.2}$$

where now P is a permeability coefficient, and A the area of the membrane.

We shall consider the kinetic analysis of facilitated diffusion at a number of levels. First we shall consider the kinetic consequences of Eqs. (4.1) and (4.2), accepting these equations as empirical descriptions of permeation behavior. Thereafter, we shall consider certain experimental data which require for their adequate description a more complex formalism than these equations allow. Finally, we shall attempt the difficult task of seeking to discover the molecular basis of permeation by setting up models for permeation and rejecting those models which do not accord with the required formalism.

TABLE 4.1

DISTRIBUTION OF THE FACILITATED DIFFUSION AND THE CO-TRANSPORT SYSTEMS BY SPECIES AND SUBSTRATE

Cell or tissue	Substrate	Defining criteria and references[a]	Comment
Bacteria:			
Escherichia coli	β-Galactosides (azide-poisoned cells)	(b) (inducibility), (d)–(f); (Kepes, 1960) (a) (c) (e) (f); (Koch, 1964)	Reduced rate of exit of galactoside is energy-dependent (Koch, 1964). In poisoned cells behaves as typical facilitated diffusion system (Winkler & Wilson, 1966)
	Galactose (exit)	(a) (b) (d); (Horecker *et al.*, 1960)	The whole system is an active transport, exit is facilitated
Staphylococcus aureus	Phosphate	(b)–(f); (P. Mitchell, 1954)	A phosphate exchange system
Streptococcus faecalis	Disaccharides	(a) (b) (e); (Abrams, 1960)	Accepts sucrose, lactose, and raffinose
Yeasts:			
Saccharomyces	Mono- and disaccharides	(a) (b) (d); (Burger *et al.*, 1959) (a) (b) (d); (Cirillo, 1961)	Aeration decreases rate of sugar entry
Saccharomyces	Glucosamine	(a) (c) (d); (Burger and Hejmová, 1961)	
Saccharomyces	Isomaltose, α-methyl glucoside	(a)–(d) (f); (Halvorson *et al.*, 1964)	The facilitated system is constitutive. It can actively transport following induction (Section 6.5)
Saccharomyces	Maltose	(a) (b) (e); (Rothstein, 1954; Sols and de la Fuente, 1961)	An inducible system
Candida (a psychrophile)	Monosaccharides	(a) (c) (d) (g); (Cirillo *et al.*, 1963)	Activation energy of transport low (12 kcal $mole^{-1}$) compared with *Saccharomyces* (49 kcal $mole^{-1}$)
"Sauternes" yeast	Monosaccharides	(a)–(d); (Sols, 1956b)	System has unusually high affinity for fructose
Other microorganisms			
Marine pseudomonad	α-aminoisobutyric acid	(b) (Drapeau and MacLeod, 1963)	Sodium-dependent, concentrative

Lower vertebrates:			
Squid nerve axolemma	Glucose	(a) (b); (Hoskin and Rosenberg, 1965)	Glucose enters; mannitol, sucrose, glutamate barred
Schwann cell	Glycerol	(a) (e); (Villegas and Villegas, 1962)	Copper an inhibitor
Frog skin	Chloride ions	(a) (e); (Ussing and Zerahn, 1951)	Copper an inhibitor
Frog skin	Na^+	(a) (c); (Cereijido *et al.*, 1964)	System present at outer face of epithelial layer
Toad bladder	Na^+	(a) (b) (c); (Frazier *et al.*, 1962)	System present at outer face of epithelial layer
Frog intestine	Sugars	(b); (Csáky and Thale, 1960) (c); (Csáky and Fernald, 1960)	A sodium + sugar co-transport system
Frog intestine	Amino acids	(b) (e) (sodium deprivation); (Csáky, 1961)	Na^+ + amino acid co-transport
Frog muscle	Sugar	(a) (c)–(e); (Narahara and Ozand, 1963)	Insulin increases V_{max}, not K_m
Pigeon erythrocytes	Glycine	(a) (c) (e) (g); (Vidaver, 1964a,b, and present text)	2 Na^+ + glycine co-transport
Mammalian tissues:			
Erythrocytes (Primates, rabbit, and various foetal animals)	Monosaccharides	(a)–(g); [see LeFevre, 1961a; Bowyer, 1957 (reviews); and present text]	The most completely characterized system
(Human, rabbit)	Glycerol	(a)–(e); (see Bowyer, 1957, and present text)	Low affinity for substrate
(Human)	Purines	(b) (c) (d); (Lassen, 1961; Lassen and Overgaard-Hansen, 1962a,b)	
(Human, rabbit)	Amino acids	(a) (c) (d) (f) (g); (Winter and Christensen, 1964, 1965)	Independent of sodium ion concentration
Reticulocytes (rabbit)	Amino acids	(a) (b) (sodium dependence), (c) (d) (f) (g); (Winter and Christensen, 1965)	A sodium-dependent transport system for glycine and alanine is lost on the maturation of the cell
Erythrocytes (human)	Urea	(b); (Danielli, 1954); (e) (tannic acid); (Hunter *et al.*, 1965)	No evidence for saturation or competition

(*continued*)

TABLE 4.1—*Continued*

Cell or tissue	Substrate	Defining criteria and references[a]	Comment
Lymph node cells (guinea pig)	Monosaccharides	(b) (d) (e); (Helmreich and Eisen, 1959)	
Leucocytes (guinea pig)	Monosaccharides	(a) (b); (Luzzatto and Leoncini, 1961)	Only pentoses studied
Leucocytes (human)	Glycine and α-aminoisobutyric acid	(a) (b) (Yunis *et al.*, 1962)	Sodium–dependent, concentrative
Fibroblasts (mouse)	Galactose, glucose	(b)–(e); (Rickenberg and Maio, 1961)	Studied in tissue culture
Ascites tumor cells (Ehrlich, mouse)	Hexoses	(d); (Nirenberg and Hogg, 1958); (b)–(e); (Crane *et al.*, 1957)	Showed crypticity-like effect of competitions
Ascites tumor cells (Ehrlich, mouse)	Nucleosides	(a) (c) (e); (Jacquez, 1962)	May use the hexose system
Ascites tumor cells (Ehrlich, mouse)	Glycine, alanine	(b)–(f); (see Christensen, 1962; and present text)	A sodium-dependent co-transport system (Kromphardt *et al.*, 1963)
Ascites tumor cells (Ehrlich, mouse)	Leucine, valine	(b)–(f); (Oxender and Christensen, 1963a,b)	Sodium-independent system, distinct from the above
Ascites tumor cells, Gardner lymphosarcoma (mouse)	Galactose	(d); (Nirenberg and Hogg, 1958)	
Muscle, striated and cardiac (many species)	Monosaccharides	(a)–(e) (g); (H. E. Morgan *et al.*, 1964; Park *et al.*, 1961; Henderson, 1964) (review)	Insulin-sensitive system, hormone affecting kinetic parameters
Diaphragm (rat)	Monocarboxylic acids	(a) (c)–(e); (Foulkes and Paine, 1961)	At high concentrations, penetration by simple diffusion is significant
Diaphragm (rat)	L-Tyrosine	(a) (b); (Guroff and Udenfriend, 1960)	Evidence not strong, viz., L-tyrosine taken up faster than D-isomer
Diaphragm (rat)	α-Aminoisobutyric acid	(a) (b); (Parrish and Kipnis, 1964)	Sodium-dependent, concentrative

Brain slices (guinea pig)	Sugars	(c) (d); (Gilbert, 1965)	Intracellular conc. may reach 2 × extracellular at low substrate levels; hence, concentrative?
Brain slices (rat)	L-Tyrosine	(b)–(d); (Chirigos *et al.*, 1960)	As above, levels reaching 1.5 ×
Brain slices (rat)	L-Histidine	(e); (DeAlmeida *et al.*, 1965)	A sodium-dependent system
Pancreas slices (mouse)	Amino acids	(c)–(e); (Bégin and Scholefield, 1965a,b)	Sodium-dependent concentrative systems
Kidney (dog)	Folic acid	(c) (d) (f); (Goresky *et al.*, 1963)	Studied *in vivo*
Kidney slices (rabbit)	D-Galactose	(c)–(e); (Krane and Crane, 1959)	Concentrative (3-fold), not yet shown to be true facilitated diffusion
Kidney slices (rat)	Amino acids	(c) (d); (L. E. Rosenberg *et al.*, 1961) (b); (M. Fox *et al.*, 1964)	Sodium-dependent (co-transport?)
Thyroid slices (sheep)	Inositol	(e) Hauser (1965)	Sodium-dependent (co-transport?)
Thyroid slices (sheep)	Iodide ion	(e); (Wolff, 1960, 1964)	Sodium-dependent
Adipose tissue (rat)	Glucose	(e); (Crofford and Renold, 1965)	An insulin-sensitive system
Intestine (many species)	Sugars	(a); (see Crane, 1965; and present text) (b)–(f); (see Crane, 1960, 1965) (g); (Crane, 1964)	Sodium-dependent co-transport
Intestine (many species)	Amino acids	(a); (Schultz and Zalusky, 1965) (b)–(e); (see Smyth, 1961)	Sodium-dependent co-transport
Placenta (sheep)	Glucose	(c); (Widdas, 1952)	
Placenta (sheep)	Fructose	(b); (Nixon, 1963)	Fructose faster than *meso*-inositol
Blood-cerebrospinal fluid barrier (dog)	Glucose	(b)–(d) (g); (Fishman, 1964)	
Blood-cerebrospinal fluid barrier (rabbit)	Glucose	(c) (d); (Bradbury and Davson, 1964)	

[a] The letters (a) through (g) refer to the defining criteria for facilitated diffusion, listed in Section 4.2, p. 127.

For the moment, we take V_{max} as a constant for each facilitated diffusion system, V_{max} depending under standard physical conditions only on the number of cells present, the cell type, and the class of permeant studied; K_m is a second constant which is independent of the number of cells present but which characterizes each cell type/permeant pair. We deliberately choose the symbols of Eq. (4.1) in order to link the description with the Michaelis–Menten formalism of enzyme kinetics. Indeed,

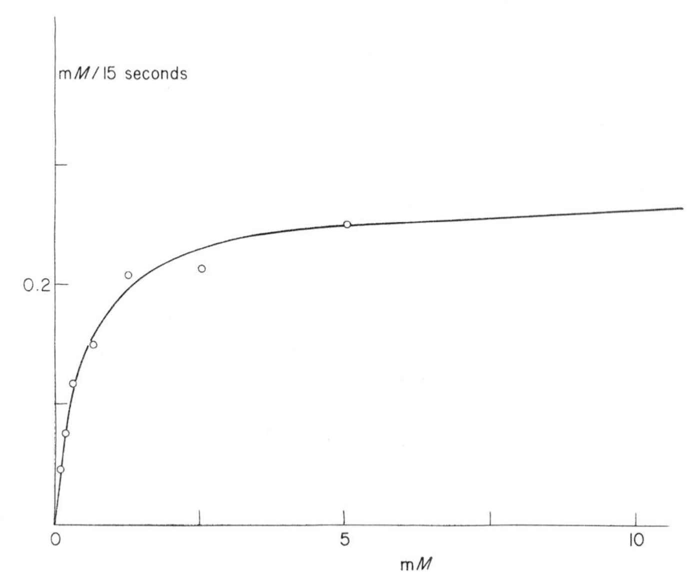

Fig. 4.1. Influx of ^{14}C glucose into human erythrocytes as a function of the glucose concentration in the medium. Abscissa: glucose concentration (m*M*). Ordinate: μmoles glucose entering 1 ml of intracellular water during 15 sec incubation at 5°C. Entrance stopped by the addition of ice-cold $HgCl_2$ solution in isotonic saline, containing potassium iodide (Stein, 1962e). ○, Experimentally determined points. Solid line is computed from the theory of Eq. (4.1), with K_m set equal to 0.5 m*M* and V_{max} set equal to 0.275 μmole ml^{-1} $(15\ sec)^{-1}$. (A determination at 20 m*M*, not shown for reasons of economy of space, gave the same uptake as the value at 5 m*M*.)

our term V_{max} expresses the maximum flux that the cell studied can demonstrate toward the permeant characterized. [This can be seen by letting the term for the substrate concentration S in Eq. (4.1) tend to infinity, when J becomes V_{max}.] Similarly, K_m is analogous to the Michaelis constant of Michaelis–Menten kinetics. Thus, K_m may in certain circumstances express some type of "affinity" of permeant with the system but formally expresses merely the substrate concentration at which the flux J is exactly one-half the limiting flux V_{max}. [This can be seen by putting $S = K_m$ in Eq. (4.1); J is now $V_{max}/2$.] The validity of the identification of K_m as an affinity is subject to the same type of ambiguity as found in classical enzyme kinetics (see for instance, Dixon and Webb, 1964), but discussion on this point is delayed until Sections 4.5 and 4.6,G.

To demonstrate the consequences of Eq. (4.1) we have plotted as the upper curve in Fig. 4.2 the unidirectional flux J against the substrate

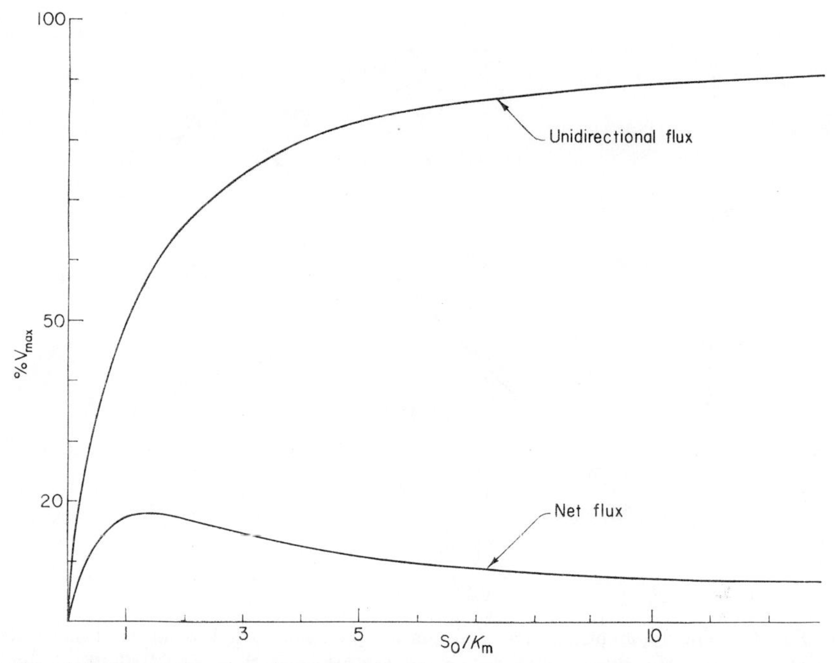

Fig. 4.2. Saturation of unidirectional flux (upper curve) and net flux (lower curve) with concentration, on the simplest formalism for facilitated diffusion, equations (4.1) and (4.3). Abscissa: the ratio of the ambient concentration of permeant (S_o) to the half-saturation constant for the system (K_m). Ordinate: flux at any S_o/K_m as a percentage of the maximum unidirectional flux. For the *net* flux, the internal concentration is chosen to be one-half the external concentration.

concentration S, computed according to Eq. (4.1). The axes chosen are, for J, the fraction of V_{max} that is reached, while S is expressed in terms of the ratio S/K_m a measure of the degree of saturation of the system. The upper curve in Fig. 4.2 thus characterizes all data of the form of Eq. (4.1) and demonstrates the saturation of the transporting system by high levels of substrate. Figure 4.2 should be compared with the experimental data of Fig. 4.1. Consider, however, Fig. 4.3 where the solid points are values computed for J from Eq. (4.1), taken only over the range $S = 0$ to $0.10K_m$. It is clear that these points will be difficult to distinguish from those fitting a straight line (for example, the line drawn, with slope 0.936), a relationship equivalent to that which Eq. (4.2) (Fick's law) would give. If the level of S is low in comparison with K_m, it will, therefore, always be rash to attempt to distinguish between behavior according to, on the one hand, Eq. (4.2)

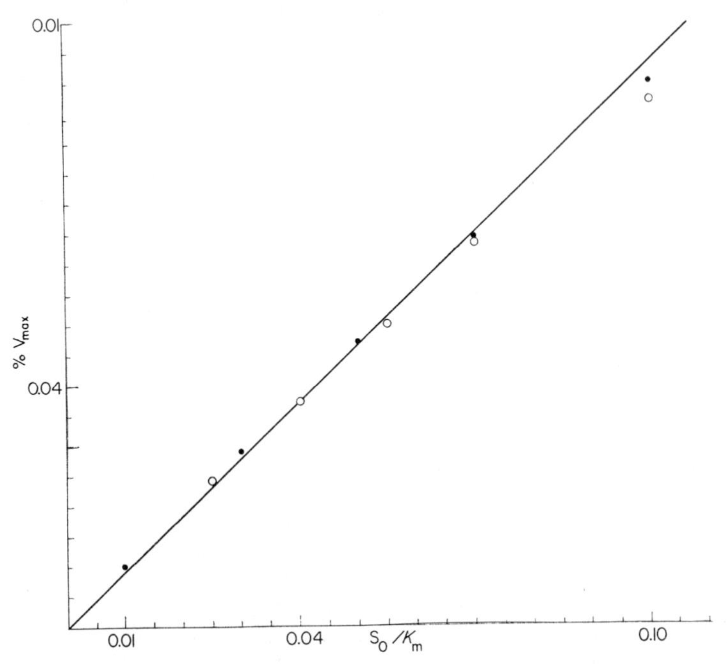

Fig. 4.3. The apparent absence of saturation phenomena at low levels of substrate relative to K_m. The straight line drawn has the arbitrary slope of 0.936. ●, Values computed for the unidirectional flux, using Eq. (4.1), over the range $S = 0$ to $0.1K_m$. ○, Data obtained by the author (Stein, 1962,e) on the uptake of radioactively labeled sorbose by human erythrocytes at 20°C. For the purpose of plotting the data, K_m had been chosen to be 500 mM, when V_{max} becomes 13 μmoles sorbose (ml cells)$^{-1}$ min^{-1}.

or, on the other, Eq. (4.1), using experimental data of the degree of accuracy generally achieved.

Some representative experimental data are given in the open circles in Fig. 4.3 where the influx of, in this case, sorbose into human erythrocytes is plotted against the concentration of sorbose. The linear relationship found here suggests either that this system is not one of facilitated diffusion or that the operative K_m is at least some tenfold greater than the highest substrate concentration reached here (compare the solid circles in this figure). Other evidence mentioned below supports the view that sorbose certainly enters this cell by a facilitated diffusion system.

The similarity between the formalism of enzyme kinetics and the formal description of transport by Eq. (4.1) can be extended to cover the action of inhibitors of transport. Competitive inhibitors will be those which have the effect of apparently increasing the term K_m in Eq. (4.1). By competing for the available sites on the transport system, these inhibitors demand an increase in the concentration of substrate which would be necessary to ensure that one-half of the transport sites are occupied by the substrate in question. Formally, K_m is replaced by the term $(1 + I/K_i)K_m$, where now I is the concentration of the competitive inhibitor and K_i is the "inhibitor constant," which can be defined as the concentration of inhibitor required to double the apparent value of K_m. A competitive inhibitor can itself be a substrate, but many cases are known in which an apparently strictly competitive inhibition is found between molecules which do not appear to be structurally analogous. (The comparative study of competitive inhibitors is discussed in Sections 8.1 and 8.2.)

In noncompetitive inhibition, the number of transport sites is diminished. This is formally expressed by a decrease in the apparent value of V_{max}; V_{max} is divided by the quantity $(1 + I/K_i)$ where I is again the concentration of inhibitor and K_i is again the "inhibitor constant," now the concentration of inhibitor required to halve the apparent value of V_{max}. Noncompetitive inhibitors can be either reversible or irreversible in their effects. We will have occasion to consider these in some detail in Section 8.3.

The determination of the inhibitor constants K_i from experimental data proceeds along the lines of classical enzyme kinetics (LeFevre, 1954; Bowyer and Widdas, 1958).

B. The Net Flow Equation

Our discussion thus far has introduced nothing beyond the most elementary enzyme kinetics. We must consider now the more complicated situation that arises when we have a *net* flow of permeant resulting from the algebraic summation of an inward flux J_{in} and an outward flux J_{out}, both being given by an equation of the form of (4.1). This case has been clearly analyzed by Wilbrandt and Rosenberg (1961) as follows:

We have

$$\text{Net flux} = J_{\text{in}} - J_{\text{out}}$$

$$= \frac{S_e V_{\max}}{K_m + S_e} - \frac{S_i V_{\max}}{K_m + S_i} \tag{4.3}$$

where S_e is the concentration of permeant at the outer face of the membrane, S_i that at the inner face. At this stage, we shall make the assumption that the terms V_{max} and K_m are the same at each surface of the membrane, that is, that the membrane is symmetric. Rearranging Eq. (4.3) we can easily obtain the result that

$$\text{Net flux} = \frac{(S_e - S_i) K_m V_{\max}}{(K_m + S_e)(K_m + S_i)} \tag{4.4}$$

Now if S_e and S_i are both small in comparison with K_m, Eq. (4.4) transforms into

$$\text{Net flux} = (S_e - S_i) V_{\max}/K_m \tag{4.5}$$

identical in form with that which would be derived directly from Eq. (4.2), that is, Fick's law.

In both Eqs. (4.4) and (4.5) the net flux will, of course, be zero for any value of S_e, if S_e is equal to S_i. But a very noteworthy distinction between these two cases arises if we take a ratio of S_e/S_i unequal to unity. Let us take, for example, the situation at the "half-equilibration time," that is, the time at which the cell is filled to 50% of its capacity after exposure to external substrate. Here $S_i = S_e/2$ or $S_i/S_e = \frac{1}{2}$. Now, if the system is behaving according to Fick's law [Eq. (4.5)], the net flux will be given by $(S_e/2) \times (V_{max}/K_m)$ and will increase without limit as S_e increases. If, however, the system is saturating [that is, behaving according to Eq. (4.4)], the net flux at the half-equilibration time will tend to zero as S_e increases. Because the system is saturated with permeant at both interfaces and is hence operating at approximately maximum velocity in both directions the flux in either direction is effectively canceled out by the similar flux in the opposite direction and, effectively, no *net* transport occurs. The lower curve of Fig. 4.2 shows this variation in the rate of net flux as a function of the ratio S_e/K_m, the degree of saturation of the system, for this case where S_i, the internal substrate concentration, is one-half of S_e. Clearly, the net flux reaches a maximum value as S_e is increased with respect to K_m. The value of the maximum net flux here is only 17.2% of the maximum unidirectional flux, while the maximum occurs in this particular case at a value of S_e equal to $1.414K_m$. Note that in the absence of saturation, that is, if the system behaves according to Fick's law [Eq. (4.2)], the net flux in the example chosen is always 50% of the unidirectional flux, the return flow being one-half of the outward flow by our chosen condition $S_i = 0.5S_e$. If we had chosen a value of S_i other than $0.5S_e$ the position and height of the maximum in the net flux would be altered. By taking the first derivative of the net flux equation with respect to the ratio S_e/K_m, we can readily show that the maximum in net flux occurs at a value of S_e/K_m which depends only on the ratio of S_i/S_e and is given by

$$(S_e/K_m) \text{ for maximum net flux} = (S_e/S_i)^{1/2} \tag{4.6}$$

On substituting this value for S_e/K_m in Eq. (4.4), we obtain a formula for the value of the net flux at this maximum

$$\text{Maximum net flux} = \frac{(1 - S_i/S_e)V_{\max}}{[1 + (S_i/S_e)^{1/2}]^2} \tag{4.7}$$

It is clear from the lower curve of Fig. 4.2 that a permeant which under the conditions of this figure exhibits a net flux of $0.1V_{\max}$ can do so at two different concentrations of permeant, one for which $S_e/K_m = 6.5$ (that is, at high S_e) and also one for which $S_e/K_m =$ approximately 0.3 (that is, at low S_e). Furthermore, if two different permeants are studied at the same substrate concentration S_e, they may manifest the same value of the net flux in spite of widely differing values of K_m. For the first permeant, a net flux of $0.1V_{\max}$ is given when the ratio of S_e/K_m is 6.5; that is, the system saturates at a low level of substrate compared with the operative level. The same value of net flux can be attained by the second substrate at a ratio of S_e/K_m of less than 0.3 since, for this second substrate, the system saturates at a high level of substrate compared with that prevailing. Such ambiguities can be resolved, of course, if K_m is known or if different values of S_i/S_e are studied. It can therefore happen that a permeant saturating at high levels of S will demonstrate a higher net flux than a permeant for which K_m is small. This paradox is in apparent contrast with the findings of simple enzyme kinetics (high affinity, high rate at a particular S and $V_{\max}$) but is a direct consequence of Eq. (4.3) and the physical situation that a high influx will be accompanied by a high efflux, resulting in a low net flux. [Similar kinetic relations are obeyed by enzyme systems catalyzing an easily reversible chemical reaction (see Dixon and Webb, 1964).]

We may appear to have spent too long on the development of this simple aspect of transfer kinetics, but it is essential that the point behind Fig. 4.2 and Eqs. (4.3) through (4.6) be thoroughly grasped, being crucial to the subsequent discussion.

C. Net Flux Versus Unidirectional Flux

A consideration of Fig. 4.2 will reemphasize that at levels of substrate which are high compared with the values of the Michaelis constant K_m, the net flux at any particular ratio of internal to external concentration will be low in comparison with the unidirectional flux at these prevailing concentrations. Formally, by dividing Eq. (4.2) by Eq. (4.4), we obtain the result that

$$\frac{\text{Unidirectional flux}}{\text{Net flux}} = \frac{S_e(K_m + S_i)}{K_m(S_e - S_i)} \tag{4.8}$$

this ratio tending to infinity with increasing S_e for any ratio of S_i/S_e. Now, it is possible to measure both the net flux and the unidirectional flux on the same sample of cells. The net flux can be determined by measuring the entrance of substrate by, for example, a chemical assay, while for the unidirectional flux one measures the rate of entry of isotopically labeled substrate when this is added to a cell suspension already equilibrated with the unlabeled permeant. One measures, in fact, the exchange of the labeled sugar entering the cell with the unlabeled sugar already present. Since the labeled sugar will be greatly diluted by the unlabeled sugar within the cell, the return flux of label is vanishingly small in the early stages so that it is effectively a unidirectional flux that is being measured. LeFevre in particular has pioneered this type of approach to the study of the transfer of sugars across the membrane of human erythrocytes and has provided definitive evidence for the inclusion of this system among the facilitated diffusion systems, according to criterion (f) of Section 4.2. Figure 4.4 depicts data obtained by LeFevre and McGinniss (1960) on the net movement of glucose from an external concentration of 184 m*M* as measured by a chemical method (open triangles and circles)—compared with the movement of radioactive tracer glucose added to cells equilibrated with unlabeled glucose also at 184 m*M*—effectively a unidirectional flux (solid triangles and circles). Note especially that the two sets of data have separate time scales, the scale for total uptake being expanded 12 times over that for tracer flux. From these data the time for half-equilibration of glucose by net transfer is some 85 times longer than the half-equilibration time for isotopic transfer.

This finding is a direct consequence of the applicability of Eqs. (4.1) and (4.3) to the transfer of glucose in this cell, for the situation where the prevailing substrate concentration S_e is high in comparison with the value of K_m. The data cannot be accounted for on any model which seeks to explain glucose transfer as occurring by simple diffusion but is easily explained by the formalism of facilitated diffusion. From Eq. (4.8), the ratio of the instantaneous rates of unidirectional flux to net flux when $S_i = 0.5S_e$ is given by the expression $\frac{1}{2} + S_e/K_m$. As a very approximate estimate we can take the time to half-equilibration as being (inversely) proportional to the instantaneous flux rate when $S_i = 0.5S_e$. Since the ratio of these two is 85 to 1 (Fig. 4.4) and S_e is 184 m*M*, K_m is here of the order of 2 to 3 m*M*. The full theory, which involves integrating the rate equations for this case, is given in the paper by LeFevre and McGinniss (1960).

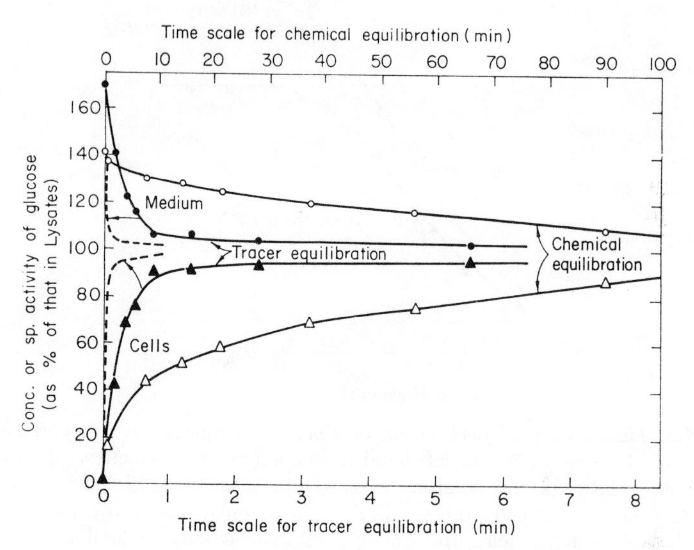

Fig. 4.4. A comparison of the rate of equilibration of tracer glucose and net glucose entry in the same cell suspension. ●, ○—Assays on incubation medium; △, ▲—cells. ○, △—Chemical equilibration; ●, ▲—tracer experiment. Note the different time scales used for tracer (bottom scale) and chemical equilibration (top scale). Broken curves are the tracer equilibration curves replotted on the reduced time scale of the chemical equilibration. (Taken with kind permission from LeFevre and McGinniss, 1960.)

On the other hand, Eq. (4.8) and Fig. 4.2 predict that if the value of K_m is *high* in comparison with the value of S_e and S_i used, then the ratio of rates of unidirectional flux and of net flow should not differ greatly from the value $S_e/(S_e - S_i)$ predicted by Fick's law. Figure 4.5 presents the data of LeFevre (1963) on the transfer of labeled and unlabeled ribose into human erythrocytes studied in a similar manner to that in Fig. 4.4. Here the ratio of the rates of net and unidirectional flux are comparable with the predictions of Fick's law. Since by other evidence (LeFevre and Marshall, 1958) it is considered that ribose does indeed enter the red cell by facilitated diffusion (and in fact uses the glucose system) it would appear from Fig. 4.5 that K_m for ribose is far larger than the operative substrate level.

We must note an extremely important consequence of Fig. 4.4 and its demonstration of the *simultaneous* applicability of Eqs. (4.1) and (4.3): it is that the inward and outward fluxes of necessity do not interfere with each other. The paths of influx and efflux are physically separate, an

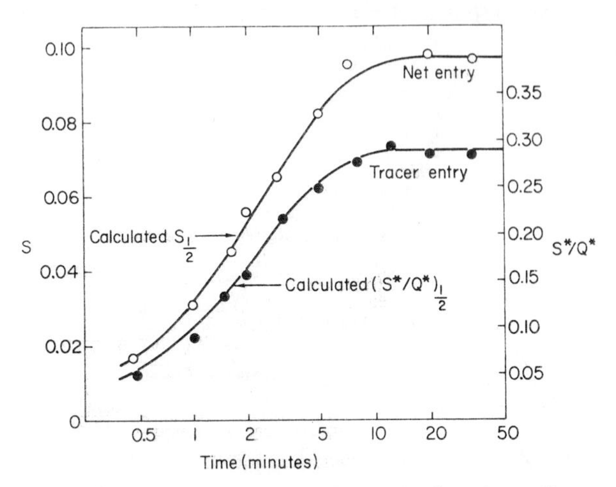

Fig. 4.5. Time course of penetration of ribose into human erythrocytes measured by a chemical method [○ and left-hand ordinate (in units which would constitute an isosmotic solution in the total suspension volume of water)] and by radioactive tracer assay on cells already pre-equilibrated with unlabeled ribose [● and right-hand ordinate (in units being the fraction cell radioactivity to total radioactivity)]. The time is plotted on the abscissa. Clearly the half-time of tracer uptake is of the same order as that of net uptake, in great contrast to Fig. 4.4. The solid lines in the present figure are theoretical curves computed on the basis of integrated forms of Eqs. (4.1) and (4.3) with $K_m = 1.5M$ and $V_{max} = 0.42M$ min^{-1}. (Taken with kind permission from LeFevre, 1963.)

assumption implicit in the derivation of Eq. (4.3) from Eq. (4.1). Were this not so, saturation of the influx path would simultaneously saturate the efflux path, and the unidirectional flux as well as the net fluxes would tend to zero with increasing concentration.

The ideas of this section will readily allow us to formulate the criteria for Ussing's flux ratio test for mediated transfer described in Section 2.8. We have seen that the unidirectional flux J_S of a permeant S that moves solely by a facilitated diffusion system is given by

$$J_{S_{\mathrm{I}\to\mathrm{II}}} = \frac{S_{\mathrm{I}}}{K_m + S_{\mathrm{I}}} V_{max}$$

with a similar expression for

$$J_{S_{\mathrm{II}\to\mathrm{I}}}$$

The ratio of the fluxes is, therefore,

$$\frac{J_{S_{\mathrm{I}\to\mathrm{II}}}}{J_{S_{\mathrm{II}\to\mathrm{I}}}} = \frac{S_{\mathrm{I}}}{S_{\mathrm{II}}} \times \frac{K_m + S_{\mathrm{II}}}{K_m + S_{\mathrm{I}}} \tag{4.9}$$

If, now, S_{I} is greater than S_{II}, the expression on the right-hand side of the equation is always less than the value $S_{\mathrm{I}}/S_{\mathrm{II}}$ predicted by Fick's law and becomes increasingly small (tending to unity) as S_{II} and S_{I} increase with respect to K_m, that is, as the system saturates. At very high substrate levels, with the system saturated at both sides of the membrane, the two fluxes will be almost identical although S_{I} may be manyfold greater than S_{II}. Substrate will exchange rapidly across the membrane, but net flow will be vanishingly small.

D. Competitive Exchange Diffusion

In LeFevre's experiments (Fig. 4.4) unidirectional flux was distinguished from net flux by the use of isotopically labeled sugar. If it is possible to distinguish analytically between different species of sugar molecules various experimental situations can be constructed in which phenomena superficially distinct from those of Fig. 4.4, but in principle identical in mechanism with them, can be found. Consider, for instance, Fig. 4.6. This figure shows data obtained by Lacko and Burger (1961) on the efflux of glucose from human erythrocytes at 0°C when various sugars were present at the external face of the cell membrane.

The efflux of glucose being measured by a chemical method is thus the net flux. This is the unidirectional flux only when no glucose is present in the external medium. The system described in Fig. 4.6 is, however, always effectively saturated with glucose at the external face; it can be calculated from Lacko and Burger's data that by the time the first measurement of the glucose concentration in the cell is made, the external glucose level is some 5 m*M*, while K_m is less than 0.5 m*M* at 0°C (Sen and Widdas, 1962a). If under these conditions, a second sugar possessing a low K_m for the system is present externally, this sugar can saturate the system at the external face of the membrane, when it will prevent the uptake of glucose from this face. This will therefore ensure that a unidirectional efflux of glucose really is being measured. External sugar, therefore, converts a net flux into a unidirectional flux. (It must be noted that Lacko and Burger have another interpretation of the data of Fig. 4.6, an important point to which we return in Section 4.5,B.)

Consider more particularly (in Fig. 4.6) the data obtained at approximately 7 min. If we compare the amount of glucose lost when D-galactose is present externally, with the control where saline only is added, D-galactose apparently pulls out a certain amount of glucose from the cell. (We know, however, that this effect is, for the most part, due to the blocking of glucose uptake that would otherwise occur in the absence of galactose.) When measurements of galactose uptake were made, Lacko and Burger found that galactose entered in an amount equivalent

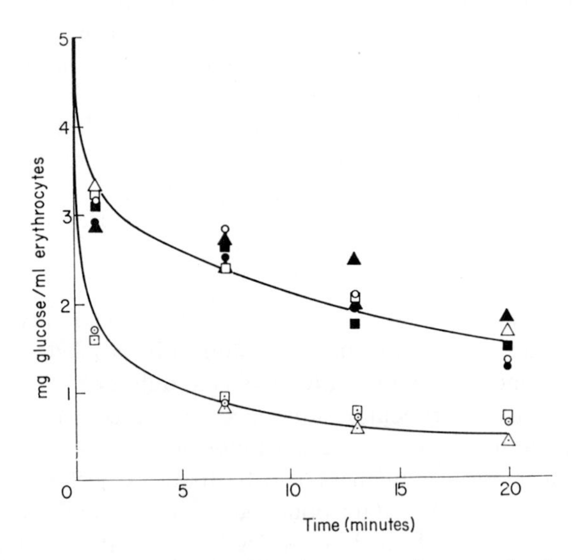

Fig. 4.6. Efflux of glucose at 0°C from human erythrocytes (preloaded with glucose to some 160 m*M* at 37°C) in the presence of various sugars (at 28 m*M*) added to the efflux medium. Upper curve—nonexchanging sugars: ○, □, saline control; △, L-sorbose; ●, D-fructose; ■, D-arabinose; ▲, D-ribose. Lower curve—exchanging sugars: ⊙, D-galactose; ⊡, D-mannose; ◬, D-xylose. (Taken with kind permission from Lacko and Burger, 1961.)

to the extra loss of glucose caused by the presence of galactose. There is apparently a mole for mole exchange of galactose for glucose. In fact, Lacko and Burger use the term "exchange diffusion" to describe this process. This term was, however, originally introduced in another circumstance as we shall consider in detail in Section 4.5,C. We shall, therefore, be more specific and use the term "competitive exchange" to describe results such as those of Fig. 4.6. Here, the major part of the apparent increase in efflux results from a decrease in the parallel influx, this decrease being brought about by the presence of the competing substrate. The phenomenon of competitive exchange will be found in any situation where a facilitated diffusion system is originally at or near saturation at both faces of the membrane and a second substrate sharing the same system is added (in substantial concentration in relation to its K_m) to one face of the membrane. The original low *net* flux will be converted into a high *unidirectional* flux. The second substrate will enter by the facilitated diffusion system and, mole for mole as it enters, will by competition with the first substrate prevent entry of that first substrate and will by an equal amount increase the net flux of the first substrate. The phenomenon of competitive exchange is a direct consequence of, and a test of, the simultaneous applicability of Eqs. (4.1) and (4.3).

E. Counterflow

There is another phenomenon, termed variously "counter-transport" (Wilbrandt and Rosenberg, 1961), "flow driven by counterflow," or more simply "counterflow" (T. Rosenberg and Wilbrandt, 1957b), that is another extremely important consequence of the saturation equations (4.1) and (4.3). Figure 4.7 (left-hand fig.), presents the data of H. E. Morgan *et al.* (1964) on the uptake of a nonmetabolizable sugar derivative, 3-*O*-methyl-D-glucose, by an oxygenated, isolated, perfused rat heart at 37°C. The upper curve depicts the uptake of the ^{14}C-labeled 3-*O*-methyl-D-glucose over a 45-min period when present alone at a concentration of 0.75 m*M*. In a second experiment (middle curve of Fig. 4.7), after 10 min of incubation with 0.75 m*M* 3-*O*-methyl glucose, the perfusion solution was substituted by one containing in addition 19 m*M*

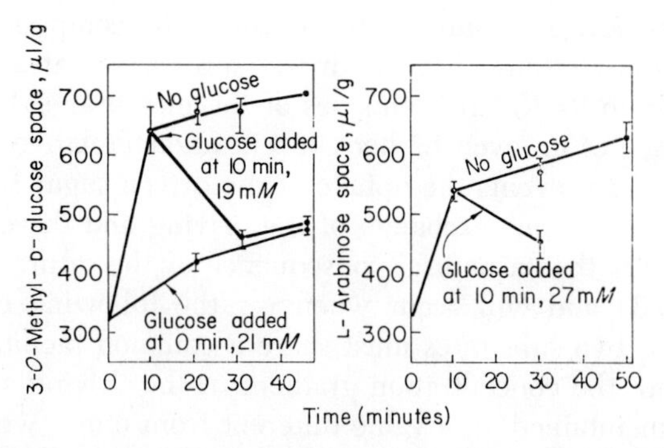

Fig. 4.7. (left) Countertransport of 3-*O*-methyl-D-glucose by D-glucose in the isolated, perfused rat heart. At zero time, hearts were perfused with buffer containing 3-*O*-methyl glucose at 0.75 m*M*. In the upper curve, no additions were made, in the middle curve 19 m*M* glucose was added after 10 min of incubation. In the lower curve 21 m*M* glucose was added at zero time. (right) Data for L-arabinose. (Taken with kind permission from H. E. Morgan *et al.*, 1964.)

glucose. The addition of glucose to the outside of the cells caused a flow of 3-*O*-methyl glucose out of the muscle until it reached a level corresponding to that reached when the 3-*O*-methyl glucose and glucose were present together during the entire course of incubation (lower curve of Fig. 4.7). Thereafter, the level of 3-*O*-methyl glucose slowly rose until an equilibrium level was reached substantially below that found in the absence of external glucose. We see that the flow of *O*-methyl glucose out of the cell after glucose is added—the counterflow—is against the prevailing concentration gradient of *O*-methyl glucose. The right-hand fig. of Fig. 4.7 presents similar data for the uptake of L-arabinose.

The difference between the phenomenon of counterflow (Fig. 4.7) and that of competitive exchange diffusion (Figs. 4.4 and 4.6) is a question of differing experimental conditions and involves no difference of mechanism. To demonstrate counterflow we arrange the experimental conditions so that in the absence of the driving substrate there is initially no net flux of the driven substrate (its concentration gradient is zero), or else the net flux is directed away from the face A to which we shall add the driving substrate. We then add the driving substrate to face A at a high concentration relative to its K_m. Then the resulting competition between the two substrates will prevent the movement of the driven substrate from face A but, for as long as the driving substrate remains at a low concentration at face B, movement of the driven substrate from face B will be largely unaffected. There will thus be a net movement of the driven substrate in the direction from face B to face A, that is, against the prevailing chemical gradient of this substrate. The phenomenon is most obvious when the concentration of the driven substrate is low relative to its K_m. In contrast, to demonstrate competitive exchange diffusion we must ensure that the driven substrate is at a high concentration relative to its K_m, at both sides of the membrane. Then the addition to one face of an even higher concentration (relative to *its* K_m) of the driving sugar prevents the uptake of the driven sugar from that face, and we see a one-to-one exchange of the driving and driven sugars.

Counterflow is thus a natural consequence of the saturation equations (4.1) and (4.3) and will occur whenever the following conditions are satisfied: First, two substrates must share a common facilitated diffusion system; second, the concentration gradient of the driving substrate must be made, or maintained at, a value different from unity; while, third, the concentration of this substrate at one or the other face of the membrane must be high in relation to its Michaelis constant. A fourth condition, of course, must also be satisfied if either counterflow or competitive exchange diffusion is to be found—the efflux and influx pathways should not interfere with one another—if a molecule moves out by using the facilitated diffusion system it must not simultaneously prevent another molecule from entering by the same system.

A fifth condition until recently has been held to be uniquely demonstrated by counter-transport (T. Rosenberg and Wilbrandt, 1957b). It was that the site or component combining with the permeants must itself move from one side of the membrane to the other during a cycle of exchange of the two permeants; that is, the membrane component must be a movable "carrier." The concentration gradient of the driving permeant would produce a concentration gradient of the carrier which would then produce a concentration gradient of the driven permeant.

Britton (1963) in a trenchant analysis has shown, however, that (as we saw above) counter-transport is kinetically equivalent to competitive exchange diffusion or to the demonstration of the occurrence of ratios of unidirectional flux to net flux greater than would be expected from Fick's law. All three phenomena require only that the pathways of entry and exit be physically separate (that is, that the inward and outward fluxes do not interfere with one another), and this condition can be as well achieved by a mobile site mechanism as by a system of unidirectional pores reserved specifically for entry or exit (see Sections 4.5 and 9.3).

The free energy required to drive out methyl glucose against its concentration gradient (Fig. 4.7) derives from the free energy present in the concentration gradient of glucose, the necessary coupling of the two flows occurring as a result of their sharing the same facilitated diffusion system. Wilbrandt and Rosenberg (1961) have derived on a simple model of competitive inhibition a general expression applicable to counterflow, which predicts the steady state distribution ratio of the driven substrate S to be

$$\frac{S_i}{S_e} = \frac{1 + G_i/K_m}{1 + G_e/K_m} \tag{4.10}$$

where G_i and G_e are the internal and external concentrations of the sugar that is driving the counterflow, and K_m is the Michaelis constant for that sugar. Equation (4.10) predicts that the distribution ratio reached by the sugar driven by counterflow is independent of the value of K_m for that sugar, and the same ratio will be reached by all sugars using the same facilitated diffusion system.

The example quoted here of Fig. 4.7 is an extreme case where the sugar driving the counterflow is rapidly metabolized and thus disappears almost entirely from one (here, the inner) face of the membrane. There is, therefore, at all times a concentration gradient present to drive a counterflow. In most cases, the sugar driving the counterflow enters the cell and accumulates. As its concentration rises within the cell, its ability to drive counterflow decreases with its decreasing concentration gradient, until an equilibrium state will be reached when the substrate driving the counterflow, and hence all other substrates, have the same concentration on either side of the membrane.

By measuring the ratio of S_i/S_e for 3-*O*-methyl glucose at various concentrations of glucose, H. E. Morgan *et al.* (1964) have derived values of K_m applicable to the transfer of glucose by the facilitated diffusion system of rat heart under various experimental conditions (see Table 4.3 of Section 4.6).

4.5 The Mobile Carrier Hypothesis

We have seen that all of the experiments described so far in this chapter can be accounted for by any model which allows for the physical separation of entry and exit pathways. Among such models is one developed particularly by Osterhout (1935), Widdas (1952), and T. Rosenberg and Wilbrandt (1955) in which facilitation of diffusion occurs by the attachment of the substrate to a component of the membrane, a "carrier" which shuttles between opposite faces of the membrane. Movement of carrier from inside to outside (exit) does not interfere with its movement from outside to inside (entry). An alternative model also described by Eq. (4.3) is one in which separate "pores" or channels are reserved for the entrance and exit of the permeant. These models are representative examples of two classes—the mobile site and fixed site models, respectively. We want now to present some experimental evidence which, when taken together with the experiments described previously in this chapter, suggest rather strongly that a fixed site model is untenable.

A. Nonpermeating Inhibitors

Consider first experiments in which the effects of nonpenetrating inhibitors of facilitated diffusion have been studied. Photometric records (Section 2.6) of the efflux of glucose from cells preloaded with sugar when efflux occurs: (1) into a simple saline medium, or (2) into the same medium with the addition of mercuric chloride, show clearly that no significant egress of glucose occurs in the latter conditions. It is important to note that the photometric record can be studied after only a few seconds of interaction between mercury and its receptors on the cell membrane. During this time little or no mercury can have entered the cell. [There is, however, a slow entry of mercuric ions into these cells as the chloride complex (Weed *et al.*, 1962).] Experiments with nonpenetrating sugars (disaccharides) (Lacko and Burger, 1962) and with other nonpenetrating competitive inhibitors (Wilbrandt, 1954) of glucose transport show that this finding is of wide occurrence. Exit as well as entrance of glucose (or of glycerol by the glycerol-specific system) can be blocked by action confined to the external face of the membrane. Interestingly, as Bowyer and Widdas (1958) have shown (Fig. 4.8), the action of externally applied nonpenetrating competitive inhibitors is, at first sight, apparently more effective on reducing exit than entrance. This is because competition between permeant and inhibitor can occur only at the external face, so that if the permeant is initially internal the maximum effect of the competitive inhibitor is displayed for the exit process.

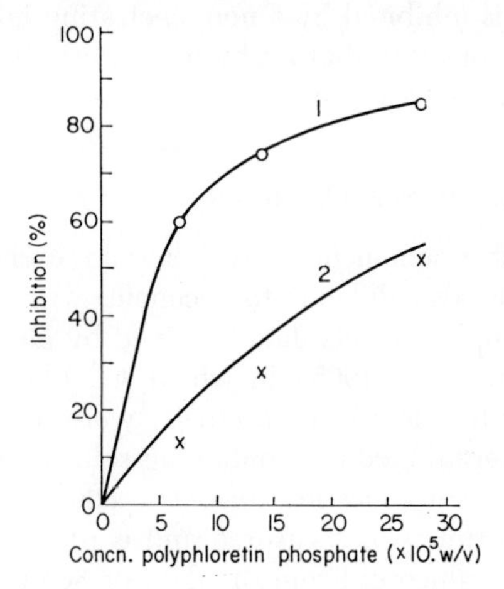

Fig. 4.8. The inhibition of glucose exit (curve 1) and glucose entry (curve 2) across the human erythrocyte membrane as a function of the concentration of the nonpenetrating inhibitor, polyphloretin phosphate, applied externally to the cell. Abscissa: concentration of inhibitor ($\times 10^5$, w/v). Ordinate: per cent inhibition of exit or entrance. (Taken with kind permission from Bowyer and Widdas, 1958.)

In no case of a facilitated diffusion system has it proved possible to demonstrate the preferential inhibition of entry over exit or vice versa. Yet the possibility of this type of inhibition would be expected if, as on a fixed site model, the entry and exit pathways were physically different portions of the membrane. Indeed, it is possible to argue that since the preferential inhibition of one direction of such a two-path system would yield a cell which could act as a valve, such a cell would contain a "Maxwell demon" and would allow the transport of substrate against a concentration gradient *without coupling to any free energy source.* On thermodynamic grounds this possibility can be dismissed.

There does not appear to be any good physical reason for assuming that we would not be able to find some chemical treatment which would block preferentially one direction of a two-path system. Thus there appear to be good grounds for rejecting any model based on unidirectional pores and hence for rejecting the fixed-site models.

The experiments on nonpenetrating inhibitors show then that the efflux and influx paths must be equally accessible to inhibitor at the external face of the cell. Yet the experiments on, for example, counter-transport establish that the carrier cannot be freely accessible simultaneously to the driving and driven permeant if these are confined to *opposite* faces of the membrane. Clearly, combination between the transport system and a permeant on the *cis* side is unaffected by any permeant present at the *trans* side, but is inhibited by a nonpenetrating inhibitor at that *trans* side. We can only conclude that a physical translocation of some part of the transport system must occur.

B. Accelerative Exchange Diffusion

There is another phenomenon involving an exchange between two permeants which is also difficult to reconcile with a fixed site model. Figure 4.9, for example, depicts data obtained by Dr. D. L. Oxender and the author (Levine *et al.*, 1965) in which red blood cells loaded with radioactive glucose to a low level internally were transferred at 0°C to a *large* volume of external medium containing saline, together with glucose at a series of increasing concentrations. The efflux of labeled glucose in a 30-sec incubation time was measured and is plotted in Fig. 4.9 against the level of external glucose. From the data of Sen and Widdas (1962a) (see Section 4.6) the value of K_m for this system is known. Thus the experimental conditions could be so chosen that with no glucose present externally the amount of glucose transferred in the extracellular fluid of the packed erythrocytes, together with that leaving the cell during the time of incubation, produced a concentration of external glucose far below K_m. A true unidirectional flux was therefore being measured. The presence of external unlabeled glucose could not (unlike Fig. 4.6) increase the net efflux by decreasing the antiparallel influx of labeled glucose, and the effect of external glucose in Fig. 4.9 appears to be a more direct one. Other experiments (Mawe and Hempling, 1965; Levine *et al.*, 1965) confirm this finding: The rate of unidirectional transfer of glucose depends on whether transfer is occurring into a glucose-rich or a glucose-free medium. Equations (4.1) and hence (4.3) do not now account for the experimental data. We term this new phenomenon of the direct acceleration of efflux by exchange "accelerative exchange diffusion." P. Mitchell (1961) clearly formulated the basis for this phenomenon, stating that here "the carried molecule tend[s] to facilitate the passage of the carrier (as the carrier-carried complex)" and Heinz and Walsh (1958) have provided definitive evidence for the operation of this phenomenon in the system which actively transports glycine into ascites tumor cells (see Chapter 5). For the modification of Eqs. (4.1) and (4.3), in order to account for Fig. 4.9, we will find it more natural to

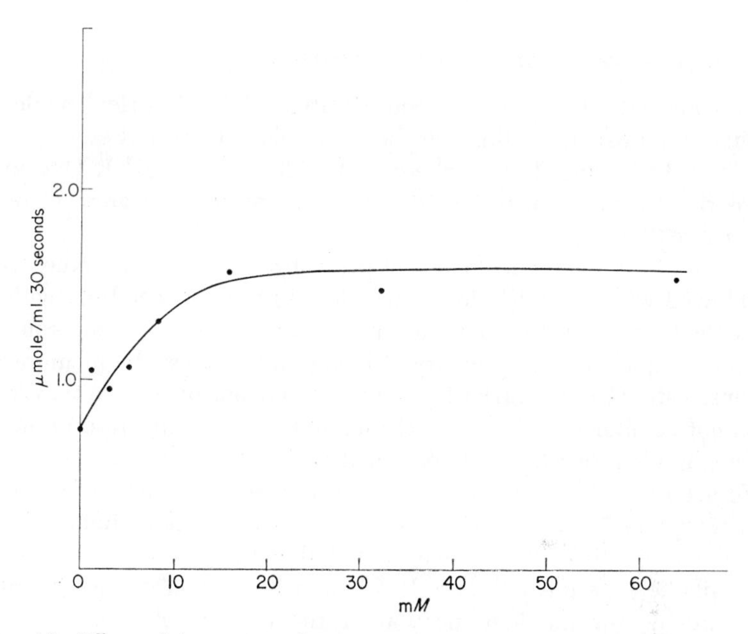

Fig. 4.9. Effect of increasing glucose concentration in external medium on the efflux at 0°C of labeled glucose from human erythrocytes preloaded to 2.6 m*M*. Abscissa: concentration (m*M*) of glucose in external medium. Ordinate: μmoles glucose lost in 30 sec per ml of cells incubated. (Taken from Levine *et al.,* 1965.)

proceed from the "carrier" model as conventionally held by most workers in this field, rather than to continue our formal analysis without using this concept.

As we shall see in Section 4.5,C, a formal description of the data of Fig. 4.9 is that the equation for the unidirectional flux out of the membrane contains a (saturation) term for the concentration of substrate at the outer face of the membrane. The data are consistent with a model in which a complex formed between the substrate and some component of the membrane moves through the membrane at a faster rate when so combined than when this membrane component is free—accelerative exchange diffusion. Lacko and Burger (1961), in the study from which Fig. 4.6 is taken, felt that their data provided an unequivocal demonstration of this point. We have seen that this demonstration will be provided only if it can be established that a unidirectional and not a net flux is being measured. Any *net* flux will, under saturation conditions on any model in which the entry and exit paths are separated, necessarily be affected by the substrate concentration on the *trans* side.

C. A Statement of the Carrier Hypothesis

We come now to a detailed consideration of the "carrier" model for facilitated diffusion, as this has been developed by LeFevre (1948), Widdas (1952), and T. Rosenberg and Wilbrandt (1957b). One might summarize the views of these authors by the following statement of the carrier hypothesis.

A carrier is a hypothetical element present in the cell membrane, which is endowed with the following properties: (1) It can combine with the molecule to be transferred to form a more or less transient complex; (2) the complex has the property of being able to cross the membrane—to translocate; (3) the carrier itself may or may not be able to translocate when not combined with the substrate; and (4) the substrate permeates at a negligible rate when not combined with carrier.

We need not here assess the reality of the carrier model. We consider the concept of "carrier" merely as a convenient fiction. Being *defined* with the properties listed above, it will thereby effectively account for the available experimental data. In Chapter 8 we shall discuss the methods available for the identification of the carrier substances and the progress that has been achieved on these lines. Meanwhile, let us explore the kinetic properties that these "substances" may have.

We shall write the equilibrium between the substrate S and the carrier E in the conventional form $E + S \underset{k_{-1}}{\overset{k_1}{\rightleftharpoons}} ES$. If we now also define rate constants k_2, k_{-2} for the translocation of the complex ES and rate constants k_3, k_{-3}, for the translocation of the free carrier E, we will have the scheme of Fig. 4.10, the simplest form required to account for the data available at present. The scheme of Fig. 4.10 carries implicitly the assumption that the membrane is symmetrical, that is, that the rate constant k_1 on one side of the membrane is equal to k_1 at the opposite face, with a similar assumption for the constant k_{-1}. We assume also that the oppositely directed rates of transit k_{-3}, k_3, on the one hand, and of k_{-2} and k_2, on the other hand, are identical. [Kinetic treatments which do not make these simplifying assumptions have been provided by

Jacquez (1961), Regen and Morgan (1964), and Britton (1966).] We shall not assume, however, that the k_2's are equal to the k_3's, that is, that the rates of transit of free and loaded carrier are the same. (That these rates are different is, as we shall see, an explanation of the data of Fig. 4.9.) We shall assume that the equilibria governed by the k_1's take place outside of the membrane, that is, that S does not enter the membrane

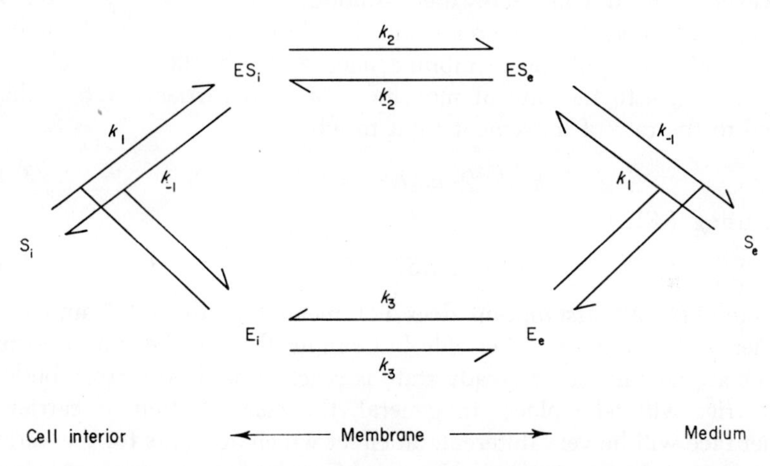

Fig. 4.10. A simple, symmetrical model for the "carrier" hypothesis, where E is the carrier (an enzyme?) and S the permeant (substrate?). The subscripts e refer to the external phase, i to the internal phase. The movements governed by the rate constants k_2, k_{-2}, k_3, k_{-3} take place within the membrane. The equilibria governed by the constants k_1, k_{-1} take place within the aqueous phases.

while the E and ES forms cannot leave the membrane. We will define—as in enzyme kinetics—a dissociation constant K_s of the carrier-substrate complex by the equation

$$K_s = k_{-1}/k_1 \tag{4.11}$$

and finally we will define a constant **r** which is the ratio of the rate of movement of loaded to free carrier, that is,

$$\mathbf{r} = k_2/k_3 = k_{-2}/k_{-3} \tag{4.12}$$

We denote components present on the external face of the membrane by the subscript e, those on the internal face by subscript i. We omit the square brackets in denoting concentrations. The total amount of carrier present per unit area of membrane, free or combined, we represent by Tot E. We have then the following relations:

$$K_s = k_{-1}/k_1 = (E_e \times S_e)/ES_e = (E_i \times S_i)/ES_i \tag{4.13}$$

from the dissociation equilibria, and

$$\text{Tot } E = E_e + E_i + ES_e + ES_i \tag{4.14}$$

a conservation equation.

These three equations alone would not be sufficient to provide an exact solution for the four unknowns E_e, E_i, ES_e, and ES_i. We have, however, a third condition: the assumption of the steady state. We assume that there is no increase (no buildup) of carrier with time on one or other side of the membrane once a steady state is reached, so that in Fig. 4.10 the rate of movement of total carrier left to right is equal to the rate of movement right to left, or

$$k_{-3}E_e + k_{-2}ES_e = k_3E_i + k_2ES_i \tag{4.15}$$

and using (4.12),

$$E_e + \mathbf{r}\, ES_e = E_i + \mathbf{r}\, ES_i \tag{4.16}$$

Note that this assumption does not mean that the total amount of carrier is the *same* on either side but simply that by the time measurements are commenced a steady state is reached, and no further buildup of carrier will take place. In general, the concentration of carrier at either face will be very different, an effect which accounts for the various phenomena of competitive exchange diffusion and of counter-transport (see Section 4.4) and many of the active transports of Chapter 6.

We should note here, too, the very important assumption implicit in our statement of Eq. (4.13): We have assumed that the movement of carrier and carrier-substrate complex through the membrane is rate-limiting (that is, it is slow) as compared with the rates of the reactions governing the formation of the carrier-substrate complex at the membrane surface. Widdas (quoted by Bowyer, 1957) has treated this assumption explicitly for the simple situation where $\mathbf{r} = 1$ (and also where $\mathbf{r}$ is less than 1). Britton (1966) has also provided a full kinetic treatment of this problem. We return to this point in Section 4.6,G.

It is now possible to solve Eqs. (4.13), (4.14), and (4.15). We obtain, on substitution, the very useful results that

$$E_i = \frac{(K_s + \mathbf{r}\, S_e)K_s \text{ Tot } E}{(K_s + \mathbf{r}\, S_i)(K_s + S_e) + (K_s + \mathbf{r}\, S_e)(K_s + S_i)} \tag{4.17}$$

with a corresponding equation for E_e to be obtained by interchanging the subscripts e and i in Eq. (4.17). Also we have from Eq. (4.13)

$$ES_i = \frac{(K_s + \mathbf{r}\, S_e)S_i \text{ Tot } E}{(K_s + \mathbf{r}\, S_i)(K_s + S_e) + (K_s + \mathbf{r}\, S_e)(K_s + S_i)} \tag{4.18}$$

with again a corresponding equation for ES_e, when e and i are transposed. From Eq. (4.18) we can easily obtain the unidirectional flux of substrate from right to left, this being identical [by assumption (4) of the carrier hypothesis] with k_2ES_i, the rate of transfer of the carrier-substrate complex from right to left. We obtain, therefore, the unidirectional flux J as follows, by substituting into Eq. (4.18),

$$J_{\text{right to left}} = k_2ES_i = \frac{k_2 \text{ Tot } E\ S_i(K_s + \mathbf{r}\ S_e)}{(K_s + \mathbf{r}\ S_i)(K_s + S_e) + (K_s + \mathbf{r}\ S_e)(K_s + S_i)} \tag{4.19}$$

Note now that the rate of flux right to left depends on both S_i and S_e, that is, it depends on the concentration of substrate at the external face, as the experiment of Fig. 4.9 demands. If, however, we assume that the rate of transfer of loaded and unloaded carrier is the same, that is, that $k_2 = k_3$, so that $\mathbf{r} = 1$, and we make this substitution in Eq. (4.19), we find an equation formally identical with Eq. (4.1):

$$J_{\text{right to left}\ \mathbf{r}=1} = \frac{k_2 \text{ Tot } E\ S_i}{2(K_s + S_i)} \tag{4.20}$$

The unidirectional flux is now no longer dependent on the *trans* concentration.

We should attempt to see the intuitive basis for Eq. (4.19) on the carrier hypothesis. It is as follows: Consider once again the experiment of Fig. 4.9, that is, where the efflux from the cell of labeled sugar occurs into a large volume of external medium free of sugar. Then as the loaded carriers traverse the membrane and reach the exterior, they will off-load their sugar and become free carrier. For a *continuing* efflux to occur, the unloaded carrier must now cross the membrane in the reverse direction to be, once more, reloaded. If we take as an extreme case, one in which the unloaded carrier cannot cross the membrane (that is, $k_3 = 0$, $\mathbf{r} = \infty$), only a movement of sugar stoichiometric with the number of carrier molecules will occur and then this once-and-for-all efflux will cease.

In a more general situation, unloaded carrier will be able to cross the membrane. Let us consider now the possibility that it crosses more slowly than the loaded form. Then the rate of efflux will be definite, but it will be limited by the rate of return of free carrier. Compare, now, the rate of efflux of label in the situation when no sugar is present externally, with the situation of Fig. 4.9 when we do have (unlabeled) sugar present at the outer face. The rate of efflux now depends on the rate of return of carrier and depends, therefore, on the degree to which the carrier is loaded, that is, on the degree of saturation of the carrier with sugar. The rate of unidirectional efflux of label then depends on the *trans* concentration of sugar, as in Eq. (4.19). If the free and loaded carriers cross the membrane at the same rate, there will be no increase in the rate of unidirectional efflux as the *trans* concentration of sugar is increased. If the loaded carrier moves more slowly than the free form, the rate of efflux will decrease as the concentration of permeant at the *trans* face increases. This then is the basis of Eq. (4.19).

It is of interest to consider further the case where the free carrier cannot cross the membrane, that is, when $\mathbf{r} = \infty$. Substituting this value for $\mathbf{r}$ in Eq. (4.19) and reducing, we have

$$J_{\text{right to left}} = \frac{k_2 \text{ Tot } E \; S_e S_i}{K_s(S_e + S_i) + S_e S_i} \tag{4.21}$$

This expression is symmetrical with respect to e and i. Thus the flux left to right is given by an identical expression, with the result that the net flux is everywhere and always zero. The flux ratio is, of course, unity for all values of the ratio S_i/S_e. A moment's consideration will convince one that if the free carrier cannot move, no net transport can occur, but a rapid exchange of tracer for unlabeled permeant or of two permeants sharing the same transport system will take place. It was the finding of this phenomenon during his studies on ion fluxes across the frog skin that led Ussing (1949a,b) to the concept of exchange diffusion. A tightly coupled exchange of permeants in the absence of any net flux, at all substrate levels, was the criterion proposed by Ussing for the phenomenon of exchange diffusion. Here we term this phenomenon "compulsory exchange diffusion," when $\mathbf{r} = \infty$. We contrast this with the exchange of permeants in the face of a low net flux, when this low flux is occasioned by a high saturation at both sides of the membrane—a situation which we have termed "competitive exchange diffusion."

"Competitive exchange diffusion" refers to a property of all transport systems for which the exit and entry paths are separate and requires only that the system be studied under conditions in which it is saturated at both sides of the membrane. "Accelerative exchange diffusion," in contrast, is more restrictive and will be found only for those systems for which the free carrier moves less rapidly than the bound carrier. The distinction between the two can be made if, when it is arranged that one face of the membrane be exposed to very low (nonsaturating) levels of permeant (so that competitive exchange is excluded), exchange still occurs. Accelerative exchange is an additional molecular property of the transporting system for those systems where it has been identified. The particular case first studied by Ussing where the free carrier cannot move at all and no net flux is ever found, is the limiting case of accelerative exchange when the parameter **r** becomes infinite. We have here still a further restriction on the molecular properties of the transporting system. This is "compulsory exchange diffusion."

In Chapter 8 we will discuss the available evidence on the nature and structure of the carriers. Meanwhile we can briefly summarize the information on the properties of carriers that kinetic analysis has revealed, which is as follows:

(1) The carriers are present in cell membranes in limited amounts.
(2) Carriers combine with permeants by undefined bonds but to an extent which depends on the nature of carrier and permeant.
(3) The carrier-permeant complex can cross the membrane and must be physically movable in order to be available to permeant first from one side of the membrane and then from the other.
(4) The rates of transit of free carrier and carrier-permeant complex can differ.

4.6 The Direct Determination of the Parameters K_m and V_{max}, Using the Flux Equations

A. The Integrated Rate Equation for Transfer by Facilitated Diffusion

Equations (4.1) and (4.3), although of value in leading to an understanding of the properties of the facilitated diffusion systems, are instantaneous rate equations and are difficult to use for interpreting experimental data, invariably obtained over a period of time. It is, however, relatively simple to integrate these equations with respect to time and thus provide forms amenable to experimental testing.

From the net flux equation (4.3) we can obtain the expression

$$dS_i/dt = \left(\frac{S_e}{K_m + S_e} - \frac{S_i}{K_m + S_i}\right) V_{max} \tag{4.22}$$

where dS_i/dt is now the rate of increase of concentration of permeant on the inside of the cell, and V_{max} is a new constant incorporating a term for the volume of the cell studied. We assume that the conditions of the experiment are such that the volume of the cell remains constant at all times. These conditions are fulfilled when tracer equilibration is studied or when the movement of substrates from concentrations relatively low compared with that isotonic for the cell is considered. For permeability measurements performed by following volume changes using the photometric method (Section 2.6) equations similar in principle to those of the present section can be derived and can be found in the original literature (Widdas, 1954b). Equation (4.22) integrates to give

$$(K_m + S_e)S_i + (K_m + S_e)^2 \ln(1 - S_i/S_e) = -K_m V_{\max} t \quad (4.23)$$

a form fundamental to nearly all the experimental methods used. S_i is now the concentration of permeant within the cell at time t. We will quote without proof various transformations of Eq. (4.23) applicable to particular experimental circumstances.

Equation (4.23) describes the predicted levels of substrate within the cell S_i at various times of assay, t, at different levels of external substrate S_e. No simple method of obtaining the desired constants K_m and $V_{\max}$ from experimental values of S_i, S_e, and t appears to be available, but in the author's laboratory (Levine, 1965) a successful method of fitting the observed data to a set of pairs of K_m and $V_{\max}$ chosen from a wide range has been derived on the basis of computer analysis. Similar procedures have been published by LeFevre (1966), Miller (1966), and for the analog computer by Hempling (1967).

B. The Half-Equilibrium Time Method

A rather easier method mathematically but one which requires more experimental data is to obtain from Eq. (4.22) an equation relating the time taken for cells to reach a distribution ratio of one-half when incubated with external substrate, that is, the half-equilibration time. This equation is obtained by making the substitution that $t = t_{1/2}$ when $S_i = S_e/2$ in Eq. (4.23). The derived equation is now (Levine *et al.*, 1965):

$$K_m V_{\max} t_{1/2} = [K_m + S_e][0.193 S_e + 0.693 K_m] \quad (4.24)$$

This equation contains the two unknowns K_m and $V_{\max}$. To obtain these two constants explicitly, we need the values at $t_{1/2}$ at two substrate concentrations at least, say, S_{I} and S_{II}. Then, from Eq. (4.14) by division,

$$\frac{t_{1/2} \text{ at } S_e = S_{\mathrm{I}}}{t_{1/2} \text{ at } S_e = S_{\mathrm{II}}} = \frac{[K_m + S_{\mathrm{I}}][0.193 S_{\mathrm{I}} + 0.693 K_m]}{[K_m + S_{\mathrm{II}}][0.193 S_{\mathrm{II}} + 0.693 K_m]} \quad (4.25)$$

Now, K_m is the single unknown in Eq. (4.25) and can be obtained by substitution in this equation. It is as well to choose widely separated values of S_{I} and S_{II} for an accurate value of K_m. With K_m known, $V_{\max}$ can be obtained from Eq. (4.24). LeFevre (1962) has solved Eq. (4.24) by a method again involving the use of a computer. For each experimental value of $t_{1/2}$ at a given concentration S_e, a range of values of K_m are taken, and V_{max} computed by Eq. (4.24). The computer will now plot the derived values of V_{max} against K_m for each $t_{1/2}/S_e$. Such a series of plots is given in Fig. 4.11 taken from the work of A. R. Kolber and LeFevre (1967). Since the acceptable values of K_m and V_{max} are those which satisfy Eq. (4.24) for all $t_{1/2}/S_e$ pairs, the point of intersection of the plots of K_m against V_{max} defines the acceptable values of K_m and V_{max}. In practice, since the value of $t_{1/2}$ is subject to experimental error, a range of acceptable values of K_m and V_{max} are obtained.

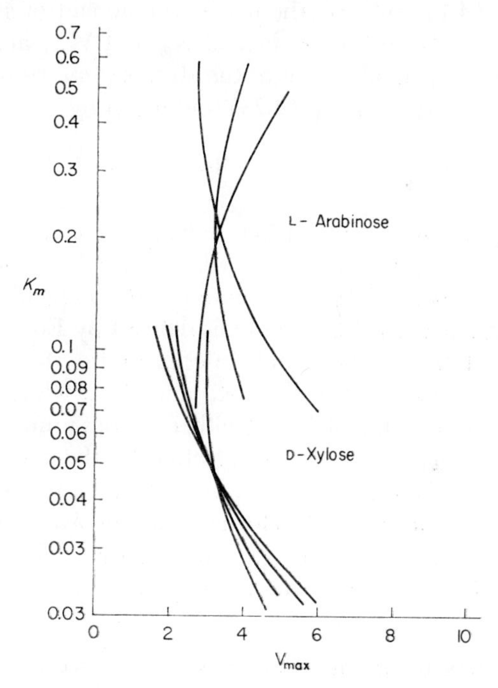

Fig. 4.11. Computer plots of the variation of K_m (ordinate) with V_{max} (abscissa) from experimental data on the penetration of L-arabinose and D-xylose into Ehrlich ascites tumor cells. Each experimental point, a pair of values of the half-time of uptake, $t_{1/2}$, at a particular concentration of sugar, S_e, is inserted into Eq. (4.24) (see Section 4.6,B) and the range of acceptable values of K_m and V_{max} are the curved lines in the figure. Where these lines intersect, the uniquely defined values of K_m and V_{max} are located. (Taken with kind permission from A. R. Kolber and LeFevre, 1967.)

C. Use of Limiting Forms of the Integrated Rate Equations

Equation (4.23) determines the internal concentration S_i reached at any time t from any level of external concentration S_e. It is of interest to develop two limiting forms of Eq. (4.23) (Stein, 1964c). The first is obtained by letting S_e tend to infinity in Eq. (4.23), and we obtain a value for the limit of S_i as S_e tends to infinity of

$$\lim_{S_e \to \infty} S_i = -K_m + \sqrt{K_m^2 + 2K_m V_{\max} t} \tag{4.26}$$

on expanding the logarithmic term in Eq. (4.23) and letting terms such as $(K_m + S_e)/S_e$ tend to unity, while terms such as $S_i^2/3S_e$ tend to zero.

This is a form similar to the limiting form of the Michaelis–Menten equation or Eq. (4.1), but now the maximum amount of substrate entering at time t depends on the product of K_m and V_{max} and not on V_{max} alone. Similarly, as S_e tends to zero, the distribution ratio S_i/S_e reaches an upper limit derived from Eq. (4.23) and given by

$$\lim_{S_e \to 0} S_i/S_e = 1 - \exp(-V_{\max} t/K_m) \tag{4.27}$$

obtained on rearranging Eq. (4.23), taking exponentials, and letting terms such as $(K_m + S_e)^2$ tend to K_m^2 when S_e tends to zero. Once again the limiting value, here of the distribution ratio, depends on both V_{max} and K_m at a particular t. If both the limit defined by Eq. (4.26) and that defined by Eq. (4.27) can be found experimentally, they will determine a single acceptable pair of values of K_m and V_{max}, which can be found by solving Eqs. (4.26) and (4.27). If either of these limits can be found and if $t_{1/2}$ is determined at one concentration S_e, the combination of the relevant two of the three equations (4.24), (4.26), and (4.27) will allow K_m and V_{max} to be determined. The hazards involved in extrapolating Eqs. (4.26) and (4.27) to the limits can be reduced by taking account of terms dropped in the expansion of Eq. (4.23).

D. A Method Based on the Balancing of Influx against Efflux

An ingenious and valuable method for determining K_m and V_{max} has been introduced by Sen and Widdas (1962a) based on the net flux Eq. (4.3). If cells are loaded by pre-equilibration with permeant, to a high level, the transfer system will be fully saturated on the inner face of the membrane if K_m is sufficiently small, and the efflux will be given essentially by V_{max}. Aliquots of such loaded cells are transferred to a number of different saline media containing various concentrations of the same permeant, and the net flux is measured. If Eq. (4.3) is obeyed, then at one particular external concentration of permeant, the influx will be exactly one-half the efflux so that the net flux will be one-half of V_{max}.

Thus where the net flux is one-half of V_{max} (the flux into zero external permeant) the external concentration must be K_m. In Fig. 4.12, taken from Sen and Widdas (1962a), the net outflow of glucose from cells loaded to a high internal level is recorded as a function of time for various levels of the glucose concentration in the external medium. That value of glucose concentration, for which the efflux is reduced to one-half of that when no external glucose is present, is K_m—and in the same experiment, of course, V_{max} can be found. Using this technique, Sen and

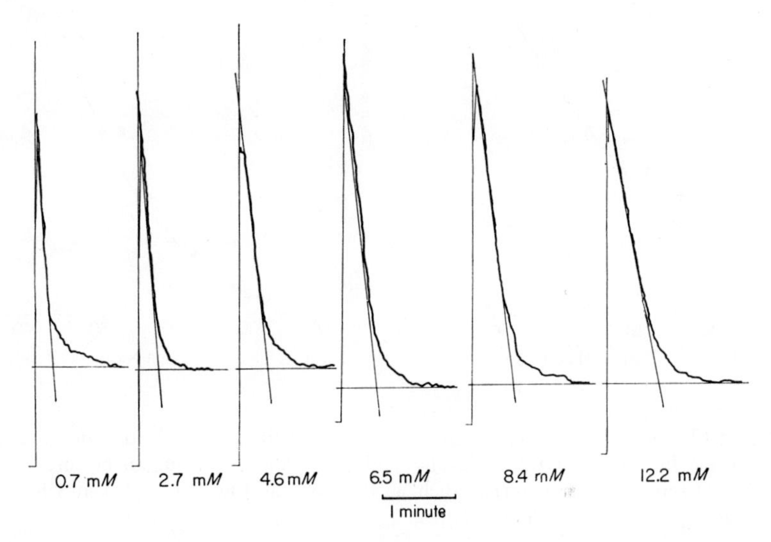

Fig. 4.12. Photometric records of the efflux of glucose from human erythrocytes at 37°C. The cells were loaded to 76 m*M* with glucose and were then added to 100 volumes of saline media containing increasing concentrations of glucose (at the concentrations shown below each figure). The linear part of each record is produced to cut the base line, and the time taken to reach this point is the measure of the rate of net efflux. (Taken with kind permission from Sen and Widdas, 1962a.)

Widdas have published a most comprehensive series of measurements of K_m and V_{max} for glucose transfer into human erythrocytes, the variation of these parameters as a function of temperature being given in Fig. 4.13. This method, originally based on the photometric recording of the volume changes following glucose egress (Section 2.6), has been confirmed using isotopic tracer techniques by E. J. Harris (1964) and Miller (1965a) and confirmed also in the author's laboratory (Levine, 1965) by a chemical method. It appears to be the method of choice in those circumstances where preloading of the cells with permeant presents no difficulty.

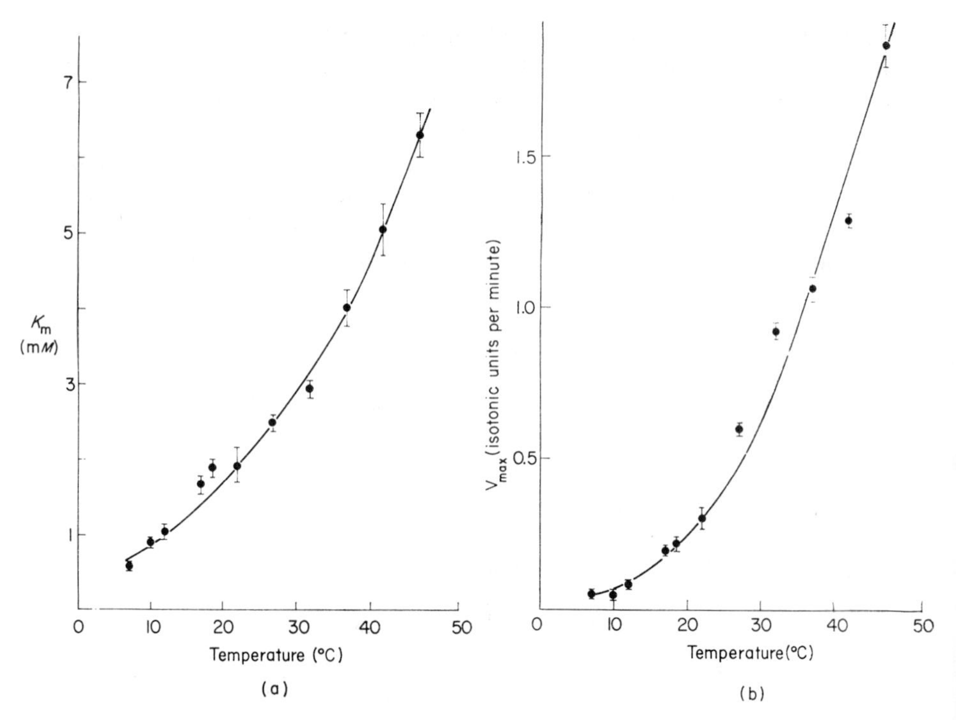

Fig. 4.13. The effect of temperature on (a) the half-saturation constant (K_m) and (b) the maximal transfer rate (V_{max}) for glucose efflux from human erythrocytes at pH 7.4. Determined from curves such as depicted in Fig. 4.12. (Taken with kind permission from Sen and Widdas, 1962a.)

E. Indirect Methods for Determining K_m and V_{max}

A study of the inhibition of permeant transfer by competitive inhibitors can lead to values for K_m and V_{max}, as can the study of counter-transport, referred to in Section 4.4,E. The available procedures are collected in Table 4.2. Values of the parameters K_m and V_{max} for a number of systems are collected in Table 4.3.

F. The Determination of the Parameter r

It will be of interest to consider briefly some of the consequences for kinetic analysis of Eq. (4.19). [A detailed treatment is available in Regen and Morgan (1964) and in Wilbrandt and Kotyk (1964).] We shall consider how the rate of unidirectional efflux J depends on the internal concentration of permeants S_i for the following two limiting conditions:

(1) The case *where there is no external permeant,* that is, where $S_e = 0$. Making this substitution in Eq. (4.19) we have

$$J_{out,\, S_e=0} = \frac{k_2 \text{ Tot } E\, S_i}{[2K_s + (1 + \mathbf{r})S_i]} = \frac{[k_2 \text{ Tot } E/(1 + \mathbf{r})]S_i}{[2K_s/(1 + \mathbf{r})] + S_i} \quad (4.28$$

Comparing this now with Eq. (4.1), we see that the term K_m, that is, the permeant concentration at which one-half the maximum velocity of efflux occurs, is now $(2/1 + \mathbf{r}) \times K_s$ and the term V_{max} is now $(1/1 + \mathbf{r}) \times k_2$ Tot E. Clearly, K_m is not a true dissociation constant but depends on the ratio $\mathbf{r}$ of the rate constants for transfer across the membrane.

TABLE 4.2

PROCEDURES USED TO OBTAIN THE KINETIC CONSTANTS K_m AND V_{max} FOR FACILITATED DIFFUSION SYSTEMS

Description	References
Balancing of entry against exit:	
Photometric	Sen and Widdas (1962a)
Isotopic	E. J. Harris (1964)
Isotopic	Miller (1965a)
Half-equilibration time for uptake of substrate:	
Photometric	Stein and Danielli (1956)
Isotopic, net uptake	LeFevre (1962)
Tracer exchange	Britton (1964)
Integrated rate equation:	
For exit and entry at brief times	Levine and Stein (1966)
Simultaneous entry of two sugars	Miller (1966)
Comparison of time course of net *vs.* tracer uptake	LeFevre and McGinnis (1960) LeFevre (1963)
Counter-transport	Regen and Morgan (1964)
Counter-transport, computer analysis	H. E. Morgan *et al.* (1964) Miller (1965b)
Competition methods:	
Substrate *vs.* competitive inhibitor at fixed inhibitor concentration	LeFevre (1954)
Substrate as a competitor (yields K_i rather than K_m)	Widdas (1954b) Levine *et al.* (1965)

TABLE 4.3

KINETIC PARAMETERS OF THE FACILITATED DIFFUSION SYSTEMS[a]

System	Substrate	Temp. °C	References[b]	K_m (mM)	K_m (isotones liter^{-1})	V_{max} (mM min^{-1})	V_{max} (isotones liter^{-1} min^{-1})
Part I—Sodium-Independent Systems							
	Sugars:[c]						
Erythrocyte (human)[d]	Glucose	37	(1)	10–13*	0.032–0.042*	198	0.64
			(2)	7.5–10	0.024–0.032	—	—
			(3)	6.5	0.021	620	2.0
			(4)	4.0	0.013	328	1.06
			(5)	4.0*	0.013*	—	—
			(6)	—	—	930	3.0
			(7)	5.6	0.018	590	1.9
		25	(4)	2.75	0.0089	130	0.43
			(8)	13*	0.042*	180	0.58
			(9)	3–5	0.0097–0.016	165	0.53
		20	(10)	~1	0.0032	300	0.97
			(4)	1.7	0.0055	66	0.21
			(11)	1.86	0.0060	210	0.68
		13	(4)	1.1	0.0036	32	0.1
			(9)	19*	0.061*	—	—
		5	(12)[e]	0.56	0.0018	9.7	0.03
			(13)	0.50	0.0016	1.1	0.0035
	Galactose	37	(2)	47	0.15	—	—
			(14)	110*	0.37*	—	—
			(15)	39, 61	0.12, 0.19	710	2.2
			(7)	21	0.064	680	2.1
		25	(9)	40–50	0.13–0.16	170	0.55
			(9)	95*	0.36*	—	—
		20	(7)	12	0.037	120	0.37

	Mannose	37	(14)	30*	0.097	—	—
			(5)	13	0.042	350	1.07
			(15)	14	0.044	680	2.1
			(7)	13	0.040	710	2.2
		20	(7)	6.8	0.021	120	0.37
	Xylose	37	(14)	110*	0.36*	—	—
			(15)	50, 71	0.16, 0.23	650	2.1
	Ribose	37	(14)	140*	0.45*	—	—
			(15)	1900, 2800	6.2, 9.0	560	1.8
			(16)	2500	8	770	2.5
	L-Arabinose	37	(15)	220, 250	0.71, 0.82	710	2.3
	Arabinose	37	(15)	5500	17	620	2.0
	L-Sorbose	25	(17)	3100	10	124	0.4
	Fructose	25	(17)	9300	30	124	0.4
Erythrocyte (foetal guinea pig)[f]	Glucose	37	(18)	4.5	0.015	520	1.68
			(19)	6.0–6.4	0.019–0.021	—	—
		27	(18)	3.12	0.010	350	1.12
		17	(18)	3.03	0.0098	94	0.292
		7	(18)	2.77	0.0090	14	0.046

[a] The table is divided into two parts. The first part records much of the available data for those systems which do not show a cation dependence. The data on K_m are given in both millimolar, and isotones per liter, units where 1 isotone = 310 mmoles; the data on V_{max}, in these same units, per minute. Numbers in parentheses refer to footnote *b* (p. 171) and give literature references or some necessary comment. The asterisk indicates that the recorded value under the heading K_m is, in fact, an inhibition constant K_i (see Section 4.4,A). The second part of the table records some of the available data on the sodium-dependent transport systems. Since either K_m or V_{max}, or both, for these systems depends on the concentration of sodium ion (see Chapter 5), the value at some arbitrary cation composition—in most cases that of some physiological saline—is of less absolute significance and the data are, therefore, treated more cursorily. The units for K_m are here millimolar, for V_{max}, either millimolar min^{-1} or that cited in the original reference. Optically active amino acids are the L-enantiomorphs and sugars, the D-enantiomorphs, except where otherwise stated. Conventional abbreviations for the amino acids are used; AIB is α-amino-isobutyric acid.

TABLE 4.3—(*Continued*)

System	Substrate	Temp. °C	References[b]	K_m (mM)	K_m (isotones liter^{-1})	V_{max} (mM min^{-1})	V_{max} (isotones liter^{-1} min^{-1})
Erythrocyte (rabbit)	Glucose	37	(20)	6	0.019	0.33	0.00011
	3-*O*-Methyl glucose	37	(20)	3.5–4.8	0.011–0.015	0.5–0.7	0.00016–0.00023
			(20)	5.3*	0.017*	—	—
	Ribose	37	(54)	1000	3.2	33	0.011
Sartorius muscle (frog)	3-*O*-Methyl glucose						
No insulin		19	(21)	3.3–4.2	0.011–0.014	0.023–0.09	0.000074–0.00029
With insulin		19	(21)	2.8–4.3	0.009–0.014	0.13–0.26	0.00042–0.00084
With insulin		29	(21)	5.2	0.017	0.27	0.00087
	Glucose	19	(21)	7.3*	0.024*	—	—
Heart muscle (rat)	Glucose						
No insulin		37	(22)	8.8	0.028	2.5	0.0081
With insulin		37	(22)	28	0.091	12.5	0.041
Diabetic		37	(22)	6.0	0.019	0.94	0.0030
No insulin		37	(23)	10–12*	0.032–0.039*	—	—
With insulin		37	(23)	20–36*	0.065–0.120*	—	—
Anoxia		37	(23)	12–18*	0.039–0.058*	—	—
Heart muscle (rat)							
No insulin	L-Arabinose	37	(24)	0.06	0.00016	0.71	0.0023
With insulin		37	(24)	25	0.081	50.3	0.16
No insulin	Xylose	37	(24)	0.21	0.00068	1.71	0.0055
With insulin		37	(24)	6.6	0.021	22.8	0.074

Brain slices (guinea pig)	Xylose	37	(25)	330	1.06	—	—
	Glucose	37	(25)	37*	0.120*	—	—
	2-Deoxyglucose	37	(25)	5 to 10*	0.016–0.032*	—	—
*Fibroblasts (mouse, in tissue culture, L-cells)	Glucose	37	(26)	1	0.0032	—	—
				13*	0.042*	—	—
	Galactose	37	(26)	0.51	0.0016	—	—
				22*	0.071*	—	—
	2-Deoxyglucose	37	(26)	2.4*	0.0078*	—	—
				7.2*	0.023*	—	—
Ascites tumor cells, mouse (Ehrlich)	3-*O*-Methyl glucose	10	(27)	2.1	0.0068	3.6	0.012
	Galactose	10	(27)	5.5	0.018	6.1	0.020
	Ribose	10	(27)	125	0.40	3	0.01
	L-Sorbose	10	(27)	330	1.10	11	0.035
	Glucose	37	(28)	20	0.065	1900	6
	Xylose	37	(28)	50	0.16	3000	10
	Galactose	37	(28)	80	0.26	2800	9
	L-Arabinose	37	(28)	200	0.6	3100	10
	Arabinose	37	(28)	900	3.0	2900	9
	Ribose	37	(28)	900	3.0	3200	10
Yeast, *Candida*	L-Sorbose	30	(29)	110	0.360	—	—
	Glucose	30	(29)	5*	0.016*	—	—
Amino acids:							
Erythrocytes, human	Leu	37	(30)	1.8	0.0055	0.52	0.0017
	Phe	37	(30)	4.3	0.014	1.5	0.0048
	Met	37	(30)	5.2	0.017	0.56	0.0018
	Val	37	(30)	7.0	0.023	1.0	0.0032
	Ala	37	(30)	0.34	0.0011	0.0068	0.000022
	Gly	37	(30)	0.30	0.0010	0.0012	0.0000039

(continued)

TABLE 4.3—(*Continued*)

System	Substrate	Temp. °C	References[b]	K_m (mM)	K_m (isotones liter^{-1})	V_{max} (mM min^{-1})	V_{max} (isotones liter^{-1} min^{-1})
Erythrocytes, rabbit	Leu	20	(31)	1.1	0.0036	1.0	0.0032
	Leu			0.9*	0.0029*	—	—
	Val	20	(31)	2.8	0.0091	1.3	0.0042
	Val			2.6*	0.0084*	—	—
	D-Val	20	(31)	20*	0.065*	—	—
	Purines:						
Erythrocytes, human	Uric acid	25	(32)	3.2	0.010	0.063	0.00020
	Hypoxanthine	25	(33)	0.1*	0.0003*	—	—
	Cations:						
Toad bladder, outer membrane	Sodium		(51)	20			
Frog skin, outer membrane	Sodium		(52)	10			

System	Temp. (°C)	References[b]	Substrate	K_m (mM)	V_{max} (mM min^{-1})	Substrate	K_m (mM)	V_{max} (mM min^{-1})
Part II—Sodium-Dependent Co-transport Systems								
			Amino acids:					
Reticulocytes (rabbit)	37	(31)	Ala	0.22	1.2	—	—	—
	37	(34)	Gly I	6.1	1.2	Gly II	0.039	0.085
Erythrocytes (pigeon)	39	(35)	Gly	0.24 (130 mM Na_e^+)	0.08	Gly	23 (6 mM Na_e^+)	0.08

Ascites tumor cells, mouse (Ehrlich)	37	(36)	Ala	0.6	5.6	AIB	1.2	5
			Betaine	40	5.0	Gly	3.5	5.5
			N-methyl AIB	0.3	4	*N*-methyl phe	12	5
			Met	1.5	4.2	Pro	1.2	5
(Note near constancy of V_{max})			Sarcosine	4.5	6	Ser	1.0	4
	37	(37)	D-Ala	15	5.5	Leu	1.5	3.0
			Val	3.5	4.5	D-Val	100	5
	37	(53)	Glutamic acid	200	70[a]			
Pancreas slices (mouse)	37	(38)	Ethionine	2.8	—	D-Ethionine	3.0	—
			Gly	4.2	—	Met	2.9	—
			D-Met	2–3	—	Val	7.0	—
	37	(39)	Pro	0.8	—			
Kidney cortex slices (rat)	37	(40)	AIB	1.25	0.07	Gly	1.67	0.07
			Phe	0.42	0.007			
	37	(41)	Arg	0.9	0.045	Lys	2.0	0.08
			Ornithine	2.2*				
Intestine (rat)	37	(42)	Ala	5.0	—	Asp	80	—
			Asp (NH_2)	17	—	Glu	26	—
			Glu (NH_2)	6.4	—	Gly	∞	—
Neutral amino acids			His	10.4	—	Ileu	1.2	—
			Leu	0.65	—	Met	0.91	—
			Phe	3.3	—	Pro	10	—
			Ser	18	—	Val	2.1	—
Dibasic amino acids		(42)	Arg	1.5	—	Lys	0.55	—
			Ornithine	6.0	—			
Intestine (rat)								
No insulin	37	(43)	AIB	~16	0.013			
With insulin			AIB	1.55	0.012			

(continued)

TABLE 4.3—(*Continued*)

System	Temp. (°C)	References[b]	Substrate	K_m (mM)	V_{max} (mM min^{-1})	Substrate	K_m (mM)	V_{max} (mM min^{-1})
Intestine (rat)	37	(44)	Ala	6.3	0.6[h]	AIB	∞	∞
Neutral amino acids			Gly	10	0.5[h]	His	6.0	0.2[h]
			Ileu	1.6	0.1[h]	Leu	2.2	0.4[h]
(Note that K_m and V_{max} increase in parallel)			Met	5.3	0.7[h]	Phe	1.4	0.2[h]
			D,L-Pro	6.2	0.5[h]	Thre	13	0.7[h]
			D,L-Try	0.25	0.03[h]	Iodo-tyr	0.4	0.2[h]
			Tyr	4.0	0.4[h]	Val	3.3	0.3[h]
Dibasic amino acids		(44)	Arg	1.2	0.1[h]	Lys	0.7	0.07[h]
			D,L-Ornithine	0.7	0.03[h]			
N-methylated amino acids		(44)	N-methyl sarcosine	1.9	0.9[h]	Betaine	1.5	4.0[h]
Intestine (hamster)	37	(45)	Met	1.3	—	D-Met	15	—
Neutral amino acids	37	(46)	Met	1.3	—	Pro	4.0	—
			Val	3.6	—			
N-methylated amino acids	37	(46)	Betaine	2.5	0.6[i]			
			N,N-dimethyl glycine	14	1.1[i]			
			Sarcosine	6.4	14[i]			
Neutral amino acids	37	(47)	Ala	7.5	28[i]			
			AIB	80	24[i]			
			Gly	43	43[i]			
			Leu	1.55	81[i]			
			Val	1.95	16[i]			
			Sugars:					
Kidney cortex slices (rabbit)	25	(48)	Galactose	1.2–4	0.1	Glucose	3–25*	—

Intestine[j] (hamster)	37	(49)	1-Deoxy glucose	14	
			6-Deoxy glucose	2	
			Arbutin	2	0.6–1.0
			Anions:		
Thyroid gland		(55)	Iodide	0.03	
			Bromide	20*	
			Cyanide	10–20*	
			Nitrite	4*	
			Nitrate	1–2*	
			Thiocyanate	0.02–0.03*	

[b] *References:* (1) Widdas (1954b). (2) LeFevre (1954). (3) Wilbrandt (1959). (4) Sen and Widdas (1962a). (5) Sen and Widdas (1962b). (6) Britton (1964). (7) Miller (1965a). (8) Levine *et al.* (1965). (9) Levine and Stein (1966). (10) LeFevre and McGinniss (1960). (11) E. J. Harris (1964). (12) By extrapolation from (4). (13) Fig. 4.1 (14) Widdas, quoted in Bowyer (1957). (15) LeFevre (1962). (16) LeFevre (1963). (17) Miller (1966). (18) Dawson and Widdas (1964). (19) *Ibid.*, cells partially irreversibly inhibited to slow transport. (20) Regen and Morgan (1964). (21) Narahara and Özand (1963). (22) Post *et al.* (1961). (23) H. E. Morgan *et al.* (1964). (24) Fisher and Zachariah (1961). (25) Gilbert (1965). (26) Rickenberg and Maio (1961). (27) Crane *et al.* (1957). (28) Kolber and LeFevre (1967). (29) Cirillo *et al.* (1963). (30) Winter and Christensen (1964). (31) Winter and Christensen (1965). (32) Lassen (1961). (33) Lassen and Overgaard-Hansen (1962b). (34) Glycine I and II, two separate sites for glycine uptake—(31). (35) Vidaver (1964a). (36) Inui and Christensen (1966). (37) Oxender and Christensen (1963b). (38) Bégin and Scholefield (1965a). (39) Bégin and Scholefield (1965b). (40) L. E. Rosenberg *et al.* (1961). (41) L. E. Rosenberg *et al.* (1962). (42) Finch and Hird (1960). (43) Akedo and Christensen (1962b). (44) Larsen *et al.* (1964). (45) Lin *et al.* (1962). (46) Hagihira *et al.* (1962). (47) Matthews and Laster (1965a). (48) Krane and Crane (1959). (49) Alvarado and Crane (1964). (50) Mawe and Hempling (1965). (51) Frazier *et al.* (1962). (52) Cereijido *et al.* (1964). (53) Heinz *et al.* (1965). (54) Steinbrecht and Hofmann (1964). (55) Wolff (1964).

[c] Note that V_{max} is, within error, a constant for the system for any species of sugar, at a particular temperature.

[d] Values of K_m for a number of other sugars as determined by LeFevre (1961a) are reproduced in Fig. 8.2.

[e] The variation of K_m and V_{max} over the range 7° to 47°C as determined by Sen and Widdas (1962a) is reproduced as Fig. 4.12.

[f] Some intermediate temperatures also given in (18).

[g] In mmoles (kg dry wt)$^{-1}$ min^{-1}.

[h] In μmoles (kg tissue)$^{-1}$ min^{-1}.

[i] In mmoles (kg tissue)$^{-1}$ min^{-1}.

[j] See also Table 5.5.

(2) The case *where there is a saturating concentration of permeant externally,* that is, $S_e = \infty$. Making this substitution now in Eq. (4.19) we obtain after reduction

$$J_{\text{out}, S_e=\infty} = \frac{(k_2/2) \text{ Tot } E \, S_i}{[(1 + \mathbf{r})/2\mathbf{r}]K_s + S_i} \quad (4.29)$$

Under these conditions, K_m is now $[(1+\mathbf{r})/2\mathbf{r}]K_s$ and $V_{\max}$ is $(k_2 \text{ Tot } E)/2$. If $\mathbf{r}$ is unity, that is, the rate constant for transfer of the free and loaded forms of the carrier are identical, then K_m is the same in the two experimental conditions defined in Eqs. (4.28) and (4.29) and is equal to K_s, a true dissociation constant, if $k_1 \gg k_2$ [see Eq. (4.30)]. Similarly, under the assumption that $\mathbf{r} = 1$, $V_{\max}$ is the same for the two conditions. If we find, therefore, that the K_m and $V_{\max}$ terms differ under the two experimental situations defined by Eqs. (4.28) and (4.29) we can interpret the data as showing that $\mathbf{r}$ is unequal to unity; in fact, we can use the comparison of Eqs. (4.28) and (4.29) to obtain a value for $\mathbf{r}$. Thus for glucose transfer across the human erythrocyte membrane at 25°C such a study by Levine and the author (Levine *et al.*, 1965) suggested a value of approximately 3 for $\mathbf{r}$. Equations for the net flux of permeant at values of S_i and S_e intermediate between zero and infinity have been developed (Britton, 1966) and also the integrated rate equations for transfer under these conditions (Levine and Stein, 1966). We record these results in Table 4.4. Two commonly used experimental procedures included in Table 4.4 need further comment. First, the K_m as determined using the procedure of Sen and Widdas (1962a) is found to be given by $K_m = [(1+\mathbf{r})/2\mathbf{r}]K_s$. Once again if $\mathbf{r} = 1$, we recover $K_m = K_s$ as originally assumed by Sen and Widdas. But even if $\mathbf{r}$ becomes infinitely large, K_m does not fall below one-half of K_s. (Contrast here the variation of K_m with $\mathbf{r}$ in the case where we measure efflux into zero external permeant.) Here k_2 [Tot $E/(1+\mathbf{r})$] is the relevant $V_{\max}$ which depends strongly on $\mathbf{r}$. Thus, Sen and Widdas determine a K_m which is that given by the conditions of Eq. (4.29) (efflux into saturating permeant) and, in contrast, a $V_{\max}$ given by Eq. (4.28) (efflux into zero permeant).

Second, we consider the situation where we measure the influx or efflux of labeled permeant added to cells equilibrated in unlabeled permeant at the same concentration (Miller, 1965b). Here the concentration of permeant at either face of the membrane is the same and we put $S_e = S_i$ in Eq. (4.19). From Table 4.4 we see that the derived equation for the dependence of flux rate on concentration is, for the first time, quite independent of $\mathbf{r}$; K_m is the true K_s (provided $k_1 \gg k_2$) and $V_{\max}$ is $\frac{1}{2}k_2$ Tot E.

TABLE 4.4

THE SIGNIFICANCE OF THE DERIVED PARAMETERS K_m AND V_{max} IN TERMS OF THE MOLECULAR PARAMETERS OF THE CARRIER MODEL (ON THE ASSUMPTION THAT $k_1, k_{-1} \gg k_2$)[a]

Conditions of experiment	Significance of K_m	Significance of V_{max}	References
Steady state (tracer added to cells, pre-equilibrated with unlabeled permeant)	K_s	$\frac{1}{2}k_2$ Tot E	(6, 9, 10)
Efflux from loaded cells into low external concentrations of permeant	$[(1 + \mathbf{r})/2\mathbf{r}]K_s$	$[1/(1 + \mathbf{r})]k_2$ Tot E	(4, 7, 11)
Flux into zero permeant at *trans* face	$[2/(1 + \mathbf{r})]K_s$	$[1/(1 + \mathbf{r})]k_2$ Tot E	(8, 20, 50)
Flux into infinite concentration of permeant at *trans* face	$[(1 + \mathbf{r})/2\mathbf{r}]K_s$	$\frac{1}{2}k_2$ Tot E	(8, 20, 50)
Permeant used as competitive inhibitor of a second permeant	$K_i \equiv K_s$	—	(1, 8)
Uptake of permeant at any S_i, S_0, using:			
Instantaneous rate equation $\frac{d\,S_i}{dt} = \frac{K_s k_2 \text{ Tot } E(S_0 - S_i)}{2K_s^2 + K_s(S_i + S_0)(1 + \mathbf{r}) + 2\mathbf{r}\, S_i S_0}$	Complex, requires "best-fit" computer analysis	Complex, requires computer analysis	
Integrated rate equation $[2K_s + 2(1 + \mathbf{r})S_0 K_s + 2\mathbf{r}\, S_0^2] \ln(1 - S_i/S_0) + [(1 + \mathbf{r})K_s + 2\mathbf{r}\, S_0]S_i = -K_s k_2 \text{ Tot } Et$	Complex, requires "best-fit" computer analysis	Complex, requires computer analysis	(7, 9, 11)
Half-equilibration time $t_{1/2} = \frac{1.386K_s^2 + 0.886(1 + \mathbf{r})K_s S_0 + 0.4\mathbf{r}\, S_0^2}{k_2 \text{ Tot } E\, K_s}$	Complex, requires "best-fit" computer analysis	Complex, requires computer analysis	(10, 15, 28)

[a] K_m is everywhere "that substrate concentration which produces half the maximum velocity of transport in the particular experimental situation used." V_{max} is this maximum velocity of transport. Other symbols are defined in the text. The derivation of these equations where not given in the text will be found in Levine and Stein (1966). The numbers in parentheses refer to the bibliography as given in footnote *b* of Table 4.3.

Finally, it is to be noted that while the parameters K_s, **r**, and k_2 Tot E are presumably constants for any permeant-carrier pair, the measured values of K_m and V_{max} will vary depending on the experimental method used for the analysis and according to Table 4.4. Discrepancies that arise, therefore, between the kinetic constants K_m and V_{max} reported by various authors [for example, for glucose transfer across the human erythrocyte membrane (Table 4.3)] are in part predicted by the analysis of this section. Table 4.5 records some of these data on the sugar-human erythrocyte system, where the measured values of K_m and V_{max} are recorded as well as the derived values of K_s and k_2 Tot E, using the value of **r** given in column five of the table. Many of the discrepancies between the published data vanish during this analysis.

TABLE 4.5

REDUCTION OF THE MEASURED PARAMETERS K_m AND V_{max} FOR THE SUGAR TRANSPORT SYSTEM OF THE HUMAN ERYTHROCYTE TO THE MOLECULAR PARAMETERS **r**, K_s, AND k_2 Tot E[a]

Temp. (°C)	Sugar	Measured K_m, or K_i if *, (mM)	Measured V_{max} (mM min^{-1})	Chosen value of **r**	K_s	$\frac{1}{2}k_2$ Tot E
37	Glucose	12* (1)	—	—	12	—
		5.6 (7)	590 (7)	3	8.4	1180
		—	930 (6)	—	—	930
37	Galactose	110* (14)	—	—	110	—
		21 (7)	680 (7)	3	32	1360
37	Mannose	30* (14)	—	—	30	—
		13 (7)	710 (7)	3	20	1420
25	Glucose	13* (8)	—	—	13	—
		—	150 (8)	4	—	375
		—	285 (8)	—	—	285
		2.75 (4)	130 (4)	4	4.4	335
		3–3.5 (9)	160–170 (9)	4	11	380
25	Galactose	95* (9)	—	—	95	—
		40–50 (9)	170 (9)	4	95	390

[a] The numbers in parentheses are literature references as given in footnote *b* of Table 4.3, where the measured parameters are collected. The value of **r** listed in column 5 is chosen to ensure good concordance of the values of K_s and $\frac{1}{2}k_2$ Tot E in columns 6 and 7. More recent data on the glucose system (Sen and Widdas, 1962b) suggest that some of the values listed for the inhibition constants K_i need to be revised downwards.

G. Nonequilibrium Kinetics

We have mentioned briefly (Section 4.5) a kinetic treatment due to Widdas in which, by analogy with the Briggs–Haldane treatment of enzyme kinetics (see Dixon and Webb, 1964), one does not have to make the assumption that the enzyme and substrate are always in equilibrium with the complex. Thus we can have the scheme

$$E + S \underset{k_{-1}}{\overset{k_1}{\rightleftharpoons}} ES_{\mathrm{I}} \overset{k_2}{\rightarrow} ES_{\mathrm{II}}$$

in which we strictly preserve the formalism of enzyme kinetics (but ignore the equilibria at the *trans* face of the membrane, as well as the possibility that unloaded and loaded carriers move at different rates). Then the term K_m applicable to the enzyme kinetics and to our transport kinetics is given by

$$K_m = \frac{k_{-1} + k_2}{k_1} = K_s + \frac{k_2}{k_1} \tag{4.30}$$

where K_s is the true dissociation constant of the complex. It can be shown (see Bowyer, 1957) that the V_{max} term for transport is given similarly by

$$V_{max} = \frac{k_2 k_{-1}}{(k_2 + k_{-1})} \frac{\text{Tot } E}{2} \tag{4.31}$$

Clearly and correctly in both equations, if k_{-1} and k_1 are large in comparison with k_2, that is, the rate-limiting step is the slow transfer of complex between the faces of the membrane, we recover the original expression [Eq. (4.1) in Section 4.4] for K_m and V_{max} with $\mathbf{r} = 1$. Dawson and Widdas (1964) have been encouraged to propose these equations to account for some data on the temperature dependence of the terms V_{max} and K_m for glucose transfer into human erythrocytes and into those of fetal guinea pigs. The availability of three temperature-dependent terms k_1, k_{-1}, and k_2 to describe the behavior of V_{max} and K_m rather than a single term k_2 for V_{max} and only two terms k_{-1} and k_1 for K_m, on the naive assumptions used previously, enable the peculiar temperature dependence of these parameters to be understood. The results are consistent with the step characterized by k_{-1}, the dissociation of ES, becoming rate-limiting at low temperatures. We should note that Eq. (4.30) predicts that K_m is always bigger than K_s. Yet in the experiments recorded in Tables 4.3 and 4.5, K_m for both glucose and galactose transport was less than the value of K_i (presumably equal to K_s) for these substances acting as inhibitors. The explanation for these results appears to lie rather in the phenomenon of the unequal rate of transfer of loaded and unloaded carriers.

4.7 The Dimerizer Hypothesis for Facilitated Diffusion

Some years ago, the author (Stein, 1962c,d), on the basis of photometric studies of the rate of entry of sugars into human erythrocytes, from various mixtures of sugar species (for example, the pair glucose and sorbose, where the rate of entry of total sugar in the mixture was apparently less than the rate of entry from solutions of the pure sugars) was led to propose the dimerizer model for facilitated diffusion. In this model permeation occurred as a result of pairs of sugar molecules interacting with a membrane component followed by the association of the sugars into dimers, a form in which a reduced number of free hydrogen-bond-forming groups would be available. This model predicted that the rate of entry of sugar would, at low concentrations of sugar, be proportional to the square of the sugar concentration. The data of Figs. 4.1 and 4.3 do not support this prediction, although studies by Luzzato and Leoncini (1961) on tumor cells and on erythrocytes (Luzzato, 1961) and by Wilbrandt and Kotyk (1964) on erythrocytes provided evidence in support of such a second-order dependence of the rate of entry on the sugar concentration. Recently, Miller (1966) and LeFevre (1966) have shown that Stein's photometric studies can be largely accounted for on the classical monomer kinetics developed in the preceding sections. In his experiments, Stein obtained the *initial* rate of sugar entry by an extrapolation procedure which both Miller and LeFevre show to be invalid for a sugar entering as rapidly as does glucose. The dimerizer hypothesis must therefore be withdrawn as far as sugar transport is concerned. We might note, however, that Bégin and Scholefield (1965b), studying the (active) transport of proline into mouse pancreas cells, find a second-order dependence of the steady state level of this amino acid on the external concentration of amino acid, while Vidaver (1964a) finds similar evidence for the bivalency of sodium to the glycine active transport system of ascites tumor cells. We return to the discussion of these systems in Chapter 5, and comment on the possible significance of bivalency in Sections 8.4 and 9.3.

CHAPTER 5

The Coupling of Active Transport and Facilitated Diffusion

5.1 The Development of the View that Sodium Transport and Metabolite Transport Are Coupled

During the last five years a most significant advance has been made in our understanding of membrane function, with the realization that sugar and amino acid transport may depend on simultaneous transport of sodium ions. This work has transferred the problem of the active transport of these amino acids and sugars (see Section 2.8) into the province of facilitated diffusion, resulting in a dramatic simplification of the conceptual framework in which these problems have been studied. In this chapter we shall discuss the evidence for this new view of metabolite transport and the consequences of a kinetic analysis of such systems.

An early indication of a possible interrelationship of metabolite and cation transport was found when the transport of amino acids into duck erythrocytes was studied (Table 5.1) (Christensen *et al.*, 1952c). Replacing half of the sodium by potassium in the medium in which the cells were incubated led to a reduction in the level to which glycine and alanine were accumulated by these cells. Similarly, the accumulation of, among other amino acids, the unnatural amino acid α,γ-diaminobutyric acid by Ehrlich mouse ascites tumor cells led to a loss of potassium ion from the cell (Christensen *et al.*, 1952b) and once again the replacement of Na^+ by K^+ (or choline) led to a reduction in the rate of accumulation of the amino acids. Christensen and his group at first considered that the presence of K^+ rather than the absence of Na^+ was responsible for these inhibitions and saw clearly that the accumulation of amino acid within the cell could result from "the potential energy inherent in the asymmetric distribution of potassium" (Christensen *et al.*, 1952a).

Definitive evidence that the presence of *sodium* ions was required for sugar transport was provided by Csáky and Thale (1960) who studied the transfer of sugars across the intestinal wall of the toad, from the mucosal side (the inside) to the serosal (the side in contact with the capillaries carrying digested food away from the gut). Table 5.2 lists representative data from these experiments. Replacing sodium by lithium, magnesium, or choline reversibly abolished transport. To be effective, the sodium ions had to be present on the mucosal side of the

TABLE 5.1

EFFECT OF SUBSTITUTING POTASSIUM FOR SODIUM IONS ON THE CONCENTRATIVE ABILITY FOR AMINO ACIDS OF DUCK ERYTHROCYTES[a]

Amino acid studied	Potassium ion added to incubation medium (m*M*)	Increase in cellular level of amino acid divided by decrease in extracellular level
Glycine	0	1.69–2.56
	80	1.12
	120	0.28
	140	0.0
Alanine	0	0.85–1.10
	105	0.71
	140	0.12

[a] Data taken from Christensen *et al.*, 1952c.

membrane. Similar observations were made on the rat intestine (Csáky and Zollicoffer, 1960). Later studies (Csáky *et al.*, 1961; Crane *et al.*, 1961) showed that not only was it the presence of sodium ion that was required, but that an intact active transport system for sodium was necessary if the accumulation of sugar was to be achieved. If active sodium transport was abolished by the use of specific inhibitors (see Chapter 6) such as the cardiac glycosides ouabain and thevetin, the active accumulation of, for example, 3-*O*-methyl glucose was inhibited (Table 5.3). Since these cardiac glycosides are presumed to be specific inhibitors of the active transport of sodium, they are clearly unlikely to affect sugar transport directly.

A crucial observation was made by Crane's group (Table 5.4) who showed that, provided sodium ions were present, an exchange occurred between labeled sugar added externally to the mucosa and sugar al-

TABLE 5.2

CATION REQUIREMENTS FOR THE ACTIVE TRANSPORT OF 3-*O*-METHYL GLUCOSE ACROSS SURVIVING TOAD INTESTINE[a,b]

Expt.[c]	Principal salt bathing mucosal surface	Salt at serosal surface if different from mucosa	Sugar transported (μmole hr^{-1}; mean ± S.D.)
1	NaCl	—	1.03 ± 0.38
	Na_2SO_4	—	0.41 ± 0.19
	LiCl	—	0.0
	$MgSO_4$	—	0.0
	Choline Cl	—	0.0
2	Li_2SO_4	Na_2SO_4	0.0
	Na_2SO_4	Li_2SO_4	0.63 ± 0.36
3[d]	LiCl	—	0.0
	NaCl	—	0.79–0.92

[a] Data taken from Csáky and Thale, 1960.

[b] Measured at 19 ± 2°C; 3-*O*-methyl glucose at 2.5 m*M*.

[c] Experiment 1 shows that Na^+ cannot be replaced by Li^+, Mg^{2+}, or choline$^+$; experiment 2 shows that Na^+ must be present at the mucosal surface; while experiment 3 shows that the effect of Na^+ deprivation can be readily reversed.

[d] Here, Na^+ replaced Li^+ after initial two hours of experiment.

TABLE 5.3

INHIBITION OF THE ACTIVE TRANSPORT OF 3-*O*-METHYL GLUCOSE ON TREATMENT WITH INHIBITORS OF ACTIVE SODIUM TRANSPORT[a,b]

Drug	Concentration (m*M*)				
	0	10^{-8}	10^{-7}	10^{-6}	10^{-5}
Ouabain	+15	—	+7	−1	−19
Thevetin	+15	+7	+1	−8	—
Acetyl strophanthidin	+15	+14	+7	+4	+5

[a] Loops of frog intestine containing a 10m*M* solution of 3-*O*-methyl glucose in a sodium sulfate Ringer solution were incubated at 30°C for 6 hr, together with the drugs listed, at the concentrations indicated at the head of each column. The change in concentration of sugar (+ indicating transfer from the mucosal to the serosal compartment), as a percentage of the original concentration within the loop, is recorded in the table. Each datum is the mean of several determinations.

[b] Data from Csáky *et al.*, 1961.

ready accumulated by strips of hamster intestine. This exchange was strictly dependent on the concentration of sodium ions, was quite unaffected by the addition of cardiac glycosides, but was inhibited by phlorizin, a potent inhibitor of sugar uptake. Under anaerobic conditions when sugars can enter the intestinal wall but will not accumulate (Fig. 5.1), this facilitated diffusion of sugar is dependent on sodium but is now not inhibited by inhibitors of the active transport

TABLE 5.4

THE INFLUENCE OF THE SODIUM ION CONCENTRATION ON THE RATE OF EXCHANGE OF ^{3}H-LABELED 1,5-ANHYDRO-D-GLUCITOL WITH ALREADY ACCUMULATED SUGAR[a,b]

		Aerobic conditions		Anaerobic conditions	
Na^+ conc. in medium (mM)	Tissue sugar level (mM)	Exchange rate	$\frac{\text{Rate}}{Na^+} \times 100$[c]	Exchange rate	$\frac{\text{Rate}}{Na^+} \times 100$[c]
145	18.0	0.69	0.48	0.29	0.20
108	13.8	0.51	0.47	0.22	0.20
72	9.6	0.37	0.51	0.18	0.25
36	4.0	0.14	0.39	0.07	0.19
0	2.1	0.07	—	0.06	—

[a] Strips of hamster small intestine were incubated at 37°C with unlabeled 1,5-anhydro-D-glucitol at 5 mM concentration for 40 min in the aerobic system, or for 15 min in anaerobiasis, until a steady state level had been reached. The concentration of Na^+ in the bathing medium was set at the levels recorded in the first column of the table. The tissue level reached in aerobiasis is recorded in the second column. (The level is probably everywhere 5 mM in the anaerobic system.) At the steady state, ^{3}H-labeled 1,5-anhydro-D-glucitol was added and the rate of exchange of label determined. These data are recorded (as mmole liter^{-1} min^{-1}) in columns (3) and (5), and the rate divided by the prevailing sodium ion level is given in columns four and six.

[b] Data of Bihler and Crane (1962) and Bihler *et al.* (1962).

[c] Note that the ratio (rate/Na^+) under anaerobic conditions is approximately one-half that in the aerobic system.

of sodium (Crane *et al.*, 1961). Similarly, Jacquez and Sherman (1965) have found that the net transport, but not the exchange, of amino acids in Ehrlich ascites tumor cells is inhibited by the metabolic inhibitor cyanide and by 2-deoxyglucose.

All these observations and many others have been succinctly explained by the admirable model put forward by Crane (1965) and depicted in Fig. 5.2. In this model, a facilitated diffusion system E_1, which can transport a sugar (G) only in combination with a sodium ion (Na), is located at the mucosal surface of the intestinal epithelial cells. A complex between sugar (or amino acid in the case of the amino acid accumulation systems), a sodium ion and the carrier is the diffusing element. We can term this "co-transport" of sugar and sodium—as distinct from counter-transport. Within the cell, the sugar and ion dissociate freely from the carrier as in any facilitated diffusion system. At the opposite end of the cell, that is, at the serosal surface, is located a system E_2 which can actively extrude sodium from the cell—a sodium

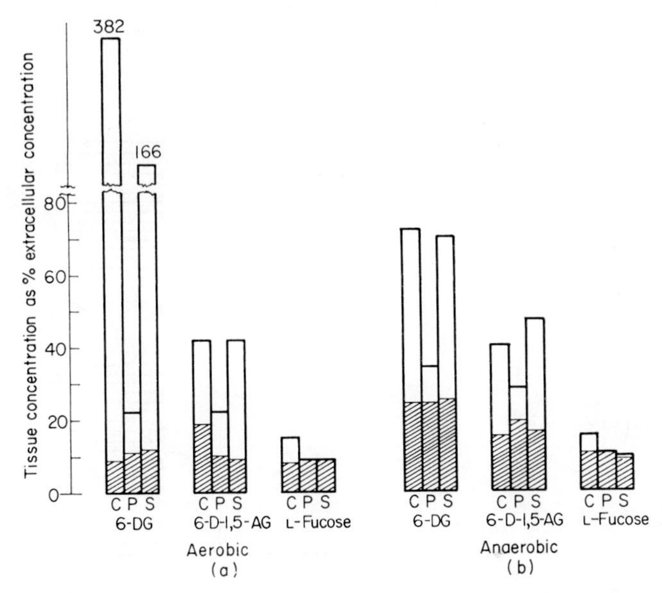

Fig. 5.1. The effect of phlorizin, strophanthidin, and Na^+ deprivation on the aerobic active sugar transport and the anaerobic entrance of sugars into strips of hamster small intestine. 6-D-G = 6-deoxy-D-glucose; 6-D-1,5-AG = 6-deoxy-1,5-anhydro-D-glucitol. Total height of column = uptake in the presence of Na^+ (143 mM). Shaded portion = Na^+ replaced by K^+. Ten minute incubation at 37°C with 0.005M sugar alone (C), with sugar + $10^{-4}M$ phlorizin (P), sugar + $10^{-5}M$ strophanthidin (S). (a) Under aerobic conditions (active transport); (b) under anaerobiasis (facilitated diffusion). (Data taken with kind permission from Crane *et al.*, 1961.)

"pump." (We shall discuss such transport systems in more detail in Chapter 6.) If this pump has a sufficiently high capacity, it will be able to maintain within the cell a low level of sodium relative to that present externally. There will thus be an electrochemical gradient of sodium ions directed inward from the cell exterior, across the mucosal interface (as well as across the serosal interface). This gradient will, if sugar or amino acid is present, drive a flow of sodium together with sugar or amino acid into the cell by the facilitated diffusion system—

this being co-transport. Kinetic analysis (Section 5.3) shows that under these conditions sugar will accumulate within the cell to reach an equilibrium distribution ratio which depends on the concentration gradient of sodium ions. Thus the active extrusion of sodium ions at the serosal surface (by a mechanism sensitive, like sodium pumps gen-

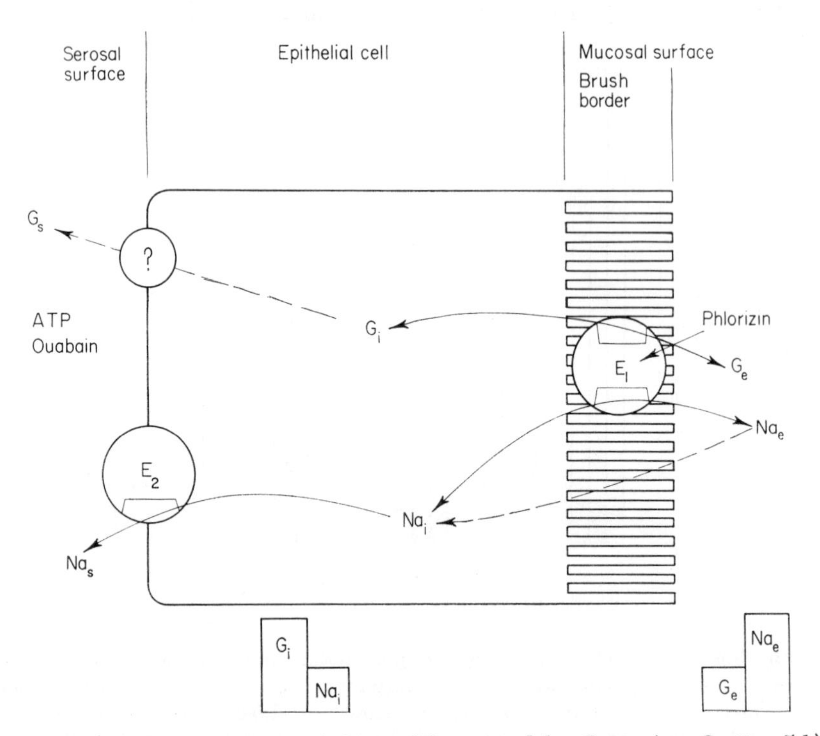

Fig. 5.2. Schematic diagram of the model proposed by Crane (see Section 5.1) for the interrelations of sodium and glucose transport in intestine; E_1 is a co-transporting sodium- and glucose-requiring facilitated diffusion system, sensitive to phlorizin, and E_2 is the sodium pump, sensitive to ouabain and requiring ATP. The height of the blocks at the foot of the figure represent the prevailing levels of glucose (G) and sodium (Na) within (subscript *i*) and at the external surface (subscript *e*) of the epithelial cell.

erally, to cardiac glycosides) maintains a low level of sodium within the cell, and the co-transport of sodium and sugar across the mucosal surface of the cell ensures that there is, therefore, a resulting concentration of sugar within the cell. If sodium extrusion is inhibited, active transport of sugar within the cell to a distribution ratio greater than unity cannot occur, but facilitated diffusion and exchange of labeled and unlabeled sugar will still take place.

5.2 Experimental Tests of the Co-transport Model

An analysis of the co-transport model of Fig. 5.2 leads to a number of predictions, an experimental test of which will then test the proposed model. These predictions are as follows:

(1) A stoichiometric relationship must exist between the number of moles of sugar (or amino acid) transferred across the mucosa and the number of moles of sodium co-transported. The stoichiometry is unity in Fig. 5.2, but this is not a necessary feature of the model.
(2) A sodium flux will drive a sugar flux but likewise a sugar flux should drive a sodium flux. Also, by driving a *net* sodium flux, an electrochemical gradient of sodium will drive a *net* sugar flux, and a gradient of sugar should drive a *net* sodium flux.
(3) The direction of net sugar flow must reverse if the electrochemical gradient of sodium is reversed, and likewise for the net sodium flow driven by a sugar gradient.
(4) The steady state concentration ratio of the metabolite must bear a definite relationship to the concentration ratio existing for the ion co-transported. The type of relationship will depend on the details of the model considered, but a reciprocity between metabolite and ion distribution ratios will be expected.
(5) Sugar uptake will be expected to obey all of the criteria of a facilitated diffusion system (see Section 4.2). In particular, saturation phenomena of the unidirectional fluxes with respect to the concentrations of both sugar and sodium ions may be expected. Tracer exchange and net flow of sugar may occur at different rates.
(6) Counter-transport of sugar will be expected to occur.

While not all of these above criteria have been tested on any single tissue, we can draw evidence from a number of sources to show that these principles apply.

Consider, as a test of criterion (2) for example, the data shown in Fig. 5.3, taken from the work of Schultz and Zalusky (1964b). Isolated strips of rabbit ileum (small intestine) are able to maintain for some time an electric potential across the tissue, directed so that the mucosal surface is negative with respect to the serosal surface. If a short-circuiting potential is applied across this preparation a continuing flow of current will be observed. (Compare this with the frog skin preparation which we shall consider in Chapter 6.) The flow of current results from the active transport of sodium ions through the tissue from the mucosal to the serosal surface (Schultz and Zalusky, 1964a), and the magnitude of the current flow is an indication of the rate of sodium ion transport through the tissue. Figure 5.3 shows that the addition of 3-*O*-methyl glucose (a sugar that can be actively transported but is not metabolizable) to the mucosal surface stimulates this current flow by increasing the rate of active transport of sodium. It is known that the drug phlorizin, if added to the mucosal surface, prevents the active accumulation of the sugar. In Fig. 5.3 we can see that the addition of

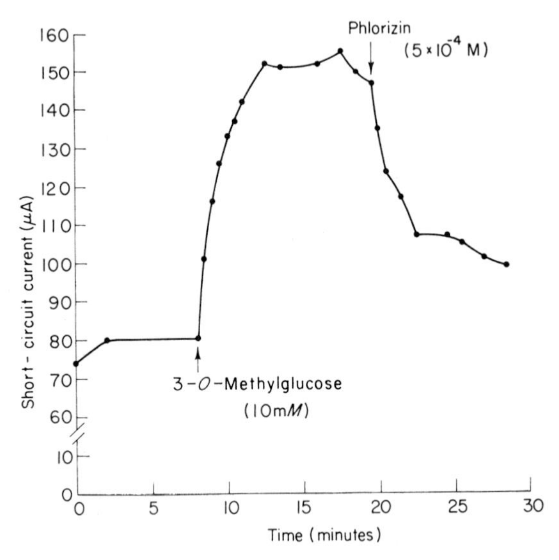

Fig. 5.3. The time course of the increase in short-circuit current following the addition of 3-*O*-methyl glucose and, subsequently, phlorizin, to a sugar-free medium perfusing a segment of a rabbit ileum. (Taken with kind permission from Schultz and Zalusky, 1964b.)

phlorizin here inhibits the additional current flow induced by the presence of sugar. When a number of sugars were similarly studied —some of these sugars being metabolizable but not transported, while others were transported but not metabolizable—in every case the only necessary condition for the stimulation of the sodium flux was that the sugar in question [or the amino acid in later experiments (Schultz and Zalusky, 1965)] be actively transported by the intestine. Consider next Fig. 5.4 also taken from Schultz and Zalusky (1964b) which shows how the increment in current flow (which, as we have seen, is equivalent to the flux of sodium through the tissue) depends strictly on the concentration of glucose in the medium bathing the mucosal surface. The sugar concentration required for half-maximal stimulation of the sodium flux is of the same order of magnitude as the concentration found by other investigators (Table 5.5) as being required for half-maximal stimulation of active sugar transport [criterion (5) of the list above]. Furthermore, Schultz and Zalusky compare the value obtained (from the upper curve of Fig. 5.4) for the maximal flux of sodium ion across the mucosa (some 3–5 μmoles/cm^2 surface. hr, depending on the sugar present) with the data found by other workers (Table 5.5) for the maximal sugar flux under similar conditions. From Table 5.5, these values range from 4–6 μmoles/cm^2 surface. hr, depending in the same way on the

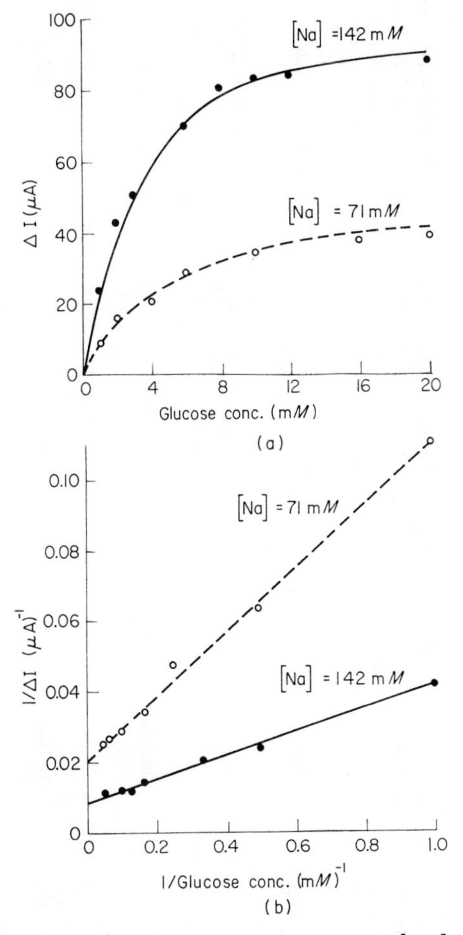

Fig. 5.4. The increase in short-circuit current across isolated rabbit ileum, as a function of glucose concentration: (a) raw data at two concentrations of sodium ion, (b) reciprocal plots of these data. (Taken with kind permission from Schultz and Zalusky, 1964b.)

TABLE 5.5

COMPARISON OF KINETIC CONSTANTS FOR THE SUGAR- OR AMINO ACID-STIMULATED SODIUM FLUX AND FOR THE ACTIVE TRANSPORT OF SUGAR OR AMINO ACID ACROSS VARIOUS INTESTINAL PREPARATIONS (ISOLATED ILEUM)[a]

Sugar or amino acid	Conc. of sugar or amino acid for half-maximal stimulation of Na flux (mM)		Conc. of substrate for half-maximal active transport (mM)		Maximal velocity of sugar-stimulated Na flux (μmole cm^{-2} hr^{-1})		Maximal velocity of active transport [μmole cm^{-2} hr^{-1} for (4, 10); mM hr^{-1} (20 cm fistula)$^{-1}$ for (5)]	
Glucose	Rabbit 4 ± 1	(1)	Hamster 1.5	(2)	Rabbit 3.7 ± 0.9	(1)	Rat 4	(4)
			Guinea pig 7	(3)			Dog 3.33	(5)
			Rat 9	(4)				
			Dog 17	(5)				
			Frog 12	(6)				
Galactose	Rabbit 10 ± 2	(1)	Hamster 2.2	(2)	Rabbit 5.1 ± 1.0	(1)	Rat 6	(10)
			Rat 35	(10)				
			Dog 42	(5)			Dog 5.31	(5)
3-O-Methyl glucose	Rabbit 17 ± 3	(1)	Frog 11	(6)	Rabbit 4.8 ± 1.0	(1)	—	
L-Alanine	Rabbit 4.5	(7)	Hamster 7.5	(8)	Rabbit 2.5	(7)	—	
			Rat 6.3	(9)				

[a] *References:* (1) Schultz and Zalusky (1964b). (2) Alvarado and Crane (1964). (3) Riklis and Quastel (1958). (4) Fisher and Parsons (1953a). (5) Annegers (1964). (6) Csáky and Fernald (1960). (7) Schultz and Zalusky (1965). (8) Matthews and Laster (1965a). (9) Larsen *et al.* (1964). (10) Fisher and Parsons (1953b).

sugar present. Schultz and Zalusky suggest, plausibly, that the stoichiometry for sodium and sugar transport might be 1 to 1 [criterion (1) above]. Table 5.5 also includes data on the sodium-stimulated active transport of alanine, a system which behaves very similarly to the sugar system. The stimulations of the short-circuit current produced by alanine or by glucose are strictly additive (Schultz and Zalusky, 1965), suggesting that we have here two separate, parallel systems for active transport. [In rat intestine, there is an apparent interaction between the transport of certain sugars and amino acids (Saunders and Isselbacher, 1965). But this interaction is indirect—it is time-dependent and non-competitive.]

The most detailed set of tests of the co-transport model comes from the ingenious studies that Vidaver (1964a,b,c) has made on glycine accumulation in pigeon red blood cells. This cell, like other red blood cells, is hemolyzed if placed in a dilute saline medium (or pure water). The hemoglobin and other soluble contents of the cell leak out leaving the membranes or cell ghosts. While the cell is being so lysed, normally impermeable substances added to the hemolyzing fluid will diffuse into the cell. Thus the cell can be "loaded" to any desired ionic composition. Following lysis, the cell is "restored" by the addition to the cell suspension of sufficient salt to bring the tonicity of the medium to its original isotonic level. These "restored ghosts" demonstrate permeability properties which are substantially the same as those of the intact cells from which they were derived. They can actively transport sodium ions (Teorell, 1952), possess unchanged glycerol (Stein, 1956) and glucose (LeFevre, 1961b) facilitated diffusion systems, and have proved a very valuable experimental object (Hoffman, 1958, 1962; Whittam and Ager, 1964). ["Ghosts" prepared from a bacterium—*E. coli*—can actively accumulate proline (Kaback and Stadtman, 1966).]

Vidaver realized that the control of the internal medium of the cell that the erythrocyte ghost allows, enabled a definitive test to be made of the co-transport model for amino acid accumulation. First, he showed (Table 5.6) that, like the other active transport systems we have considered, active glycine uptake by intact pigeon erythrocytes is strictly dependent upon the concentration of sodium in the external medium [criterion (2) above]. (Note that glycine uptake in these cells appears to occur by two processes. One process obeys Michaelis–Menten kinetics; that is, it is saturable with respect to the glycine concentration and in addition depends strictly on the external sodium ion concentration. The other process appears to occur at a rate that increases linearly with the concentration of glycine. This latter component, which may be either simple diffusion or a mediated transfer possessing a very high K_m for glycine, is unaffected by variations in the sodium ion concentration. We shall consider in the following discussion only the saturable component of glycine influx.) Analysis showed that criterion (5) above was satisfied in that the saturable component of glycine influx obeyed kinetics such that the K_m for glycine in the Michaelis–Menten formalism was dependent on the sodium ion concentration, the V_{max} term being unaffected.

TABLE 5.6

CATION REQUIREMENTS FOR GLYCINE ENTRY INTO PIGEON ERYTHROCYTES[a]

Expt. no.	KCl_{out} (mM)	$NaCl_{out}$ (mM)	Other additions (mM)	Glycine influx (μmole/ml pellet water in 15 min at 39°C)
A[b]	106	40	None	0.23
	146	0	None	0.041
	12	40	LiCl, 94	0.23
B[b]	116	30	None	0.22
	146	0	None	0.050
	116	0	LiCl, 30	0.042
	116	0	Choline Cl, 30	0.040
C[b]	51	95	None	0.59
	20	95	Sucrose, 62	0.57
	6	95	Sucrose, 90	0.52

[a] Data taken from Vidaver, 1964a.

[b] From A, potassium cannot replace sodium. From B, neither lithium nor choline can replace sodium. From C (and A, third line) the absence of potassium is not deleterious.

Surprisingly, K_m for glycine uptake was dependent on the square of the sodium ion concentration rather than on the first power of this term, obeying a relation of the form

$$K_m = K_1/(\mathrm{Na}^2 + K_2)$$

This finding has been confirmed for the uptake of glycine, but does not appear to be general for amino acid uptake (Wheeler *et al.*, 1965). Thus alanine and α-aminoisobutyric acid uptake by these cells (and also by ascites tumor cells and reticulocytes) depends strictly on the first power of the sodium ion concentration. We comment further on this point in Section 5.3.

When cells were loaded with various concentrations of sodium by being lysed in the appropriate medium and the tonicity thereafter restored, the degree to which glycine was accumulated depended on the ratio of external to internal concentration of sodium. In Fig. 5.5 we have plotted, from Vidaver's data, the change in distribution ratio of glycine over a 40-min incubation period (expressed as the ratio of the final distribution ratio of glycine to the initial distribution ratio) against the ratio of external to internal sodium ion concentrations. An accumulation of glycine is shown as an upwardly directed solid arrow, a depletion as a downward arrow. Clearly whether cells accumulate or extrude glycine, and the degree to which they do so, depend on the concentration gradient of sodium [criteria (3) and (4) above]. When the ratio of internal to external sodium ion concentrations was held at unity over a forty-fold variation of the actual concentration on either side of the membrane, the

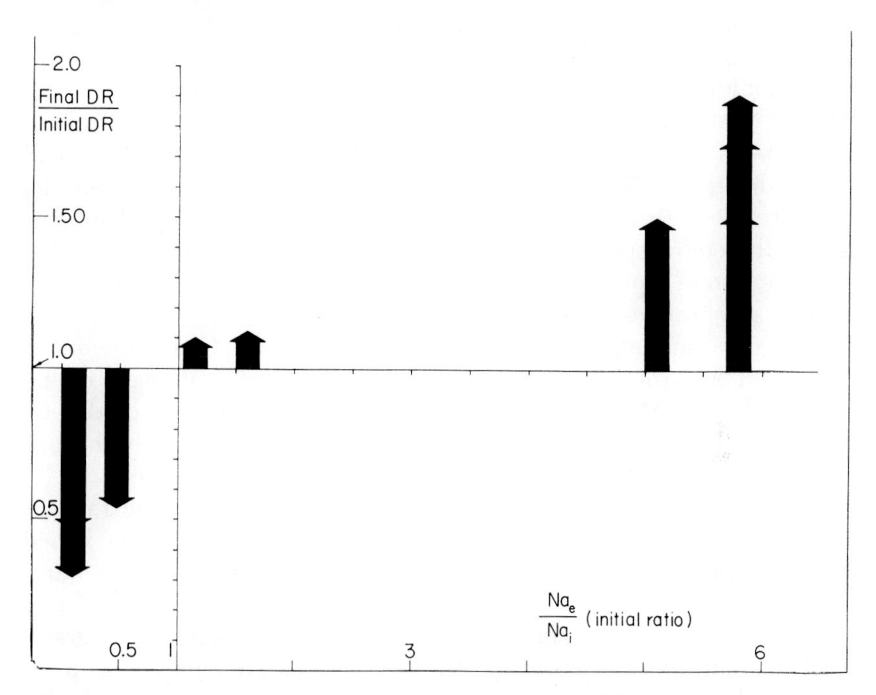

Fig. 5.5. The influence of the prevailing sodium ion levels on the accumulation of glycine by lysed and restored pigeon erythrocytes. Data of Vidaver (1964b) plotted as follows. Abscissa: ratio of external sodium ion concentration to internal sodium ion concentration at the beginning of the incubation period. Ordinate: ratio of the distribution ratio of glycine at the end of the incubation period to its value at the beginning of the period. The cells were incubated with glycine and sodium for 40 min at 39°C. Upwardly directed arrows indicate an accumulation of glycine, downward arrows, a loss of glycine by the cells.

internal flux of glycine was always equal to the external flux of glycine, that is, there was no net extrusion or accumulation, although the absolute fluxes varied by fourfold. If an electrochemical gradient of sodium ion was imposed even at zero chemical gradient (by making use of the Donnan potential imposed by a nonpenetrating anion, toluene 2,4-disulfonate), transport of glycine occurred in the direction and to a degree determined by the electrochemical gradient of sodium [criteria (3) and (4) above]. Finally, in this brilliant series of experiments, a direct measure of the sodium-dependent entry of glycine and simultaneously of the glycine-dependent entry of sodium was performed on the same set of cells to test [criterion (1) above] the stoichiometry of the co-transport.

The ratio of the amount of sodium ion entering to the amount of glycine entering in four different experiments covering a 2½-fold range of rates of entry was 1.5, 1.8, 1.9, and 2.2, while the predicted value—from the dependence of the K_m for glycine on the second power of the sodium ion concentration—was 2.0. [A difficulty associated with this test of stoichiometry is that the uptake of sodium ion in the absence of glycine is substantial and the glycine-induced entry a relatively small (approximately 40%) increment in this figure, but the accord of the derived ratio with that predicted is still impressive.] A point of interest is that the co-transport of sodium and glycine in this cell required the presence also of a penetrating anion in the medium. Chloride and nitrate were most effective but acetate was less so.

Additional support for the co-transport model comes from the work of Eddy and Mulcahy (1965). These authors showed that even in the presence of 2 m*M* sodium cyanide, when metabolism is blocked such that adenosine triphosphate (ATP) is no longer available, glycine can be accumulated by Ehrlich ascites tumor cells, provided there is an inwardly directed gradient of sodium ions. In these studies, the ratio of the concentration of glycine in the cells to that in the medium varied systematically from 0.3–8 when the ratio of the extracellular to intracellular sodium concentration was varied experimentally from 0.2–5.4. A plot of the log of the glycine gradient against the log of the Na^+ gradient led to an upper estimate of 60% for the efficiency of the coupling between these two species.

We might mention finally an autoradiographic study made by Kinter and Wilson (1965) of the uptake of ^{14}C- and ^{3}H-labeled sugars and amino acids by hamster intestine (Fig. 5.6). As judged by the blackening of the X-ray film in their preparations, Kinter and Wilson found that the concentration of permeant is highest at the mucosal end of the cell and lowest at the serosal end, a result perfectly consistent with the model of Fig. 5.2.

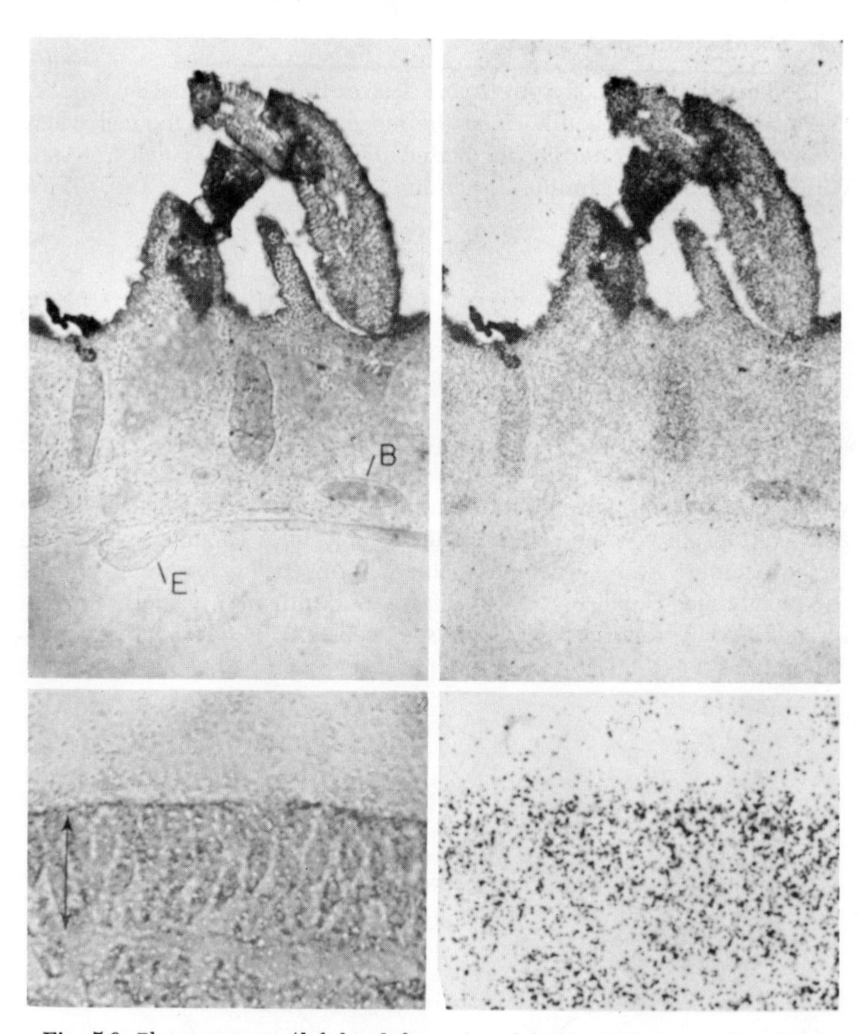

Fig. 5.6. Phase contrast (left-hand figures) and bright field views of underlying autoradiographs (right-hand figures) of strips of hamster intestine incubated with ^{3}H-labeled L-methionine. Upper figures show villi and muscularis (contracted edge at *E*). 80 ×. Lower figures: a strip of columnar absorptive cells near the tip of a villus. Magnification, 600 ×. The line of greatest autoradiographic grain density coincides with the brush border surface. (Taken with kind permission from Kinter and Wilson, 1965.)

There is thus very substantial evidence that the co-transport model of Fig. 5.2 is a valid description of the active uptake of sugar and amino acid in a variety of tissues. A full kinetic analysis of the model is, therefore, of some interest.

5.3 The Kinetics of Co-transport

The kinetic analysis of co-transport derives from the model of Fig. 5.2. We consider first the equilibria at the mucosal surface of the cell where we have the co-transporting facilitated diffusion system. Such a system will be governed by equilibria such as

$$E + \mathrm{Na} \underset{k_{-1}}{\overset{k_1}{\rightleftharpoons}} E\mathrm{Na} \tag{5.1}$$

$$E + \mathrm{G} \underset{k_{-2}}{\overset{k_2}{\rightleftharpoons}} E\mathrm{G} \tag{5.2}$$

$$E\mathrm{Na} + \mathrm{G} \underset{k_{-3}}{\overset{k_3}{\rightleftharpoons}} E\mathrm{GNa} \tag{5.3}$$

$$E\mathrm{G} + \mathrm{Na} \underset{k_{-4}}{\overset{k_4}{\rightleftharpoons}} E\mathrm{GNa} \tag{5.4}$$

where E is the free carrier, Na the sodium ion, and G the sugar or amino acid co-transported. We assume that the only form capable of diffusing is the complex EGNa, the intermediate forms being anchored outside the membrane; E will be free to diffuse and, although it *may* diffuse at a rate different from that of EGNa (see Section 4.5b), we shall assume—

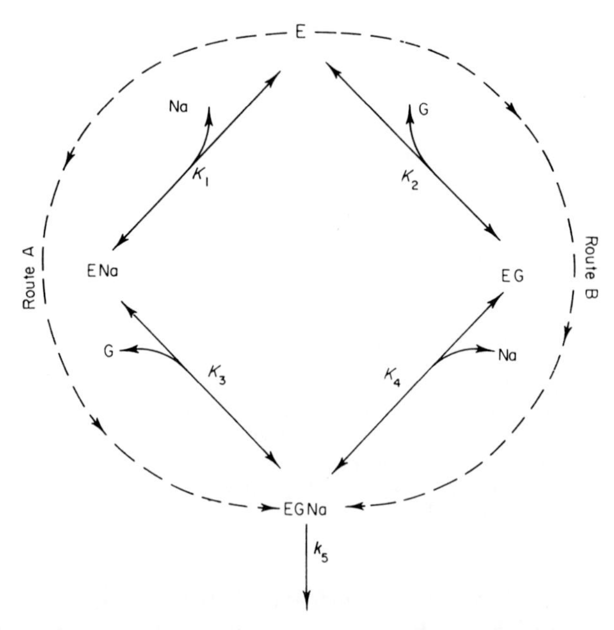

Fig. 5.7. Receptor-substrate interrelationships in a system which co-transports sodium, Na and a permeant G; K_1, K_2, K_3, and K_4 are equilibrium constants, k_5 is the rate constant for the transfer of the complex EGNa across the membrane.

until it is necessary to do otherwise—that the rates of transit of $EGNa$ and of E are identical. Equations (5.1) through (5.4) will be governed by equilibrium constants K_1 through K_4, respectively.

In Fig. 5.7 we depict these equilibria. It is clear that the complex $EGNa$ can be formed by two routes. In route A the formation of $EGNa$ occurs through the intermediate form ENa, that is, by the prior addition of Na. In route B, the intermediate EG is first formed; that is, the metabolite necessarily binds to the carrier before sodium is bound. We can set up kinetic tests which will allow a decision to be made as to whether route A or route B is followed in a particular case. We shall see that if route A is followed, the sodium ion effectively allows a complex to be formed between carrier and substrate. If route B is followed, the sodium ion is allowing an already formed complex to diffuse—it affects a diffusivity term. We can, therefore, distinguish between "sodium-dependent avidity" and "sodium-dependent mobility" of the carrier.

The conservation equation applied to the equilibria of Fig. 5.7 yields the result that

$$\text{Tot } E = E + EG + E\text{Na} + EG\text{Na} \tag{5.5}$$

Substituting for the terms EG, $EGNa$, and ENa, using the equilibrium equations (5.1) through (5.4), and rearranging, we find that in the general case,

$$EG\text{Na} = \frac{\text{Tot } E.\text{G}.\text{Na}\{[1/(K_1K_3)] + [1/(K_2K_4)]\}}{1 + (\text{Na}/K_1) + (\text{G}/K_2) + (\text{GNa}/K_1K_3) + (\text{GNa}/K_2K_4)} \tag{5.6}$$

Then the rate of transfer across the membrane, the unidirectional flux, is given as in Section 4.5,C by k_5 . $EGNa$ where k_5 is the rate constant for the translocation of $EGNa$ through the membrane. This rate gives the unidirectional fluxes of both G and Na, to the extent that these species are transported by the co-transport system.

If, now, we approach Eq. (5.6) in terms of Fig. 5.7, we see that we can obtain the kinetics for the formation of $EGNa$ by path B—sodium-dependent mobility—as the preferred route by letting K_3 in Eq. (5.6) become infinite. Then $EGNa$ will never be formed from EG. Equation (5.6) transforms into Eq. (5.7):

$$\text{Rate of penetration by route } B = \frac{k_5 \text{ Tot } E[\text{Na}/(K_4 + \text{Na})] \,.\, \text{G}}{[K_2K_4/(K_4 + \text{Na})] + \text{G}} \tag{5.7}$$

Similarly, for penetration by route A, we put K_2 and K_4 = infinity and we have

$$\text{Rate by route } A(K_2 \text{ and } K_4 \text{ infinite}) = \frac{k_5 \text{ Tot } E \,.\, \text{G}}{[(K_1 + \text{Na})/\text{Na}]K_3 + \text{G}} \tag{5.8}$$

Equation (5.7) predicts saturation kinetics for the sodium ion dependence and also for the co-transported species G. If, however, K_4 is large compared with Na, the V_{max} term will depend linearly on Na while the K_m term will be independent of Na. This situation has indeed been found experimentally. Representative data are given in Fig. 5.4b, which is a double-reciprocal plot of the data of Schultz and Zalusky (1964b) discussed previously (Fig. 5.4a). The intercepts on the abscissa of the two straight lines in the figure are coincident. Thus, K_m is unaffected by Na, but V_{max} is here directly proportional to the concentration of sodium ion. For the co-transport of sodium and glucose in the rabbit ileum, therefore, the experimental data are consistent with Eq. (5.7) in which transfer occurs by route *B* of Fig. 5.7 through the obligatory intermediate *EG*. In the case of the uptake of α-aminoisobutyric acid by Ehrlich ascites tumor cells (Wheeler *et al.*, 1965) both V_{max} and K_m depend on the sodium ion concentration (Fig. 5.8b), but the kinetics here are not fitted exactly by Eq. (5.7).

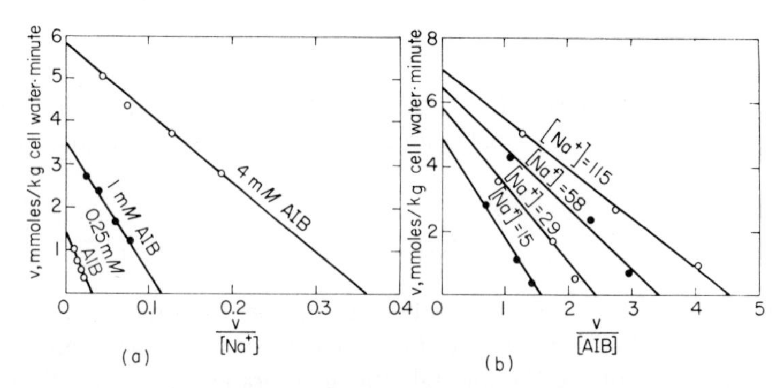

Fig. 5.8. Uptake of α-aminoisobutyric acid (AIB) by Ehrlich ascites tumor cells, during 1 min of incubation at 37°C. (a) Data plotted as a function of sodium concentration at various levels of AIB, showing a Michaelis–Menten dependence on the first power of the sodium ion concentration. (b) The same data plotted as a function of the AIB concentration at various levels of sodium ion. Both V_{max} and K_m are affected by Na^+. (Data of Inui taken with kind permission from Wheeler *et al.*, 1965.)

For Eq. (5.8) a strict dependence of K_m on concentration of Na according to the form

$$K_m = K_3(1 + K_1/\text{Na}) \tag{5.9}$$

is predicted, with no functional dependence of the V_{max} term on Na. Some experimental data are presented in Fig. 5.9. The predictions of Eq. (5.8) are clearly fulfilled. For the case of glycine accumulation by pigeon erythrocytes we have seen, however, that K_m depends on the

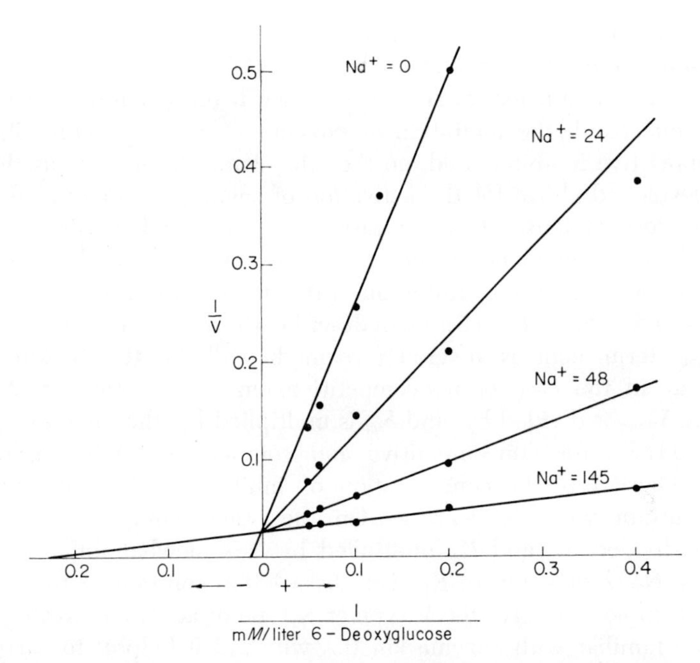

Fig. 5.9. Dependence of the rate of transport of 6-deoxyglucose into hamster intestine on the sugar and Na^+ concentrations. Ordinate: reciprocal of rate of uptake (10 min incubation at 37°C). Abscissa: reciprocal of millimolar concentration of 6-deoxyglucose in the everted intestinal sac. The concentration of Na^+ for each set of points is indicated on the figure, with K^+ replacing Na^+ in the bathing medium. The extrapolated K_m values were as follows: Na^+ 145 m*M*, K_m 4 m*M*; Na^+ 48 m*M*, K_m 15 m*M*; Na^+ 24 m*M*, K_m 40 m*M*; Na^+ 0 m*M*, K_m 100 m*M*. (Figure and data taken with kind permission from Crane *et al.*, 1965.)

square of the sodium ion concentration. This formalism is readily derived if, following Vidaver (1964a), one introduces into the equilibria (5.1) through (5.4) the additional sequences

$$E\text{Na} + \text{Na} \underset{k_{-6}}{\overset{k_6}{\rightleftharpoons}} E\text{Na}_2 \tag{5.10}$$

and

$$E\text{Na}_2 + \text{G} \underset{k_{-7}}{\overset{k_7}{\rightleftharpoons}} E\text{Na}_2\text{G} \tag{5.11}$$

We have seen above that the stoichiometry of glycine influx and sodium influx in co-transport in pigeon erythrocytes accords with Eq. (5.11) rather than Eq. (5.3). The strict dependence of the K_m term of the glycine kinetics on the sodium ion concentration (albeit to the second power) and the independence of the $V_{\max}$ term on Na suggest strongly that the formation of ENa_2 is here an obligatory intermediate in the formation of the transferred complex ENa_2G.

There is a certain formal similarity between the equations governing, on the one hand, the inhibition of enzyme systems by competitive or noncompetitive inhibitors and, on the other hand, the equations that we have developed above for the activation of sugar and amino acid transport by sodium ions. Thus the case [Eq. (5.8)] where the apparent Michaelis constant alone is affected is somewhat analogous in form to the case of competitive inhibition. (We might call this case the "*K* kinetics.") Similarly, Eq. (5.7) can describe a form of activation in which the V_{max} term alone is affected (we might call this the "*V* kinetics"), rather as in the case of noncompetitive enzyme kinetics. In enzyme kinetics, V_{max} is divided by, and K_m is multiplied by, the same expression $[(K_i + I)/K_i]$ for noncompetitive and competitive inhibition, respectively, where I is the concentration of inhibitor and K_i its dissociation constant with the enzyme. On the other hand, in co-transport V_{max} is divided by, and K_m multiplied by, a somewhat different factor $[(K_{Na} + Na)/Na]$—where K_{Na} is a dissociation constant of the sodium-carrier complex—to give the *V* type or *K* type of activation, respectively.

Those familiar with enzyme kinetics will find it helpful to carry over the methods for determining the V_{max}, K_m, and K_i terms in such equations, into co-transport kinetics, but the differences as well as the similarities of form should be noted.

We have in the above discussion considered only two limiting forms of Eq. (5.6), one obtained by letting K_1 and K_3 be infinite, and the other by similarly letting K_2 and K_4 be infinite. There are, however, a number of other limiting forms possible and these are collected in Table 5.7. Of these forms, only (2) can be neglected since it predicts saturation in Na but not in G, a situation which has not been found experimentally. The form (5) will, as we have seen, demonstrate *K* kinetics and could thus account for the data on sugar transport across hamster intestine and, in basically the same way, for Vidaver's data on pigeon erythrocytes. The forms (1), (3), and (4)—all with K_4 large in comparison with Na—and again (6)—with K_1 large in comparison with Na—will demonstrate the *V* kinetics of the data of Fig. 5.4b.

The statement that an experimental demonstration of *K* kinetics demands that K_3 be large, that is, that the path *B* be followed, cannot therefore be unequivocally made. Similar kinetics will be found if the perfectly reasonable assumptions of, for instance, case (6) are valid: Path *A* is followed but the binding of G to the "carrier" is unaffected by its prior binding to Na. [Both cases (4) and (6) are, however, "uneconomic" of carrier; they require K_1 and K_4 to be large so that very little of the carrier will ever be in the form of EGNa, the transferable complex. Case (1) is also uneconomic of carrier requiring that all three constants K_1, K_3, and K_4 be large.]

It will therefore be difficult to decide between behavior according to one or another of the several models of Table 5.7 by simple kinetic analysis. If it is possible to detect binding of G to the carrier (for example, by a protective action of G against the action of some inhibitor) then, whether the complex between E and G is formed in the absence of Na (route B) or only in its presence (route A) could be determined and a

TABLE 5.7

LIMITING FORMS OF EQ. (5.6) GIVING THE RATE OF PENETRATION OF THE COMPLEX EGNa

	K_3 large and in addition:	
K_1 is large:	K_2 is large:	$K_1 = K_4$:
$\dfrac{k_5 \text{ Tot } E \dfrac{\text{Na}}{K_4 + \text{Na}} \text{G}}{K_2 \dfrac{K_4}{K_4 + \text{Na}} + \text{G}}$	$\dfrac{k_5 \text{ Tot } E \dfrac{K_1}{K_2 K_4} \text{G} \cdot \text{Na}}{K_1 + \text{Na}}$	$\dfrac{k_5 \text{ Tot } E \cdot \text{G} \dfrac{\text{Na}}{K_4 + \text{Na}}}{K_2 + \text{G}}$
(1)	(2)	(3)
	K_4 large and in addition:	
K_1 is large:	K_2 is large:	$K_2 = K_3$:
$\dfrac{k_5 \text{ Tot } E \dfrac{K_2}{K_1 K_3} \text{G} \cdot \text{Na}}{K_2 + \text{G}}$	$\dfrac{k_5 \text{ Tot } E \cdot \text{G}}{\dfrac{K_1 + \text{Na}}{\text{Na}} K_3 + \text{G}}$	$\dfrac{k_5 \text{ Tot } E \cdot \text{G} \dfrac{\text{Na}}{K_1 + \text{Na}}}{K_3 + \text{G}}$
(4)	(5)	(6)

decision made between certain of these models. Alternatively, by working at very high sodium levels—or under conditions of temperature, pH, etc. in which K_1 and K_4 are lowered—certain of these cases might also be distinguishable. Thus, as Table 5.7 records, of the cases of V kinetics for (3), (4), and (6) the K_m term is invariant with Na but for case (1) K_m tends to zero with increasing Na. Since in the study of Schultz and Zalusky (1964b) (made over a concentration range of Na up to 140 mM) K_m was strictly invariant, case (1) is, on this ground, perhaps somewhat disfavored.

Finally, in these same experiments, the nature of the sugar transported was shown to affect the V_{max} term. Unless this reflects differences in the rate constant k_5 for different sugars, a situation quite unlike that found for the facilitated diffusion of sugars, form (4) might be indicated by this finding. For form (4), the equilibrium constants for the association of the sugar, K_2 and K_3, appear in the V_{max} term.

It is clear that much precise work will still be needed to clarify the detailed molecular events occurring in co-transport.

5.4 The Distribution Ratios at the Steady State

The discussion in the previous section shows how the initial rate (that is, the unidirectional flux) of sodium or of the co-transported molecule depends on the concentration of these penetrating species. It is also of interest to consider how the distribution ratio of the metabolite, at the steady state, is affected by these concentrations. We shall consider only the two limiting forms given by Eqs. (5.7) and (5.8). The first form, V kinetics, where the concentration of sodium ion affects only the maximum velocity of co-transport, gives, when K_4 is large in comparison with Na,

$$\text{Influx} = \frac{V_{max} \cdot G_e \cdot Na_e}{K_m + G_e} \tag{5.12}$$

where V_{max} is the maximum flux at any concentration of sodium ion, K_m is the appropriate Michaelis constant, and the subscript e refers to the concentrations at the external face of the membrane. Replacing this e by i, referring to the internal concentration of these permeants, we obtain the corresponding equation for efflux by the co-transport system. Then at the steady state and, assuming that G can leave and enter the cell *only by using the co-transport system,* the influx according to Eq. (5.12) will exactly equal efflux and we have the following on equating, reducing, and simplifying for the case of V kinetics:

$$\frac{G_i}{G_e} = \frac{Na_e}{Na_i} \tag{5.13}$$

The concentrations of the two co-transported permeants are inversely related. The data of Eddy and Mulcahy (1965), quoted in Section 5.2, are largely in accord with Eq. (5.13). Note that the distribution ratio G_i/G_e is independent of both V_{max} and K_3, and so should be the same for all metabolites that use the same co-transport system—or indeed for those metabolites that use any other co-transport system of any power (in the sense of "rate of doing work") provided it has the mechanism and stoichiometry of Eq. (5.12). We shall return to this point when we consider the complexities of amino acid co-transport in Section 5.5.

Similarly if co-transport occurs by the mechanism of Eq. (5.8), where the sodium ion concentration affects only the Michaelis constant for the interaction of E and G, we have at the steady state for the case of K kinetics:

$$\frac{G_i}{G_e} = \frac{Na_e}{Na_i} \times \frac{K_1 + Na_i}{K_1 + Na_e} \tag{5.14}$$

Now, if K_1 is very large compared with Na_i and Na_e, Eq. (5.14) reduces to Eq. (5.13) but, if the equilibrium constant K_1 is small in comparison with the operative sodium ion concentrations, this system should fail to show any concentrative uptake of G, since the right-hand side of Eq. (5.14) will be approximately unity. If, as in the case of glycine uptake by pigeon erythrocytes (Vidaver, 1964a,b,c), the stoichiometry of the system is such that two sodium ions are co-transported for each glycine molecule, Eq. (5.14) is replaced by

$$\frac{G_i}{G_e} = \frac{Na_e^2}{Na_i^2} \times \frac{K_1K_6 + Na_i^2}{K_1K_6 + Na_e^2} \tag{5.15}$$

where K_6 is the equilibrium constant governing the equilibrium defined by Eq. (5.10). Again if K_1K_6 is sufficiently small we will have a distribution ratio of unity for G at all relative *trans*-membrane concentrations of sodium. If, however, constant K_1K_6 is large in comparison with the prevailing sodium ion concentrations, we obtain

$$G_i/G_e = (Na_e/Na_i)^2 \tag{5.16}$$

as the determining condition, and now the requirement for a stoichiometry of 2 has intensified the dependence of the distribution ratio of G on the distribution ratio of the sodium ion. If one takes Vidaver's data (Vidaver, 1964b) and plots these as in Fig. 5.10, this double logarithmic plot of the ratio of G_i/G_e against Na_e/Na_i shows a dependence of the glycine distribution ratio on the sodium ion concentration ratio to a power between two and three, fairly consistent with the prediction from Vidaver's model [that is, Eq. (5.15) with K_1K_6 large].

Clearly, the distribution ratio of metabolite is determined by the distribution ratio of sodium, and we must now inquire as to how this latter ratio is itself determined.

If we consider only what appears in Eqs. (5.13) through (5.16), then it would seem that the distribution ratios of sodium and of the metabolite G are mutually determined—the system is in a steady state at any distribution ratio G_i/G_e given by the relevant one of these equations. If there is no other force than the concentration gradient of the co-transported species acting on the system, and if these species can penetrate the membrane only by using the co-transport system, any of an infinite range of steady state ratios of G_i/G_e is available to the complete system. The particular steady state that will be achieved depends only on the initial state of the concentrations of the sodium ion and the co-transported molecule. Thus, for example, if we have a cell containing at zero time neither sodium nor the co-permeant in an infinite volume of

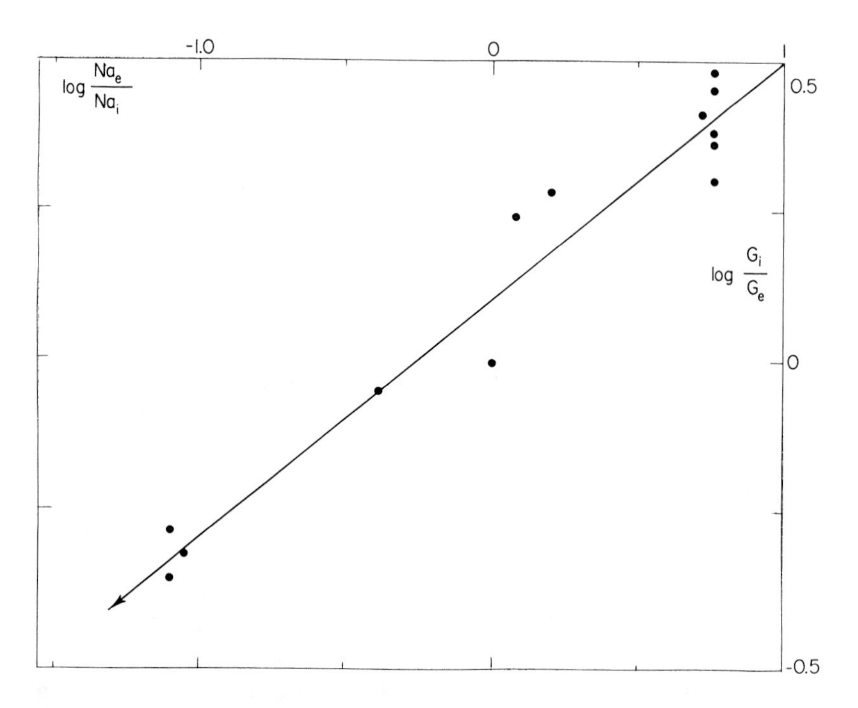

Fig. 5.10. Stoichiometry of glycine and sodium ion interrelationships for glycine uptake by lysed and restored pigeon erythrocytes. Abscissa: logarithm of the ratio of the external to internal sodium ion concentrations. Ordinate: logarithm of the ratio of internal to external glycine concentration, after incubation for 40 to 100 min at 39°C. The straight line drawn has slope of 2.5, indicating that some two to three sodium ions are concerned in the transport of each glycine molecule. (Data taken from Vidaver, 1964b.)

medium of concentration Na_e with respect to sodium and G_e with respect to the co-permeant, the influx will be (if, to simplify, we assume that the Michaelis terms are large) proportional to G_eNa_e. Efflux will be proportional to G_iNa_i at any instant. At the steady state when efflux is equal to influx, $G_iNa_i = G_eNa_e$ but now, since a molecule of G enters with each Na, we have always that $G_i = Na_i$.

Then

$$\mathrm{Na}_{i\ \text{steady state}} = \mathrm{G}_{i\ \text{steady state}} = \sqrt{\mathrm{G}_e\mathrm{Na}_e} \tag{5.17}$$

while

$$\mathrm{G}_i/\mathrm{G}_e = \mathrm{Na}_e/\mathrm{Na}_i = \sqrt{\mathrm{Na}_e/\mathrm{G}_e} \tag{5.18}$$

and the concentrations of both sodium and G internally are given by the geometric mean of the initial external sodium and G concentrations. In passing we may note that (1) if the initial external concentration of sodium ion is equal to that of the co-permeant, at the steady state there will be no concentration gradient of either G or Na, and (2) the concentration ratio for G determined by Eq. (5.18) is the maximum that would be found for the particular Na_e present. If any sodium happens to be present in the cell initially, the outward flux of G will at all times have been greater than is assumed in the derivation of Eq. (5.18), and the resulting distribution ratio of G at the steady state will be lower than that given.

In practice a situation such as that envisaged in the above paragraph will occur only if the cell or tissue studied is in an unphysiological state, for example, if it has been poisoned by cardiac glycosides. In any physiological situation the internal sodium ion concentration of the cell is determined and maintained by the sodium pump. Thus a *primary active transport system* determines the term Na_i in Eqs. (5.13) through (5.16), thereby determining through the *secondary active transport system,* that is, the co-transport system, the steady state distribution ratio for the co-transported species G.

The internal concentration of the sodium ions at the steady state is given, once again, by the balance of efflux against influx, but now sodium efflux is composed of (at least) two parts: one flux through the co-transport system, depending on the product of G_i and Na_i, and the second flux through the active sodium pump, proportional to the Na_i concentration. We shall discuss the mechanism of such active sodium transport in Chapter 6. In the meanwhile we can construct the following equation for the steady state governing the distribution of sodium ions:

Influx of sodium ions by co-transport = Efflux by co-transport + efflux by pump

or, when the Michaelis terms are large,

$$k_5 \frac{V_{\max}}{K_m} \mathrm{G}_e \,.\, \mathrm{Na}_e = k_5 \frac{V_{\max}}{K_m} \mathrm{G}_i \,.\, \mathrm{Na}_i + k_8 \mathrm{Na}_i \tag{5.19}$$

where K_m is some particular combination of the terms K_1 through K_4 in Eq. (5.6), and k_8 is the rate constant for the sodium pump. Thus,

$$\mathrm{Na}_i = \frac{k_5(V_{\max}/K_m)\,.\,\mathrm{G}_e\,.\,\mathrm{Na}_e}{k_8 + k_5(V_{\max}/K_m)\mathrm{G}_i} \tag{5.20}$$

and if we assume that k_8, the rate constant for the sodium pump, is greater than the remaining term in the denominator of Eq. (5.20)—and this is equivalent to the assumption that the sodium pump determines the internal concentration of sodium—then we have that

$$\mathrm{Na}_i = k_5/k_8 \frac{V_{\max}}{K_m} \mathrm{G}_e\,.\,\mathrm{Na}_e \tag{5.21}$$

or

$$\mathrm{Na}_i/\mathrm{Na}_e = \frac{k_5(V_{\max}/K_m)}{k_8} \mathrm{G}_e \tag{5.22}$$

This equation, except for the saturation term in G_e, which for simplicity we ignored in the derivation here, adequately accounts in the experiments on rabbit ileum (see Fig. 5.4) for the dependence of the sodium current on the external glucose and sodium concentrations. The internal sodium concentration, as would be expected, depends on the balance between influx by the co-transport system [determining the numerator of Eq. (5.21)] and efflux by the pump [the k_8 term in Eq. (5.21)]. Thus Eq. (5.21), or more exactly (5.20), will determine the internal concentration of sodium ion. The distribution ratio for sodium ion will be given by Eq. (5.22) and this will determine by Eqs. (5.13) through (5.16) the distribution ratio of the co-transported species G. If there is a parallel entry system for sodium ions other than the co-transport system, and if this parallel system has a far higher capacity for sodium influx than the co-transport system, Eqs. (5.21) and (5.22) will not apply; the internal sodium level will depend only on the relative rates of influx by this parallel route as compared with efflux by the pump. The internal level will be unaffected by variations in the external level of the species G.

A complete kinetic description of a co-transporting system, then, requires the experimental determination of a number of relevant constants. These are: (1) the Michaelis constants governing the saturation of the system by the co-transported species G and, if such saturation is found, by sodium ions; (2) the rate constant for the transfer of the complex [carrier + sodium + co-transported species]; and (3) the rate constant for the extrusion of sodium by the active transport. With these constants known, the behavior of the system at any concentration of G_e, Na_e, G_i and Na_i can be predicted and the distribution ratio of G found for any Na_e. The rate constant for sodium pumping that applies should be independent of the species that is co-transported, and the demonstration that this is in fact the case would be of much interest. A complete kinetic analysis along these lines of a co-transport system is still lacking, but its determination would be a good test of the theory developed in the preceding pages.

5.5 A Tertiary Active Transport System?

The work of Eddy and Mulcahy (1965) discussed in Section 5.2 has provided substantial evidence that the accumulation of glycine by the Ehrlich ascites tumor cell occurs by a co-transport system. Other studies (Oxender and Christensen, 1963a,b; Christensen, 1966), however, have shown that for the amino acids in general uptake is rather complex, and numerous separate systems of overlapping specificity have been shown to be concerned in amino acid uptake. Thus neutral amino acids are accumulated by the Ehrlich ascites tumor cell by at least two systems, one bearing the *L* site which has a preference for leucine, valine, cycloleucine (an unnatural amino acid), and methionine, while the other system bearing the *A* site prefers alanine, glycine, α-aminoisobutyric acid, and, once again, methionine. (The evidence for these statements is presented in Section 8.1.)

The *L* and *A* systems differ in a number of ways. For instance, *L*-site-using amino acids are accumulated to lower steady state distribution ratios than are the *A*-site-using compounds (Fig. 5.11). Second, *L*-site-using, but not *A*-site-using, amino acids stimulate a mutually enhanced rate of exchange among each other. Thus, for *L*-site amino acids the rate of transfer of carrier-substrate complex is faster than that of the free carrier (see discussion in Section 4.5,B). The fact that the two systems share an overlapping specificity—in particular, the fact that methionine is common to the two—has prompted Christensen (1964a) to make the

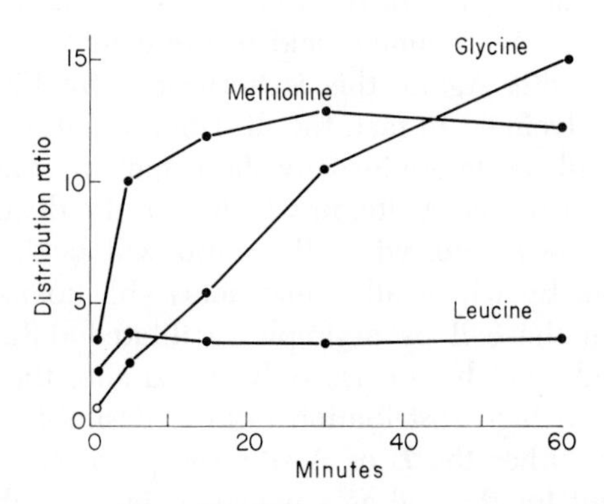

Fig. 5.11. Time course of uptake of three amino acids by Ehrlich ascites tumor cells at 37°C. Initial level of the test amino acid is 1 m*M*. (Taken with kind permission from Oxender and Christensen, 1963b.)

following very interesting suggestion: While *A*-site amino acids, including methionine, may be actively accumulated by a sodium-dependent co-transport, the resultant gradient of methionine can be used to drive *L*-site-preferring amino acids into the cell by counter-transport. Thus one might visualize a set of satellite transport systems depending less and less directly on the fundamental ion gradient. Sodium is pumped out of the cell by the primary transport system, leading to a sodium ion distribution ratio substantially less than unity. This is the primary active transport system. This distribution ratio is coupled by co-transport of sodium ion to yield an unequal distribution ratio in the *A*-site-using amino acids—secondary active transport. The resulting unequal distribution ratio of, for example, methionine, will impose by counter-transport on those amino acids sharing with it the use of the *L* site, a distribution ratio which is again more or less far from unity. This is the tertiary active transport. The distribution of leucine will be determined by the methionine gradient according to the counter-transport equation developed by T. Rosenberg and Wilbrandt (1957b) (see Section 4.4,E):

$$\frac{S_i}{S_e} = \frac{K_G + G_i}{K_G + G_e} \tag{5.23}$$

where S is the amino acid driven by the gradient of G, and K_G is the appropriate Michaelis constant for G. Inasmuch as K_G is unlikely to be vanishingly small, Eq. (5.23) shows that the distribution ratio of the driven amino acid will be lower than that of the driving amino acid, in conformity with Fig. 5.11 (compare leucine and methionine). Also, since the free energy for the uptake of *L*-site-using amino acids derives from the free energy present in the concentration gradient of the coupling amino acid methionine, the distribution ratio of methionine will be less than that of a pure *A*-site amino acid to the extent to which it, methionine, uses the *L* site. Again, this is borne out by Fig. 5.11 (compare methionine and glycine). In fact, the distribution ratio of any amino acid in this system will be determined by the extent to which it is preferentially accumulated by the *A* site, in which case the distribution ratio will be high, or by the *L* site, when this ratio will be low. The *L* site is effectively a leak by which all amino acids able to use the *L* site are drained off from the cell by a simple facilitated diffusion system. For those amino acids which can use only the *L* site, the *L* site is clearly advantageous, if a high distribution ratio is desirable. For those amino acids which use either the *L* or *A* site, the presence of the *L* site is a disadvantage, but for the cell as a whole—which is what matters—presumably the satellite systems offer some distinctive feature of value.

We should note that in Section 5.4 [Eq. (5.13)] it was pointed out that a similar distribution ratio should be reached for all co-transported systems sharing the same stoichiometry, whatever the capacity (V_{max} and K_m) of that system. That the amino acids in general do not distribute to the same ratio suggests strongly that either the L site does not have the same stoichiometry as the A site (and this could readily be tested) or that the interpretation offered here is correct, that is, that the L site is a tertiary active transport system, one not linked directly to the sodium gradient. We might recall here that the data for glycine (an A-site amino acid) uptake by Ehrlich ascites tumor cells (Eddy and Mulcahy, 1965), mentioned in Section 5.2, suggested that the distribution ratios for glycine and for sodium ions were more or less strictly reciprocal, in accordance with the view that A-site transport is a secondary system.

To conclude, it may be worthwhile to write down the general equation for the transport of a permeant that moves both by a co-transport system (A site?) of maximum velocity V_A for which it has a Michaelis constant K_A and by a facilitated diffusion system (L site?) of maximum velocity V_L for which it has a Michaelis constant K_L. In this case, K_A is determined by the sodium ion concentration on either side of the membrane according to Eq. (5.9) or perhaps by a similar equation involving the second power of Na; K_L will be determined by the concentration of those permeants sharing the facilitated diffusion system, by the formalism of true competitive inhibition. By equating the efflux and influx at the steady state, we obtain after simplification

$$\frac{(V_L + V_A)(K_{A_i}K_{L_i}) + (V_A K_{A_i} + V_L K_{L_i})G_i}{(V_L + V_A)(K_{A_e}K_{L_e}) + (V_A K_{A_e} + V_L K_{L_e})G_e} \qquad \frac{(K_{L_i} + G_i)(K_{A_i} + G_i)}{(K_{L_e} + G_e)(K_{A_e} + G_e)} \tag{5.24}$$

where the subscripts i and e attached to the Michaelis terms emphasize that these terms are different on either side of the membrane, since the co-transporting sodium ion concentrations and the counter-transporting permeant concentrations differ on these two sides. By putting either $V_A = 0$ or $V_L = 0$ in this equation we obtain the two limiting cases (1) where the amino acid does not travel by the co-transport system, its concentration being determined wholly by counter-transport; and (2) where it travels only by the co-transport system, counter-transport having no effect on its distribution ratio. The intermediate forms with $V_A \neq V_L \neq 0$ are difficult to handle. If either V_A or V_L, then, is equal to zero we obtain the general result that

$$\frac{G_i}{G_e} = \frac{K_{in}}{K_{ex}} \tag{5.25}$$

where for the case of co-transport only

$$\frac{G_i}{G_e} = \frac{K_{in}}{K_{ex}} = \frac{K_1 + Na_i}{Na_i} \times \frac{Na_e}{K_1 + Na_e} = \frac{Na_e}{Na_i} \times \frac{K_1 + Na_i}{K_1 + Na_e} \tag{5.26}$$

and for the case of counter-transport only, by the equations of competitive inhibition,

$$\frac{G_i}{G_e} = \frac{K_{in}}{K_{ex}} = \frac{1 + (A_i/K_A) + (B_i/K_B) + (C_i/K_C)}{1 + (A_e/K_A) + (B_e/K_B) + (C_e/K_C)} \tag{5.27}$$

where A, B, C . . . are the concentrations of the species of permeants sharing the same facilitated diffusion system, with Michaelis constants K_A, K_B, K_C . . . and the subscripts i and e have the above meaning. An equation identical with Eq. (5.25) is obtained from Eq. (5.24) if, by coincidence

$$K_{A_i} = K_{L_i} = K_{in} \qquad \text{and also} \qquad K_{A_e} = K_{L_e} = K_{ex}$$

These limiting cases clearly show up the complexity of the interacting transport systems in a living cell.

CHAPTER 6

The Primary Active Transport Systems

6.1 Criteria for Distinguishing between Primary and Secondary Active Transport Systems

A simple yet sufficient criterion of the existence of an active transport system is that the following condition be satisfied: A net flux of the permeant must occur in a direction opposite to that of the electrochemical gradient of the transferred species (Section 2.7). In Chapter 5 we saw that many active transports arise as a result of the co-transport of a metabolite together with sodium ions. The linking of the fluxes of the metabolite and the sodium ions results from the requirement that both these species must be attached to the mobile membrane carrier if transport is to occur. Thus the force driving the flux of sodium ions (the electrochemical gradient of sodium) indirectly drives by co-transport the flux of metabolite. We have seen, too (Section 5.5), that by counter-transport the active transport of a third species can be linked by two stages with the electrochemical gradient of sodium ions. It is conceivable that all active transport in a cell might occur by such secondary and tertiary transports, the fundamental concentration gradient arising perhaps from the continual intracellular depletion of some cell constituent by metabolic activity. Let us consider how we could distinguish between this situation and a true case of primary transport.

We begin by considering the possible forces that can drive a net flux. In Section 2.8 we saw how it was possible to consider J_i the flux of the ith component as resulting from (1) the electrochemical gradient in i itself:— $\Delta\mu_i$; or (2) an interaction with the flux J_j of some other component or (3) the action of some chemical reaction J_r. Formally, we can write

$$J_i = R_{ii}\,\Delta\mu_i + \sum_{\substack{j=0\\ j\neq i}}^{n} R_{ij}J_j + R_{ir}J_r \qquad (6.1)$$

where the interaction coefficients are the respective terms R. Simple or facilitated diffusion arises if R_{ii} is positive and both R_{ij} and R_{ir} are zero.* Co-transport in the most general sense arises if at least one term R_{ij} is positive. Primary active transport will occur if R_{ir} is nonzero. Now a net flux is a vector quantity—it has direction, "in" or "out" of the cell, as well as magnitude. Since it is a fundamental law of the physical world that any vectorial flux must be driven by a force which is also vectorial ["It is impossible for a force of a certain tensorial character to give rise to a flow of a different tensorial character"—Curie's theorem (Curie, 1908)], the force driving such a net flux must be vectorial. The coefficients R_{ii} and R_{ij} in Eq. (6.1) can be scalar since the terms J_i and J_j are vector quantities. If, however, J_r is considered as a flow of chemical reaction (for example, the consumption of glucose by the tissue) this flow is a scalar quantity, thus the coefficient R_{ir} must be a vector—the reaction in some way creates a directed flow (Kedem, 1961). If J_r is considered to be the gradient in "affinity" of a reaction across a membrane ["affinity" being the degree of advancement of a reversible chemical reaction $B \rightleftharpoons C$ (see De Groot, 1959)], R_{ir} will be scalar. The choice of the meaning to be attached to J_r will be largely a question of convenience. What is always true is that the product $R_{ir}\, J_r$ must be a vector. All active transport in a cell will, therefore, in the last analysis result from the linking of transport either to a concentration gradient in some molecular species, to an affinity gradient in some particular chemical reaction, or to directed flow brought about by a coupled chemical reaction. [A stimulating discussion on these points is given by Jardetzky (1964).]

To drive a net flux is an irreversible process and must result in the irreversible diminution in the gradient which produces the flux. Thus, if the flux depends on a concentration gradient, there must occur a flow of matter concurrent with the flow of the driven species. On the other hand, if the flux is coupled to a directed chemical reaction, there need be no flow of matter across the membrane but merely a change in the steady state of the specific chemical reactants located in a particular position with respect to the membrane and a transformation of some material substance(s) into others of lower free energy. This analysis gives us our criterion for distinguishing between primary and secondary transport systems: In a secondary transport, the net flux of the driven species A is linked mechanically to the flux of some other species B down the electrochemical gradient of B. In a primary transport, the net flux need not be accompanied by a flow of matter other than A, rather the flux is

* It might be useful in some circumstances, however, to consider the flow of a facilitated diffusion "carrier" as the flow of a component j.

linked mechanically to the participation of some molecular species *B*, *C* . . . in a chemical reaction.

We shall, in the following discussion, assume in the first place that all active transports are secondary. To disprove this assumption in any particular case we shall therefore have to provide an experimental demonstration that, while the active transport of *A* is not linked to the trans-membrane movement of any species *B*, *C* . . . , it *is* linked stoichiometrically to the transformation of some molecular species *F* into *G*, *H* . . . by a chemical reaction located asymmetrically across the membrane. The difficulty of demonstrating that there is no species *B*, the flux of which is linked to that of *A*, can be formidable. A demonstration that flux is directly linked to a vectorial chemical reaction is, in practice, easier to obtain.

6.2 Indications that Certain Active Transport Systems May Not Be Linked to Other Driving Fluxes

It has been argued most forcefully by Ling (1962), Troshin (1961), and E. J. Harris and Prankerd (1957) that active transport driven by a pump located in the cell membrane does not exist at all. These authors would hold rather that the distribution ratios higher or lower than unity that exist between the cell and the extracellular fluid are entirely a result of binding of the distributed components by some cellular constituent. The arguments of these authors have been ably dealt with by P. Mitchell (1959, 1961), Keynes (1961a), and Glynn (1959). A stimulating discussion between the protagonists of these rival theses is reported verbatim in the proceedings of the *Symposium on Membrane Transport and Metabolism* held in Prague (Kleinzeller and Kotyk, 1961). In essence, the counter-arguments are (1) that the components apparently actively transported are indeed free to move within the cytoplasm, while (2) exerting their full osmotic effect; also, (3) that there is insufficient material present within the cell with the binding capacities that would be necessary if all accumulation were by binding. We shall not enter further into this controversy here but shall attempt rather to consider what type of membrane-located system can drive these active transports.

That active transport of sodium ions occurs as a result of a primary rather than a secondary transport—and that active transport results by some mechanism other than by cytoplasmic binding—was most clearly demonstrated in a classic paper by Ussing and Zerahn (1951) in the case of frog skin. [Subsequent studies on various intestinal preparations, on toad bladder and within kidney tubules (see Table 6.1) have fully

TABLE 6.1

SOME EPITHELIAL CELL PREPARATIONS THAT ACTIVELY TRANSPORT SODIUM IONS[a]

Tissue	Net Na^+ flux (μequiv cm^{-2} hr^{-1})	Net water transport (μl cm^{-2} hr^{-1})	References
Teleost intestine (*Cottus scorpius*)	12	4.9 ± 3.0	House and Green (1963)
Frog skin	1.4		Curran *et al.* (1963)
Toad bladder	1.6		Leaf (1960)
Necturus kidney (proximal tubule)	0.22	2.8	Oken *et al.* (1963)
Rat ileum	3.6	90	Curran and Solomon (1957)
Dog ileum	7.2	108	Visscher *et al.* (1944)
Rabbit ileum:			
—No glucose	2.8		Schultz and Zalusky (1964a)
—With 11 m*M* glucose	3.9		Schultz and Zalusky (1964a)
Rabbit ileum *in situ:*			
—No glucose	4.2	220	A. H. G. Love *et al.* (1965)
—With 25 m*M* glucose	7.1	350	A. H. G. Love *et al.* (1965)
Rabbit gall bladder	8	53	Diamond (1964a)

[a] Data on net water transport are also included.

confirmed and generalized these findings.] This experiment of Ussing and Zerahn is shown diagrammatically in Fig. 6.1. The membrane S is a piece of frog skin (or other epithelial cell layer in the later studies of Table 6.1), held in place by a plastic frame, which separates the two com-

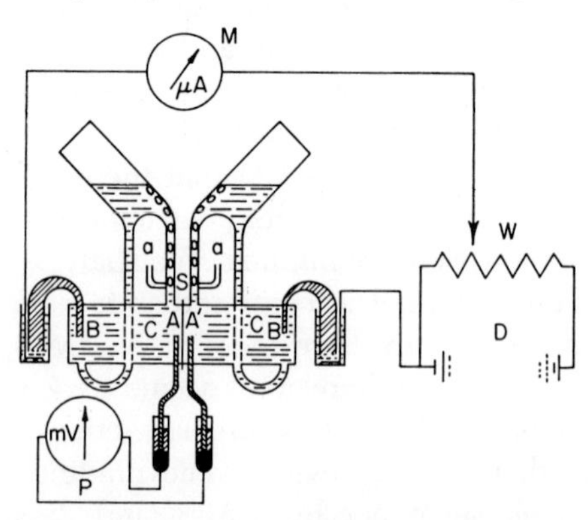

Fig. 6.1. Apparatus used for determining Na flux and short-circuit current, where *C* is the celluloid chambers holding the frog skin S, *A,A'* are agar-Ringer bridges to calomel electrodes and potentiometer *P*, and *B,B'* are agar-Ringer bridges to outside emf from battery *D*. (Taken with kind permission from Ussing, 1954.)

partments *C*. These compartments contain a synthetic salt medium, aerated and containing glucose, but identical in both sections. The electrodes *A* and *A'* measure the electrical potential difference between the two compartments. A separate pair of electrodes *B* and *B'* enable a potential from an external source *D* to be applied across the membrane in any direction but, in particular, in such a direction and of such a magnitude as to neutralize completely the inherent potential of the skin. (We consider the basis for this potential in more detail in Section 6.7.) The membrane is thus short-circuited. There is now no electrical gradient across the membrane nor, since the compositions of the bathing solution in both compartments are identical, is there any chemical gradient; yet under these conditions a steady flux of sodium ions continues to move from the outside (mucosal) surface of the skin to the inside (serosal). The magnitude of this flux can be accurately determined by chemical analysis. This transfer of positively charged sodium ions is accompanied by a flow of electric current which can be measured by the galvanometer *M*. Table 6.2, taken from Ussing (1954), shows that the magnitude of this short-circuit current is exactly given by the magnitude of the sodium flux. For our purposes this experiment demonstrates two points: First, that a binding of the sodium ions by some unknown component cannot account, in this case, for the active transport—the compositions of the bathing media are defined and identical. Second, the movement of sodium ions cannot arise by co-transport—once again, there is no gradient between the two compartments which can drive the flux of a co-transported species. We shall see later (Sections 6.7 and 7.3) that this seminal experiment of Ussing and Zerahn has led to many other advances in our understanding of membrane transport. Here we take note only of the conclusion that, since the only movement of matter which occurs in such experiments as that depicted in Fig. 6.1 is the movement of sodium ions, the over-all net transport of sodium (whatever its detailed mechanism may be) is by this experiment shown unequivocally to be a primary transport system. The net sodium flux depends absolutely on a continuing metabolism within the frog skin, and it is here that the seat of the vectorial chemical reactions driving the transport must arise. [It should be mentioned that there is indeed a small flux of hydrogen ions occurring at the beginning of such experiments (Fleming, 1957) but that the magnitude of this flux has been shown to be too small to account for the movement of sodium ions by a hydrogen ion linked exchange system.] Studies with the other epithelial cell layer preparations listed in Table 6.1 have given essentially similar results.

To locate more exactly the site of the vectorial chemical reactions we should prefer to work with single cell populations rather than with the

TABLE 6.2

SODIUM FLUX AND CURRENT FLOW ACROSS SHORT-CIRCUITED FROG SKIN[a]

Influx			Efflux		
Experiment	Na^+ Flux[b,c]	Current[c]	Experiment	Na^+ Flux[b,c]	Current[c]
3	64	63	6	0.8	102
	64	55			
	57	49	5	2.6	164
				2.4	118
1	102	99			
	93	99	3	6.0	136
				5.6	124
5	139	133			
	118	112	1	9.7	130
				10.5	139
2	177	174			
	176	162	2	5.3	111
	124	123		9.1	108
				13.0	108
				13.6	112
4	248	253			
	260	224			
	205	205	4	14.7	92
				13.2	100

[a] Data from Ussing, 1954.

[b] *Note*: On the average the influx is some 5% higher than the total current, while the efflux is itself 5% of the total current. The *net flux* is, therefore, exactly equal to the short-circuit current.

[c] In millicoulombs cm^{-2} hr^{-1}.

tissues of Table 6.1. The problem of demonstrating primary transport across the cell membrane, however, is a formidable one; it has only very recently become possible to control absolutely the composition of the fluids bathing the two sides of a cell membrane. The difficulty can best be realized by considering Fig. 6.2. This figure shows an experiment by Post and Jolly (1957) in which red blood cells were prepared (by storage at 2°C in a sodium-rich medium) in order to have the cation composition shown on the left-hand side of the figure—sodium inside some 115 m*M*, potassium inside some 15 m*M*. After two hours of incubation with a metabolizable substrate (inosine) at 37°C, in a medium containing only 150 m*M* sodium and no potassium, the suspension was divided into two parts. Enough isotonic potassium chloride (solid symbols) and sodium chloride (open symbols) were added to each pot to have these contribute 21 mM/liter to the cations of the medium. In the case of the medium containing added potassium ions, an outward flux of sodium ion and an inward flux of potassium ion commenced immediately. By the time "4 hours" on Fig. 6.2, the chemical gradients for both sodium ion and potassium ion are oppositely directed to the direction of the measured net flux. In the normal red blood cell, however, there is an electrical (Donnan) potential across the membrane, which potential (from the

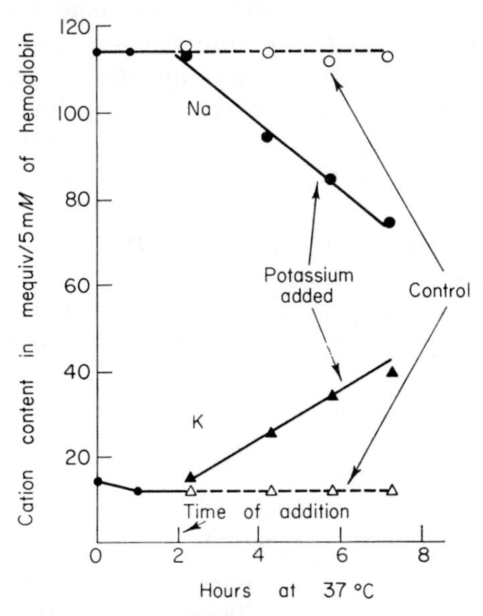

Fig. 6.2. Restoration of cation levels in cold-stored erythrocytes (refrigerated for 9 days at 2°C in a sodium medium), when incubated at 37°C. Control cells were incubated in the absence of potassium, test in the presence of 21 mequiv/liter of K^+. (Taken with kind permission from Post and Jolly, 1957.)

steady state distribution of the passively moving chloride ions) can be computed to lie between 0 and 10 mV, inside negative. This will constitute a small inward-driving force for the potassium ions but, by the time "6 hours" on Fig. 6.2—when there is clearly no significant cessation of potassium influx—both sodium and potassium ions are being driven against their prevailing electrochemical gradients. Clearly, any co-transport or counter-transport between these two cations cannot account for the active transport of *both* ions but, without following in detail the concentration changes of all other cell constituents here, we cannot be certain that both cations are not transported by a secondary transport of some other cell constituent.

We can approach a little closer to the desired proof by considering some results of Maizels (1961). Maizels has shown that human erythrocytes suspended in an isotonic lactose solution become "leaky" and can then be charged with any desired internal concentration of ion by incubation in a defined ionic medium. If calcium chloride is then added to the cell suspension, the normal cation impermeability of the cell is restored. In one experiment Maizels obtained erythrocytes with an internal cation composition of 116 m*M* of sodium and 14 m*M* of potassium. These cells were then suspended in a medium containing 114 m*M* of sodium and 10 m*M* of potassium, together with adenosine and glucose. After 24 hours of incubation the internal cation composition of the cell was 25 m*M* in sodium and 102 m*M* in potassium—both cations having moved against an electrochemical gradient at the expense of the metabolism of glucose and adenosine. The internal and external concentrations of all permeable constituents of the cell are here controlled, but it could perhaps be argued that a gradient of glucose and adenosine is still present and is driving transport by a secondary linkage.

Another example we might consider is that of the active transport of sodium and potassium across the membrane of the nerve axon, where a detailed comparison of the internal composition of the axon and of the bathing sea water can be made. An elegant procedure has been introduced by Baker *et al.* (1961) in which the axoplasm is squeezed out of the axon by the use of a miniature roller, enabling a chemically defined medium to be placed on the inside of the nerve axon membrane. With the cell immersed in a defined external fluid the fluxes occurring across the membrane can be fully documented and a direct test for primary transport applied. Indeed, such a preparation can actively extrude sodium and accumulate potassium, provided that energy-rich intermediates such as ATP and arginine phosphate are available in the inside medium.

Finally, a partial demonstration of the independence of sodium fluxes and the flux of a sugar into a bacterial cell is available. The bacterium *Escherichia coli* possesses a system which actively accumulates β-galactosides against a concentration gradient (Rickenberg *et al.*, 1956; Kepes, 1960). We have shown (J. Kolber and Stein, 1966a) that the unidirectional flux of an accumulated β-galactoside, *O*-nitrophenyl-β-D-galactopyranoside (ONPG), is not absolutely dependent on the presence of sodium ions (Fig. 6.3). While diminished in sucrose and still further diminished if potassium replaces sodium, the flux of ONPG still proceeds with an unchanged V_{max} parameter in sucrose- or potassium-containing media. The parameter K_m is affected by these replacements, but the gradient of potassium ion across the cell membrane can be shown to be wholly insufficient to account for the degree of accumulation of β-galactoside (a 2000-fold concentration for lactose) by any co-transport mechanism. In addition, "ghosts" prepared by osmotic shock from *E. coli* can actively accumulate proline (Kaback and Stadtman, 1966), if provided with energy sources. Yet these ghosts have lost 85% of the cell proteins, presumably most of the small molecular weight components of the cell.

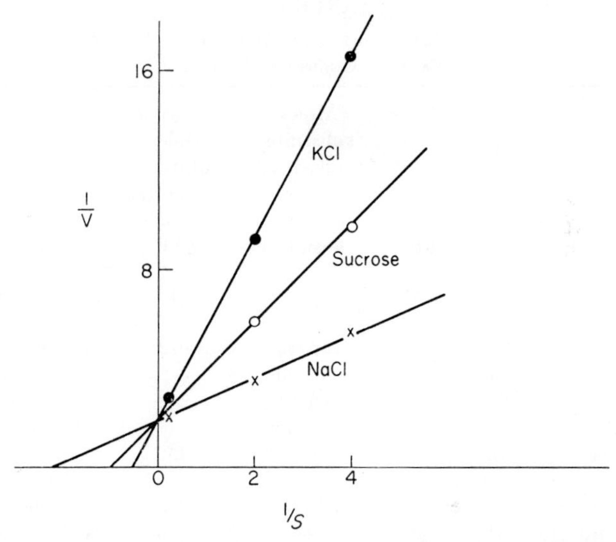

Fig. 6.3. Influence of ionic composition of medium on the uptake of *O*-nitrophenyl galactoside by lactose-induced cultures of *E. coli*. Abscissa: reciprocal of galactoside concentration in mM^{-1}. Ordinate: reciprocal of optical density reading of liberated nitrophenol, at 420 mμ. Five minutes incubation at 0°C in 0.1*M* salt or sucrose, as indicated. (Data of J. Kolber and Stein, 1966b.)

6.3 The Linkage of Active Transport to Chemical Reactions

In all the cases discussed above it can be shown that a very close relation exists between the active flux and some energy-producing metabolic activity of the cell (Table 6.3), providing excellent evidence for the existence here of primary transports. (The elucidation of the intimate details of this linkage between a directed flux and energetic catabolism is perhaps the most important task that faces those engaged in the problem of understanding the molecular basis of membrane function.) Studies on the giant nerve axons of squids provide a very direct example of the linking of active transport with a chemical reaction. Sodium extrusion against an electrochemical gradient of sodium ions can be followed in these cells by observing the rate of loss of radioactive sodium from fibers loaded by soaking in ^{22}Na-chloride artificial sea water (Keynes, 1961b). If the inherent metabolism of these fibers is blocked by applying metabolic inhibitors such as cyanide or alkaline dinitrophenol, sodium efflux ceases (Fig. 6.4) but can be restored by the microinjection of adenosine triphosphate (ATP) within the fiber. Figure 6.4 shows also that a boiled solution of ATP in which the high-energy bonds of the substrate have

TABLE 6.3

STOICHIOMETRY OF ACTIVE TRANSPORT AND ADDITIONAL ENERGY CONSUMPTION IN VARIOUS CELLS OR TISSUES

Cell or tissue	Substrate	Moles substrate transported per addl. mole O_2 consumed	Moles substrate transported per mole ATP	References
E. coli	Thiomethyl galactoside	6	1	Kepes (1964)
Frog skin	Na^+	16–20	~3	Zerahn (1961)
Toad bladder	Na^+	16–20	~3	Leaf (1961)
Toad bladder	Na^+	13.6/CO_2 evolved	2–3	Maffly and Coggins (1965)
Turtle bladder	Na^+ (anaerobic)	—	15	Klahr and Bricker (1965)
Human erythrocyte	Na^+	—	3.5	Sen and Post (1964)
Human erythrocyte	K^+	—	2.3	Sen and Post (1964)
Human erythrocyte	Na^+	—	3.2 ± 0.2	Whittam and Ager (1965)
Human erythrocyte	K^+	—	2.4 ± 0.15	Whittam and Ager (1965)
Squid nerve	Na^+	—	0.7	Caldwell *et al.* (1960)
Carrot slices	KCl, NaCl, KBr, KI	4	$1\frac{1}{3}$	R. N. Robertson (1960)

been broken down has no effect on sodium efflux. The addition of ATP to the outside of the nerve fiber also has no effect on sodium efflux—the breakdown of ATP has the spatially localized character required for a vectorial link with transport. There are, however, complications which prevent an easy interpretation of these findings. First, the original level of sodium efflux is never restored completely by injections of ATP, although it can be almost completely restored by injections of arginine phosphate or a system which will generate ATP *in situ* (Keynes, 1961b). It may be that a particular ratio of ATP to ADP is required for maximum stimulation of sodium transport. Second, in fibers using their own internal energy sources, the efflux of sodium depends rather critically on the presence of potassium externally, and an exchange occurs of sodium moving outward for potassium moving inward. (We shall discuss this more fully in Section 6.4 and Chapter 8.) If arginine phosphate is injected into poisoned nerve fibers (Fig. 6.5), this sensitivity of sodium efflux to the external potassium ion concentration is transiently restored; but in cells injected with ATP, the sodium efflux driven by ATP breakdown is insensitive to removal of external potassium. It is apparently still too early to be certain whether these microinjection experiments do actually reproduce the phenomena occurring in intact nerve fibers, but these experiments certainly suggest that sodium extrusion is directly linked to a metabolic effect, the breakdown of high-energy phosphate bonds. The stoichiometry of this relation has been partially elucidated in these nerve fibers, the number of sodium ions extruded per energy-rich phosphate bond injected being about 0.7.

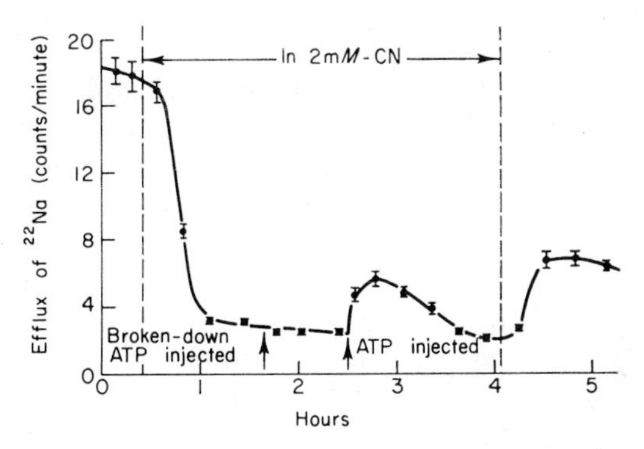

Fig. 6.4. The effect of the microinjection of high-energy phosphate compounds on the sodium efflux from cyanide-poisoned squid axon. The microinjection of a boiled ATP solution and of ATP itself is shown. (Taken with kind permission from Keynes, 1961b.)

Using red blood cell "ghosts" (see Section 5.2), Gardos (1954) has similarly shown that the active influx of potassium can be supported by the breakdown of the added ATP. The presence of arsenate which prevents ATP formation by the endogenous resources of the cell did not cut down this potassium accumulation, showing that it was the *added* ATP which was directly effective here. Hoffman (1958, 1962) has confirmed and extended these studies considerably.

Whittam (1958), using intact red cells, showed clearly that the rate of potassium influx fell in parallel with the fall in internal ATP

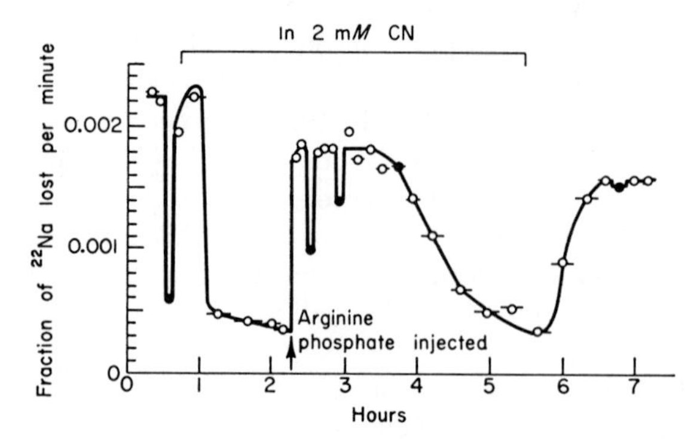

Fig. 6.5. The microinjection of arginine phosphate and the sodium efflux from cyanide-poisoned squid axon. (Taken with kind permission from Keynes, 1961b.)

concentration in cells deprived of glucose, but neither level fell significantly in cells metabolizing glucose. When the influx of potassium was inhibited by the addition of a cardiac glycoside (ouabain) to cells deprived of glucose, cessation of potassium influx was accompanied by a reduction in the rate of disappearance of ATP in these cells.

Whittam and Ager (1964, 1965) carried these studies to a more definitive conclusion. Using Maizels' technique (Maizels, 1961) of rendering erythrocytes ion-permeable by suspending them in lactose solutions, a range of internal potassium and sodium ion concentrations can be established within the cell, and glucose and adenosine can also be introduced. By measuring the rate of formation of inorganic phosphate and lactate produced by the cells when exposed to a variety of external and internal cation concentrations and in the presence and absence of ouabain, Whittam and Ager were able to test accurately the relationship between ion pumping and metabolism. Figure 6.6 shows the results obtained

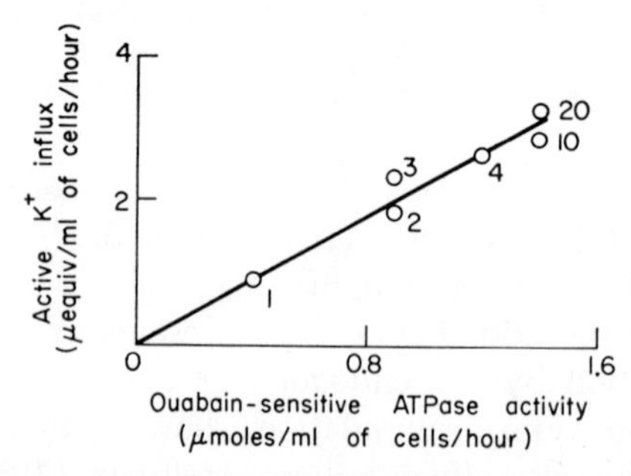

Fig. 6.6. The parallelism between active K^+ influx and the ouabain-sensitive ATPase activity for human erythrocytes incubated in media of varied K^+ concentration (given, in m*M*, beside each point on the graph). (Taken with kind permission from Whittam and Ager, 1965.)

when the total ouabain-sensitive ATPase activity (given by the sum of the inorganic phosphate and the lactate produced) was plotted against the active component of potassium ion influx, at various levels of external potassium ion concentration.

Figure 6.7 shows similarly the relation between active K^+ influx and the ouabain-sensitive ATPase activity at different levels of ouabain. In all cases, a strictly linear relation can be demonstrated between the amount of active ion transport and the amount of ouabain-sensitive (and hence transport-linked) ATPase activity. The stoichiometry was such that for every mole of ATP hydrolyzed, some 2.4 ± 0.3 moles of potassium ion

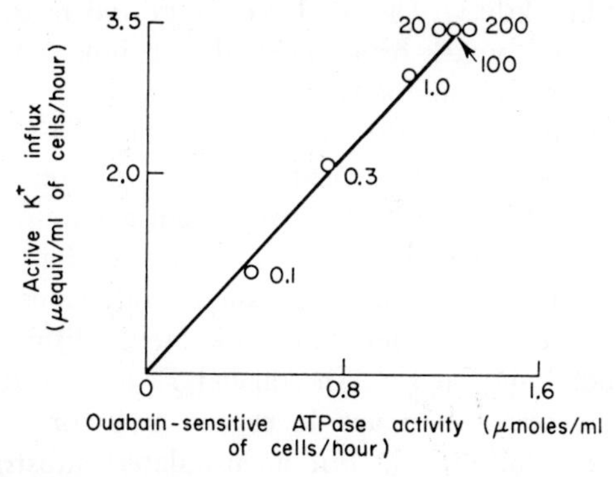

Fig. 6.7. As Fig. 6.6, but at the various levels of ouabain concentration, indicated, in μM, by the figures next to each point on the graph. (Taken with kind permission from Whittam and Ager, 1965.)

were actively transported inward while 3.2 ± 0.2 moles of sodium ion were transported outward. These are very similar figures to those obtained using the frog skin (Table 6.3). A calculation showed that the free energy liberated by the hydrolysis of one mole of ATP (estimated to be -13 kcal/mole under the conditions of these experiments) was more than sufficient to account for the free energy required to maintain the gradients of sodium and potassium ion found (9 kcal/mole as a maximum in these experiments). As we shall see in Section 6.4, there are very definite vectorial properties associated with the ATP hydrolysis by these red cell preparations, and the ATPase activity here has therefore all the characteristics required for it to be the chemical reaction driving ion transport by these cells.

Thus, where one can to a certain extent control the environment on both sides of the cell membrane, it is possible to make a very clear test of the existence of a direct linkage between active transport and a chemical reaction and to begin to identify the reaction concerned (see Section

8.4,D). For the active transport of sodium ions across frog skin (Ussing and Zerahn, 1951) and toad bladder (Leaf, 1961), for the uptake of β-galactosides by the permease of *E. coli* (Kepes, 1961) and for the active uptake of anions by plant tissue (reviewed by R. N. Robertson, 1960), it has proved relatively easy to establish a numerical relationship between the number of molecules or ions actively transported and the moles of metabolite combined (Table 6.3). It is, however, not possible to show in all these preparations that the movement of the actively transported species is directly linked to a particular metabolic reaction. In fact, as Robertson discusses in his review, there is still much controversy as to whether in plant tissue, the metabolic reaction linked to active transport is ATP breakdown (as we have suggested in the case of nerve and red blood cells above) or a more direct link with the oxidative phosphorylation step in metabolism.*

To choose, perhaps arbitrarily, one of these studies for more detailed consideration, we might further consider Kepes' work on the β-galactoside permease of *E. coli* (Kepes, 1964). This system accumulates, in addition to the normal substrate, lactose, various substrates that are not metabolized. This feature is an advantage in such work since it is the transporting of such substrates, rather than their metabolism, that can be the only link with metabolic energy. In starved *E. coli,* oxygen uptake and hence endogenous respiration are increased two- or threefold by the addition of a nonmetabolizable but accumulated substrate (Fig. 6.8). Cells that cannot accumulate substrate, that is, mutants which lack the gene for "permease," show no increased oxygen uptake when the nonmetabolizable β-galactoside is present. Oxygen consumption in competent cells continues until well after the steady state level of accumulation of β-galactoside within the cell is reached, one atom of oxygen (equivalent to three residues of high-energy phosphate) being consumed for every three molecules of thiomethyl galactoside transported. The rapid onset of oxygen consumption, its continuation through the steady state of accumulation, and the proportionality between the rate of transport and oxygen uptake all suggest that transport and energy consumption are very closely linked. In cells metabolizing succinate the level of ATP decreases some 15 to 25%, the ADP level rising by an equivalent amount, within one minute of transport commencing, again stressing the close link of active transport with metabolic reactions. In an endeavor to ascertain more exactly at which metabolic step transport and energy utilization are linked, Kepes has used the elegant spectroscopic techniques of

* MacRobbie (1965) shows, for instance, that chloride ion uptake by *Nitella translucens* is directly linked to a light-driven electron transfer reaction, but potassium ion uptake is linked to ATP utilization.

Chance (1952). A difference spectrum in the region 400–480 mμ was measured between two bacterial suspensions, one actively transporting a β-galactoside, the other free of any β-galactoside. A shift in the absorption spectrum corresponding to the oxidation of cytochrome *b* suggested that the link between transport and metabolism might occur at a point in the respiratory chain between cytochrome *b* and O_2 but the exact nature of this link is still far from clear.

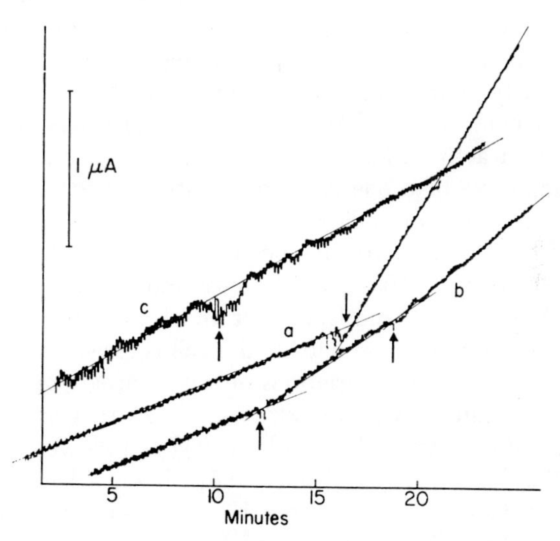

Fig. 6.8. Oxygen consumption, measured by the dropping mercury electrode in μA (ordinate), by a suspension of *E. coli* grown on mineral medium supplemented with maltose and starved for 45 min. The arrows indicate the times of addition of permeants, thiophenyl galactoside in curve *a*, thiomethyl galactoside in curves *b* and *c*. A permease-positive strain of *E. coli* (ML 308) was used in curves *a* and *b*, a permease-negative strain (ML 3) in curve *c*. (Taken with kind permission from Kepes, 1964.)

This and the similar studies on erythrocytes, epithelial cell layers, and plant tissues provide suggestive (although not in all cases direct) evidence for the primary nature of the transports concerned. It will, therefore, be worthwhile to consider further the properties of these systems on the assumption that we are here considering true primary transports.

6.4 Kinetics of the Primary Transport Systems

With the background of Chapters 4 and 5, we can deal relatively quickly with the kinetics of the primary transport systems. In brief, the rate of unidirectional flux up the electrochemical gradient is, in general, a saturable function of the concentration of the permeant on that side of the membrane. (A list of the available kinetic analyses and the results of these analyses is presented in Table 6.4.) We note that for sodium efflux from the nerve axon no evidence for saturation has yet, however, been presented. The rate of flux of, for example, sodium ions, can also depend on the presence of some other ion, for example, potassium, at the opposite side of the membrane, the side to which flow is being directed. The rate of transport of sodium ions is then stoichiometrically related to the rate of transport of potassium ions in the opposite direction. The flux of the transported species depends also on the rate of metabolic activity within the cell. Conversely, the rate of metabolic activity can depend, as we have seen in the previous section, on the prevailing rate of active transport.

As an example we take one case from Table 6.4 to discuss in greater detail: the coupled sodium efflux, potassium influx system of the human erythrocyte. The kinetics here have been studied very carefully by Glynn (reviews in Glynn, 1957, 1959) and by Post and his colleagues (Post and Jolly, 1957). Figure 6.2, taken from Post and Jolly (1957), shows clearly that the presence of potassium outside the cell is essential if sodium efflux is to occur. The influence of variations of this external *potassium* ion concentration on the rate of *sodium* extrusion shows that a true Michaelis–Menten formulation [Eqs. (4.1) and (4.19)] is followed with the V_{max} term (from the data of Post and Jolly, 1957) being 7.1 μmoles of sodium per liter cell water per hour and the K_m term—the potassium ion concentration giving a half-maximal stimulation of sodium efflux—being 2.2 m*M* at 37°C. The excellent correspondence between these figures and those of other workers will be obvious from an inspection of Table 6.4. Figure 6.6 shows how the rate of ouabain-sensitive ATPase activity depends on the level of external potassium. A K_m of 1.8 m*M* is operative in this case. Figure 6.9a shows the influence of the internal sodium concentration in these red cells on the rate of active sodium extrusion. The rate of extrusion varies linearly with the concentration up to a level of some 50 m*M* internal sodium, but apparently saturates thereafter. Similarly, the ouabain-sensitive ATPase activity is stimulated by internal sodium, with directly comparable kinetic parameters (Whittam and Ager, 1965). In contrast to this finding, Hodgkin and Keynes (1956) report that no variation could be found in the rate of extrusion of sodium ion from ^{24}Na-loaded squid axons, as the internal concentration of sodium was varied over the range 30–130 m*M*.

One of the most striking findings of these studies on cation transport in the erythrocyte was that the dependence of the rate of potassium influx on the external potassium ion concentration (Fig. 6.9b), being

TABLE 6.4

KINETIC PARAMETERS K_m AND V_{max} FOR SOME PRIMARY ACTIVE TRANSPORT SYSTEMS

System (substrate and cell)	Temp. (°C)	K_m (mM)	V_{max}[a] μmole (gm dry wt.)$^{-1}$ min^{-1}	References
Sugars:				
Galactose, *E. coli*	37	0.007	2.5	Horecker *et al.* (1960a)
β-Galactosides, *E. coli*				
Thiomethyl galactoside	26	0.5	148	Kepes and Cohen (1962)
Thiodigalactoside	26	0.02	20	Kepes and Cohen (1962)
Thiophenyl galactoside	26	0.25	>86	Kepes and Cohen (1962)
Lactose	26	0.07	158	Kepes and Cohen (1962)
O-Nitrophenyl galactoside	26	1	—	Kepes and Cohen (1962)
Iso propylthiogalactoside	28	0.24	—	Koch (1964)
O-Nitrophenyl galactoside	28	1.0–1.1	—	Koch (1964)
O-Nitrophenyl galactoside (in NaCl)	0	0.6	~30	J. Kolber and Stein (1966a)
O-Nitrophenyl galactoside (in KCl)	0	2.0	~30	J. Kolber and Stein (1966a)
O-Nitrophenyl galactoside (in sucrose)	0	1.2	~30	J. Kolber and Stein (1966a)
β-Methyl galactosides, *E. coli*				
Thiomethyl galactoside (by this system)	37	0.43	—	Rotman (1959)
Methyl galactoside	37	0.14	—	Rotman (1959)
Carbohydrates, *Staph. aureus*				
Lactose	37	0.005	—	Egan and Morse (1966)
Maltose	37	0.10	—	Egan and Morse (1966)
α-Methyl glucoside	37	0.02	—	Egan and Morse (1966)
Sucrose	37	∞	—	Egan and Morse (1966)

(*continued*)

TABLE 6.4—*Continued*

KINETIC PARAMETERS K_m AND V_{max} FOR SOME PRIMARY ACTIVE TRANSPORT SYSTEMS

System (substrate and cell)	Temp. (°C)	K_m (m*M*)	V_{max}[a] μmole (gm dry wt.)$^{-1}$ min^{-1}	References
Carbohydrates, yeast				
Glucose	30	80	—	Avigad (1960)
Maltose	30	80	—	Avigad (1960)
α-Methyl glucoside	30	80	—	Avigad (1960)
Sucrose	30	∞	—	Avigad (1960)
α-Glucosides, yeast (maltose system)				
α-Methyl glucoside	30	12	0.011	J. J. Robertson and Halvorson (1957)
Maltose (K_i)	30	24	—	J. J. Robertson and Halvorson (1957)
α-Glucosides, yeast (Isomaltose system)				
α-Thioethyl glucoside	30	1.3	2.5	Halvorson *et al.* (1964)
Carbohydrates, yeast, *Rhodotorula gracilis*				
Xylose ($+O_2$)	28	2	3210	Kotyk and Höfer (1965)
Xylose ($+N_2$)	28	0.8	2820	Kotyk and Höfer (1965)
Arabinose ($+O_2$)	28	41	4030	Kotyk and Höfer (1965)
Arabinose ($+N_2$)	28	130	4270	Kotyk and Höfer (1965)
L-Rhamnose ($+O_2$)	28	3.4	550	Kotyk and Höfer (1965)
L-Rhamnose ($+N_2$)	28	4.4	190	Kotyk and Höfer (1965)
Glucose ($+O_2$)	28	0.55	2980	Kotyk and Höfer (1965)
Amino Acids:				
Valine, *E. coli*	37	0.003	—	G. N. Cohen and Rickenberg (1956)
Histidine, *Salmonella typhimurium* (two systems)		0.11	—	Ames (quoted in Kepes, 1964)
		0.00017	—	Ames (quoted in Kepes, 1964)

Cations:				
Potassium uptake, erythrocytes (external [K$^+$] varied)				
Horse	38	2–3	1.0–1.9[a]	T. I. Shaw (1955)
Sheep	38	7–9	0.5–0.6	T. I. Shaw (1955)
Beef	38	7–8	0.7–0.8	T. I. Shaw (1955)
Human	38	2.1	2.2	Glynn (1956)
Human	37	2.2	7.1	Post and Jolly (1957)
Sheep (HK cells)	37	2.0	1.20	Tosteson (1964)
Sheep (LK cells)	37	2.0	0.25	Tosteson (1964)
Sodium efflux, erythrocytes (internal [Na$^+$] varied)				
Human	37	~18	8.5	Post and Jolly (1957)
Sheep (HK cells)	37	16	1.20	Tosteson (1964)
Sheep (LK cells)	37	16	0.25	Tosteson (1964)
Potassium uptake, yeast	25	5–10	13–17	Rothstein and Bruce (1958)
Anions:				
Orthophosphate uptake, yeast (pH 4.0)		0.4 (total phosphate)	4.3	Goodman and Rothstein (1957)
Orthophosphate uptake, *Staph. aureus* (pH 5.5–8.5)	25	0.8	1.4	P. Mitchell (1954)
Miscellaneous:				
Vitamin A, rat intestine		0.34	—	Skála and Hrubá (1964)
Biotin, hamster intestine		0.06	—	Spencer and Brody (1964)

[a] All V_{max} measurements under cation system in mmole (liter cells)$^{-1}$ hr^{-1}.

again of a Michaelis–Menten form, demonstrated a value for the K_m term for half-maximal stimulation of potassium influx of 2.2 m*M*, identical with that for half-maximal stimulation of sodium efflux and for stimulation of the ouabain-sensitive ATPase activity. Note in Fig. 6.9b that there is in addition a linear component due to the passive influx of potassium.

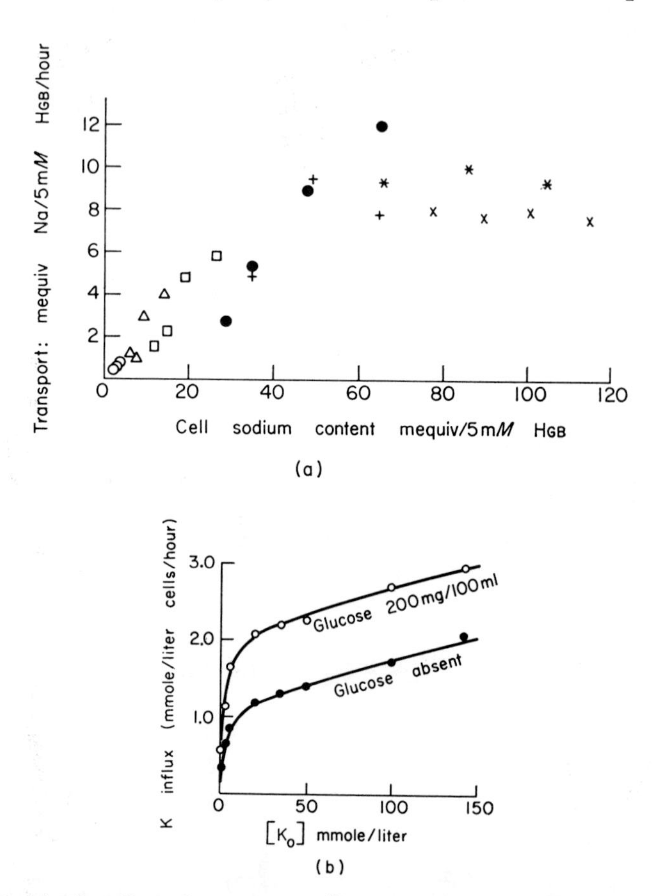

Fig. 6.9. (a). The effect of increasing sodium content on the efflux of sodium from human erythrocytes at 37°C, in the presence of high levels of external potassium. Cells were stored at 2°C for different periods of time (4–25 days) in the presence of different potassium-sodium media, to reach the levels of internal sodium indicated on the abscissa. (Taken with kind permission from Post and Jolly, 1957.) (b). The effect of increasing potassium levels in the external medium on the influx of potassium into human erythrocytes. The data are given for two cases, in the presence of glucose (upper curve) and in the absence of glucose (lower curve). The removal of glucose leaves the nonsaturating component of influx largely unaltered but decreases the saturating (and, presumably, the active) component. (Taken with kind permission from Glynn, 1956.)

These findings on the "contralateral" action of sodium and potassium ion are reminiscent of, and have the same mechanistic base as, the phenomenon of "accelerative exchange diffusion" discussed in Section 4.5,B. The explanation proposed there, presumably applicable here too, is that the membrane carrier (for sodium efflux) cannot return to the inside of the membrane unless it is bound to an alternative substrate (potassium ions in this case) at the other face of the membrane. On this assumption, the potassium ions should ride in with the return of the sodium efflux carrier and there should be a direct one-for-one link of sodium efflux with potassium influx. T. I. Shaw (1955) has proposed a carrier-type model (Fig. 6.10) which accords with these findings. Some

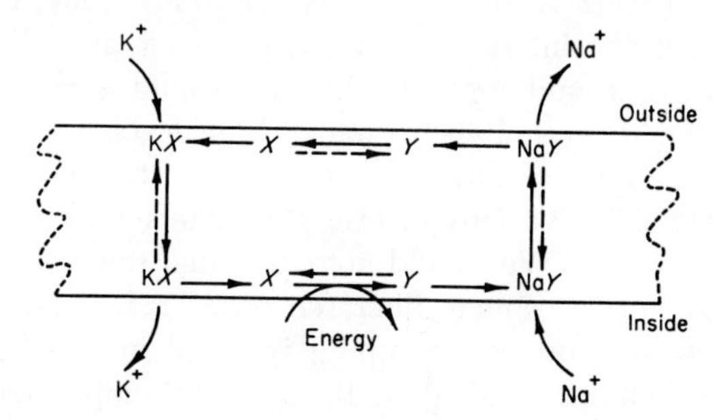

Fig. 6.10. A carrier model, proposed by T. I. Shaw (1955), to explain the link between the active sodium and potassium movements in erythrocytes. (Taken with kind permission from Glynn, 1957.)

of the available data (Glynn, 1957; Tosteson and Hoffman, 1960) support such a tightly coupled one-for-one exchange of sodium and potassium, but Post and Jolly (1957) find rather that 2 atoms of potassium exchange for three atoms of sodium. Perhaps, in this latter case, some other ion substitutes for, or neutralizes, the potassium in the return flow of carrier. We might note that Squires (1965) has considered the question of the stoichiometry of the binding of sodium and potassium ions for the cation carrier. It appears that the carrier is bivalent: Two sodium ions have to be bound to the carrier at the inner face of the membrane for efflux to occur. Likewise, two potassium ions are required to be bound at the external surface for potassium influx to occur and, finally, two molecules of ATP must interact simultaneously with the carrier on the inner face of the membrane to provide the energy source for transport. We shall discuss the implications of this finding and of the similar data on the glucose and glycerol facilitated diffusion systems when we come to discuss possible mechanisms of membrane transfer in Chapters 8 and 9.

It is the peculiar vectorial properties of cation transport here that should be reemphasized. For the complete system to be operative (1) sodium must be present on the inner face of the cell membrane, (2) adenosine triphosphate must be present at this face of the cell while (3) potassium ions must be present at the outer face of the membrane. With these three conditions satisfied, the three processes of sodium extrusion, potassium influx, and the associated ATP hydrolysis as depicted in Fig. 6.10 will occur. If one of these conditions is not satisfied, no one of these processes can take place. We should note, however, that the specificity of these cation requirements is not absolute. The external potassium ions can be replaced, as far as both inward transport and the hydrolysis of ATP are concerned, by lithium, rubidium, cesium, or ammonium ions (Maizels, 1961; Post, 1957; Post and Jolly, 1957). Lithium can substitute for the internal sodium ions, as far as ATP hydrolysis is concerned (Whittam and Ager, 1964) but lithium is not actively transported outward (Maizels, 1954). Finally, high levels of external sodium ions depress the ability of external potassium to stimulate the hydrolysis of internal ATP. (The various kinetic parameters for these systems are collected in Table 6.4.) We should note, too, that the orthophosphate ion and the adenosine diphosphate liberated on ATP hydrolysis in reconstituted ghosts carrying out ion pumping is found inside the ghosts (Sen and Post, 1964; Whittam and Ager, 1964), again emphasizing the vectorial nature of this process.

Of great importance for the elucidation of the kinetics and other properties of the cation transport systems was Schatzmann's discovery (Schatzmann, 1953) that active potassium influx can be specifically and greatly inhibited by the action of the cardiac glycosides. We have mentioned this effect several times previously in these chapters. Here we mention only the vectorial aspect of cardiac glycoside inhibition—the glycoside has to be present at the outer (potassium-binding) surface of the membrane for its action to be manifested. At this surface the effect of the cardiac glycoside can be overcome by the addition of potassium ions to high concentrations (Glynn, 1957; Bower, 1964).

6.5 Primary Active Transport Systems Uncoupled to the Energy Input

We have seen that if red blood cells are, by the technique of reversal of hemolysis, loaded with sodium and with ATP and are then placed in a medium containing potassium ions, there is a tightly coupled exchange of sodium for potassium ions and a simultaneous and stoichiometric hydrolysis of the ATP.

However, Garrahan and Glynn (1965) have shown that, if the cells are loaded with only low levels of radioactively labeled sodium (some 5–

10 m*M*) and with a low ratio of ATP to inorganic phosphate (1 m*M* ATP and 5 m*M* inorganic phosphate), there is still a substantial efflux of this sodium, but this efflux now no longer requires the presence of external potassium ions and is in fact reduced by the presence of external potassium. The efflux of ^{24}Na does require the presence of sodium on the outside of the cell, that is, it is an *exchange* of labeled and unlabeled sodium. It requires also the presence of the internal ATP but takes place without detectable hydrolysis of this ATP. The exchange of sodium ions is inhibited by the presence of cardiac glycosides.

These findings lead to a significant further understanding of the mechanism of cation transport. They suggest that for the "carrier" to transfer sodium across the membrane, it must bind to ATP—but the ATP need not be split. In the absence of external potassium, this ATP-bound form of the carrier can shuttle sodium in both directions across the membrane. But if potassium is present externally and if the level of internal ATP is sufficiently high, the bound ATP will be split and simultaneously there will be an exchange of internal sodium and external potassium. Skou (1964) has shown that in the presence of sodium, potassium, and magnesium, when ATP is split by a highly purified preparation of the membrane-bound ATPase (see Section 8.4,D), there is no binding of the liberated phosphate to the enzyme. This study together with that of Garrahan and Glynn suggests that it is the binding of ATP prior to its splitting that enables the membrane ATPase to bind sodium on the inside of the cell and presumably potassium at the outer surface. The splitting of ATP without the incorporation of the liberated inorganic phosphate into the enzyme leads to the production of a rearranged form of the enzyme in which the external potassium and the internal sodium exchange with each other. We shall discuss in Section 8.4,D some of the properties of the membrane ATPase preparation and the significance of these studies for the mechanism of cation transport.

In a similar fashion, the β-galactoside permease can be studied under conditions of restricted energy input. The entry of *O*-nitrophenyl-β-D-galactoside (ONPG) into cells of *E. coli* containing an intact β-galactosidase system is not strictly an active transport system. The uptake of β-galactoside proceeds more slowly than its hydrolysis within the cell so that uptake of ONPG always takes place down its concentration gradient. If the metabolic energy production by the cell is blocked by the presence of sodium azide (Koch, 1964), the entry of ONPG continues unabated and with unaltered kinetic parameters. The system behaves as (and indeed is) a facilitated diffusion system for the entry of β-galactosides. Strong evidence for this latter statement is presented in the valuable study by Winkler and Wilson (1966).

If the function of ATP is to modify the system so that it can no longer bind galactoside at the inner face of the membrane (that is, if K_m for galactoside is now very high on the inside) while the kinetics for entry remain unaffected, then this is an identical situation to that of a co-transport system in sodium ion imbalance, and the discussion of Section 5.4 and Eq. (5.22) will allow one to conclude that an active accumulation of galactoside will result. The system will bind galactoside at the outer face of the membrane and the galactoside will enter. At the inner face of the membrane a reaction linked with ATP hydrolysis must ensure that the affinity of the carrier for galactoside is reduced to a very small value. Little or no efflux of galactoside can thus occur, so that the free carrier is returned to the outer face of the membrane and the cycle of uptake can continue (Winkler and Wilson, 1966).

Such a model would suggest the presence, in the membrane of induced cells, of a membrane-bound permease. This permease may well be a galactoside-dependent ATPase, but in any event should bind galactoside in the absence of ATP but not in its presence. We shall see in Section 8.4,C that a component with galactoside-binding properties, associated with induction of the permease, has been isolated from *E. coli*.

A rather different situation is found for the α-methyl glucoside (isomaltose) permease of the yeast *Saccharomyces cerevisiae* (Halvorson *et al.*, 1964). In noninduced cells, that is, cells grown in the presence of glucose, a facilitated diffusion system for isomaltose is present in these cells. The system has a fairly low affinity for isomaltose ($K_m = 50$ mM at 30°C), is unaffected by interference with the energy metabolism of the cell—the action of sodium azide or dinitrophenol—and is thus a typical facilitated diffusion system. If the cells are grown in the presence of isomaltose but in the absence of glucose, an active accumulation of isomaltose occurs, an active transport system for this sugar being induced. The induced system has a high affinity for external substrate (K_m is circa 1 mM at 30°C), while the exit system cannot be shown to saturate when the internal substrate concentration is varied. Most important is the fact that the exit rates in induced and noninduced cells are nearly identical. The kinetic data are collated in Table 6.5. Mutant forms of the yeast exist in which the active transport system is absent. For all such mutants the facilitated diffusion system is also missing—the two systems appear to be controlled by the same gene. Thus the facilitated diffusion carrier must, during the process of induction, become linked to an energy-yielding reaction in the cell. In contrast to the *E. coli* system, the energy link here increases the affinity of the entry system for sugar, rather than decreasing the affinity of the exit system. Either procedure introduces an asymmetry and will convert a facilitated diffusion system into an active transport system. The particular method used by *Saccharomyces* appears to be a more economical one. If internal energy sources are low, reducing the affinity of the carrier at the inner face will cut down the loss of already accumulated substrate—there will be little leak through such a low-affinity facilitated diffusion system. In *E. coli*, breaking the link with the energy-yielding reaction will leave operative an effective and presumably undesirable facilitated diffusion system.

6.6 The Kinetics of "Pump" and "Leak" Systems

The existence of a unidirectional flux of substrate brought about by a primary active transport system would in principle lead to the unlimited accumulation of substrate within the cell. That this does not occur and that instead a high but finite level of substrate is, in general, reached within the cell at the steady state, is due to the almost universal presence

TABLE 6.5

KINETICS OF α-ETHYL THIOGLUCOSIDE (TEG) ENTRY AND EXIT IN INDUCED AND NONINDUCED CULTURES OF THE YEAST *Saccharomyces cerevisiae* AT 30°C[a,b]

			Exit rate			
			No addition		+MG	
Cells grown on	Equilibrium level G_∞ (μmole/gm cell)	Entry rate (μmole/gm cell/hr)	Initial rate (μmole/gm cell/hr)	Rate constant (hr^{-1})	Initial rate (μmole/gm cell/hr)	Rate constant (hr^{-1})
Glucose	0.44	0.36	0.09	0.22	1.58	3.6
MG	19.0	18.0	3.42	0.18	73.0	3.84

[a] Induced cells grown on α-methyl glucoside (MG), noninduced on glucose. Exits were studied in normal medium and in the presence of $10^{-2}M$ MG.

[b] Data taken from Halvorson *et al.*, 1964.

of "leaks" allowing substrate to escape from the cell. The steady state level of substrate within the cell is, as we shall demonstrate, defined by this balance between, on the one hand, accumulation by the pump and, on the other, the leakage of substrate across the membrane. One may speculate on the possibility that the leak is a physiological necessity and has had to be evolved in conjunction with the pump for any substrate. Were this not the case, and were the substrate to be osmotically active within the cell, the steady and undrained accumulation of substrate must so raise the internal osmolarity of the cell that the resulting water inflow would burst the cell membrane (Section 7.1), no matter how rigid the cell wall. It is interesting to note that the accumulation of phosphate by mitochondria, an apparently unidirectional uptake for which no leak has been reported, is accompanied by the formation of insoluble deposits of calcium phosphate (hydroxylapatite). The phosphate is thereby rendered osmotically inactive (Lehninger, 1965).

We can distinguish at least two types of leak, the two resulting in somewhat different formal descriptive equations which are worth recording.

If, *case (a) the efflux is unidirectional* (where, for example, efflux is occurring by the altered form of the active influx system), we have the following situation:

$$\text{Ingress by action of the pump} = \frac{V_{\max} S_e Q}{K_m + S_e} \qquad (6.2)$$

where S_e is the external concentration of permeant, K_m the Michaelis constant of the pump, and $V_{\max}$ a term being the maximum velocity of activity of the pump per unit concentration of some metabolite Q driving the primary transport.

$$\text{Egress by the unidirectional leak} = kS_i \qquad (6.3)$$

where k is a rate constant and S_i the substrate concentration within the cell. Then at the steady state

$$kS_i = \frac{V_{\max} S_e Q}{K_m + S_e} \qquad (6.4)$$

and the distribution ratio (DR) is given by

$$\text{DR} = \frac{S_i}{S_e} = \frac{V_{\max} Q}{k(K_m + S_e)} \qquad (6.5)$$

We will note certain consequences of Eq. (6.5). First, as S_e tends to zero, the distribution ratio tends to a maximum given by $V_{\max}Q/kK_m$ and if $V_{\max}Q$ and K_m can be found from a study of the unidirectional flux or of the rate of exchange of labeled substrate at the steady state, then k can be found from the minimum (extrapolated) distribution ratio. Second, as S_e tends to infinity, the *distribution ratio* tends to zero, while the *internal concentration* tends to a limiting value given again by $V_{\max}Q/kK_m$. If either $V_{\max}$ or Q is zero, S_i is zero and, of course, so is the distribution ratio; $V_{\max}$ will be zero if the system is absent (in a mutant or noninduced cell, in the case of transport in bacteria) or if the system is inhibited. If the metabolic energy source for transport is absent, Q will be zero. From Eqs. (6.2) and (6.4) we can see that for any level of external substrate S_e the distribution ratio should be always $1/S_ek$ times the unidirectional flux of permeant by the pump system or, in other words, the initial velocity of uptake of substrate divided by the distribution ratio should always be proportional to S_e, the coefficient of proportionality being the rate constant k. Finally, we might put Eq. (6.4) into the reciprocal form when we obtain

$$\frac{1}{S_i} = \frac{k}{V_{\max}Q} + \frac{kK_m}{V_{\max}Q} \times \frac{1}{S_e} \tag{6.6}$$

from which equation, in the usual fashion, the constant K_m and the composite term $k/V_{\max}Q$ can be found from the intercepts of a double reciprocal plot. Figure 6.11 depicts some data obtained for the β-galactoside

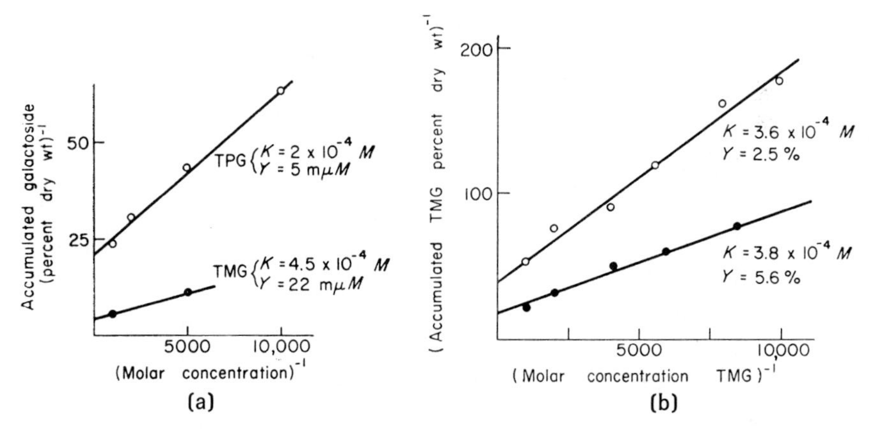

Fig. 6.11. Accumulation of thiogalactosides by the galactoside permease of *E. coli*. Lineweaver-Burk plots of the steady state level of accumulated galactoside (ordinate) against the external level of substrate (abscissa). (a) Uptake of two different galactosides, thiophenyl-β-D-galactoside (TPG) and thiomethyl-β-D-galactoside (TMG). (b) Uptake of TMG by two cultures, induced for 5 min (upper curve) and 1 hr (lower curve), respectively. (Taken with kind permission from Rickenberg *et al.*, 1956.)

permease of *E. coli* (Rickenberg *et al.*, 1956). In Fig. 6.11a the double reciprocal plot of the steady state level of galactoside is plotted against the substrate concentration for two different galactosides. The values of both K_m and the term $V_{\max}Q/k$ differ for these two substrates. In Fig. 6.11b the data for one particular galactoside (thiomethyl-β-D-galactoside) are measured for two cultures of *E. coli*, one having been grown for 5 min and the other for 1 hr in the presence of the substrate as inducer. From these data K_m is not affected by the process of induction but $V_{\max}Q/k$ is increased. Presumably, though we cannot draw such a conclusion from this study alone, it is the term V_{max}, the quantity of transporter per unit area of cell surface, that increases on induction. Clearly, the formalism of Eq. (6.6) and hence of Eq. (6.5) fits the galactoside permease data adequately, at least over the concentration ranges studied. Such a simple picture is not found for all permeases, however. A plot showing the absence of the strict relationship between the initial rate of uptake of substrate and the final distribution ratio, that Eqs. (6.2) and (6.4) would predict, is available for the system transporting glucuronides into *E. coli* (Stoeber, 1961) (Fig. 6.12). At high levels of substrate, there

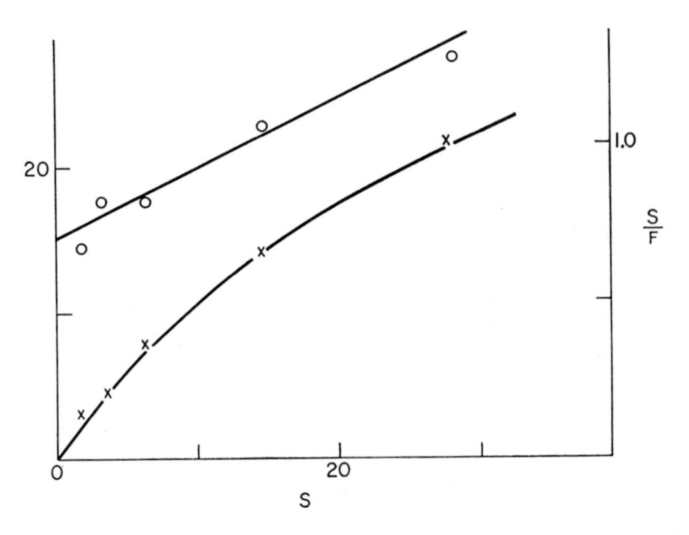

Fig. 6.12. Accumulation of thiophenyl glucuronide by the glucuronide permease of *E. coli.* Lower curve, initial velocity of uptake (left-hand scale in μmole gm^{-1} min^{-1}) against steady-state level of accumulation [S, abscissa in μmoles gm^{-1}]. Upper curve, data plotted as required by Eq. (6.10). Steady-state level divided by initial velocity (S/F, right-hand scale in minutes) against steady-state level on the abscissa. From the slope and intercept of the straight line, K_f is 32 μmoles gm^{-1}, while k is 45 μmoles gm^{-1} min^{-1}. Data due to F. Stoeber. (Taken with kind permission from Kepes and Cohen, 1962.)

is significantly more substrate present within the cell than this simple relationship would predict. We can account for at least part of this deviation by making the assumption that *case (b) efflux and the nonactive component of influx occur by a facilitated diffusion system.* We have, now,

$$\text{Egress} = \frac{kS_i}{K_f + S_i} \tag{6.7}$$

where k is a rate constant and K_f the Michaelis constant for the parallel facilitated diffusion system, while

$$\text{Ingress} = \frac{V_{\max} S_e Q}{K_m + S_e} + \frac{k S_e}{K_f + S_e} \tag{6.8}$$

Then from the steady-state condition we have

$$\frac{k S_i}{K_f + S_i} = \frac{V_{\max} S_e Q}{K_m + S_e} + \frac{k S_e}{K_f + S_e} \tag{6.9}$$

and now from Eqs. (6.8) and (6.9) the ratio

$$\frac{S_i \text{ at the steady state}}{\text{unidirectional flux}} = \frac{K_f + S_i}{k} = \frac{K_f}{k} + \frac{S_i}{k} \tag{6.10}$$

As the upper curve of Fig. 6.12 we have plotted, as Eq. (6.10) requires, the term on the left-hand side of this equation against S_i, using the data of the lower curve of Fig. 6.12. A good linear plot is obtained from which we can determine the terms K_f and k. Equation (6.9) does not yield a simple form for the distribution ratio S_i/S_e.

If K_f is very high, that is, the parallel facilitated diffusion pathway is mediated but of low affinity, we obtain a simple form for the distribution ratio as

$$\text{DR} = S_i/S_e = 1 + \frac{K_f \cdot V_{\max} \cdot Q}{k(K_m + S_e)} \tag{6.11}$$

on setting $K_f \gg S_e$, S_i in Eq. (6.9).

The limit of the distribution ratio here as S_e tends to zero is now $1 + K_f V_{\max} Q / k K_m$, similar to the result from Eq. (6.5), but as S_e tends to infinity the distribution ratio, instead of approaching zero, will tend to unity. This situation would be found for the data on potassium influx in red blood cells (Fig. 6.9) if these were plotted as distribution ratios. It is also found for the secondary transport systems accumulating amino acids in intestine (Akedo and Christensen, 1962a) and pigeon erythrocytes (Vidaver, 1964a).

6.7 Ion Pumping by Epithelial Cell Layers

We have already briefly considered the over-all picture of ion transport across epithelial cell layers (Section 6.2 and Fig. 6.1). Our discussion on the complexities of leak and pump systems in single cells should now enable us to approach the detailed mechanisms involved in epithelial cell transport. A salient observation in the analysis of this problem was the demonstration (Koefoed-Johnsen and Ussing, 1958) that, in the absence of potassium ion in the medium bathing the *inner* (serosal) surface of frog skin, the electrical potential across the skin, which was thought to be associated with the active transport of sodium from outside to inside, decreased. The presence or absence of potassium in the medium bathing the *outer* membrane had in contrast no effect on the potential and, hence, on the rate of sodium pumping. Our discussion on the linked sodium-potassium pump in the red blood cell (Section 6.4) might suggest an immediate explanation of this finding. We might postulate the existence of a linked sodium-potassium exchange pump at the inner (serosal) membrane of this epithelial cell layer. Thus Fig. 6.13 represents the model proposed by Koefoed-Johnsen and Ussing

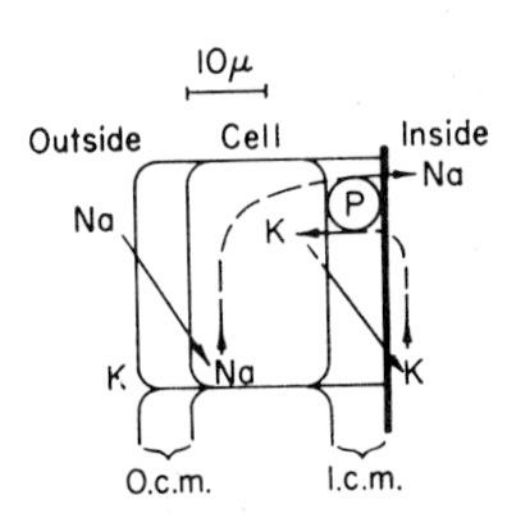

Fig. 6.13. A model representing the possible movements of cations across the outer (o.c.m.) and inner (i.c.m.) cell membranes of an epithelial cell. At the outer membrane is a specific facilitated diffusion system for sodium; *P* is a pump mechanism with a one-to-one exchange of K for Na. (Taken with kind permission from Ussing, 1959.)

(1958) to explain, in part, this finding. If the linked sodium-potassium pump is to be able to move salt through the cell layer, the entry of salt into the cell across the outer membrane must be possible. Koefoed-Johnsen and Ussing proposed that a facilitated diffusion system specifically transferring sodium was present at this membrane. If potassium is to be continually available on the inside of the inner membrane, the outer membrane must be impermeable to this ion, but the inner membrane must allow a leak of potassium. Potassium thus circulates between the cell and the inner medium—pumped in by the linked pump, leaking out by the conventional leak. In the steady state a flow of sodium ions (and chloride ions and, as we shall see in Section 7.3, in certain circumstances, of water) will occur through the cell, at the expense of the metabolic energy required to maintain the intracellular ion gradients. This model predicts (1) that the outer membrane be selectively sodium permeable, (2) that the inner membrane be selectively potassium permeable (but this selectivity need not be absolute), and (3) that the sodium pump be linked to potassium pumping. Predictions (1) and

(2) require further that the electrical potential that the sodium pump develops across the cell layer (Section 6.2 and Fig. 6.1) should be composed of two potentials—one between the inside medium and the cell interior across the inner cell membrane, the other between the cell interior and the outside medium—across the outer cell membrane. The inner membrane should behave as a potassium-selective membrane. The potential across this inner membrane should thus be a diffusion potential, dependent on the (relative) concentrations of potassium in the cell interior and in the medium bathing the inner surface of the skin, as given by Eq. (6.12).

$$E_{\mathrm{K}} = \frac{RT}{F} \ln \frac{\mathrm{K_{cell}}}{\mathrm{K_{outside}}} \tag{6.12}$$

Here R is the gas constant, F the faraday, and T the absolute temperature. The outer membrane likewise should be sodium-selective and develop a diffusion potential determined by the sodium concentration in the outer medium, given by Eq. (6.13).

$$E_{\mathrm{Na}} = \frac{RT}{F} \ln \frac{\mathrm{Na_{outside}}}{\mathrm{Na_{cell}}} \tag{6.13}$$

The potential across the membrane E_M should be the sum of these two diffusion potentials as

$$E_M = E_{\mathrm{Na}} + E_{\mathrm{K}} \tag{6.14}$$

The true diffusion potential developed by these membranes can be most easily seen if the short-circuiting flow of anions through the system is blocked by using impermeable sulfate ions in the bathing solutions. Figure 6.14 shows how the potential across the frog skin depends on the concentration of potassium ion in the inner solution (Fig. 6.14a) and sodium ion in the outer solution (Fig. 6.14b). The dashed lines are the slopes (58 mV for a tenfold change in concentration) predicted by Eqs. (6.11) through (6.14) and given by the term RT/F. The accord of these experiments with the predictions made on the basis of Fig. 6.13 and the dependence of the rate of pumping on the potassium concentration in the inner medium seemed completely to justify the model of Fig. 6.13.

It has clearly been disappointing for all concerned in such research to find in recent years that this elegant model which accorded so well with the information obtained from single cell systems is in fact an oversimplification. The first breach in the evidence came from the work of Ussing's own school (Bricker *et al.*, 1963) when it was found that the removal of potassium ion from the solution bathing the inner surface of the frog skin led to a shrinkage of the epithelial cells. (Potassium chloride was found to be leaking out of the cells down the potassium ion gradient,

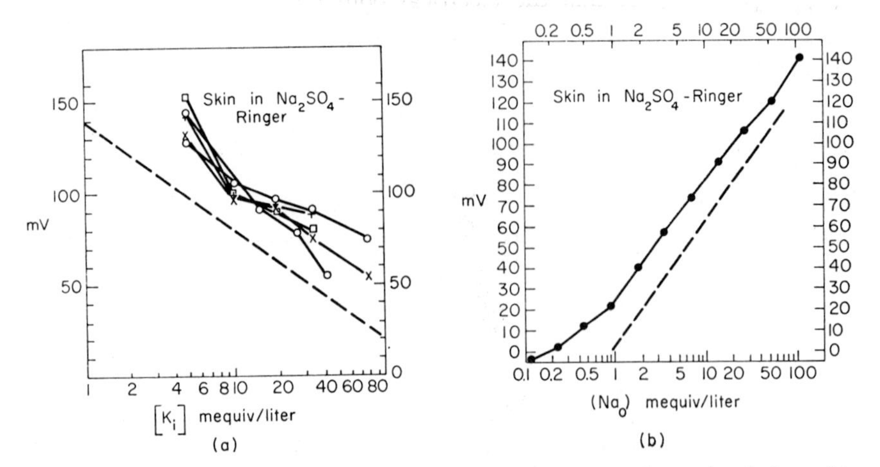

Fig. 6.14. Dependence of the potential difference across the isolated frog skin on (a) the inside K concentration and (b) the outside Na concentration. The use of sulfate avoids short circuiting by anion permeation. The dashed lines in the figures are the theoretical predictions if the cell surfaces were behaving as a potassium electrode in (a) or as a sodium electrode in (b). (Taken with kind permission from Ussing, 1959.)

and the accompanying solvent flow led to the observed cell shrinkage.) Such shrinkage, whether caused by electrolyte shifts or by an induced osmotic gradient, inhibited sodium pumping. If the shrinkage of the epithelial cell layers was prevented or reversed—for instance, by simply diluting the medium bathing the inner skin surface—active salt transport was restored and could continue for several hours in the complete absence of potassium ion from the medium bathing the inner skin surface. Unless the free movement of the extruded potassium ions from the cell interior to the medium is prevented in some way that is not understood, there can be no continued cycling of potassium ions as Fig. 6.13 would require. The sodium ion pumping is presumably accompanied, therefore, by chloride movement, since electrical neutrality must be maintained.

The other grounds for supporting the model of Fig. 6.13, the dependence of the electrical diffusion potentials on the ionic concentration of the bathing fluids, has been questioned by the experiments of Snell and Chowdhury (1965) who studied the effects of simultaneously changing both media, the internal and external solutions, on the diffusion potentials. The contralateral effects of ion concentrations were more complicated than the simple model of Fig. 6.13 would predict. Their data showing the dependence of the membrane potential on the potassium ion concentration bathing the inner surface of skin, as a function of the sodium ion concentration bathing the outer surface, are given in Fig.

6.15. Clearly the slope of the curve relating the "diffusion potential" to the concentration of the ion (which should, in the model of Fig. 6.13, determine this potential) instead of being the constant value RT/F [Eqs. (6.12) and (6.13)], depends on the ionic composition of the medium bathing the opposite face of the membrane.

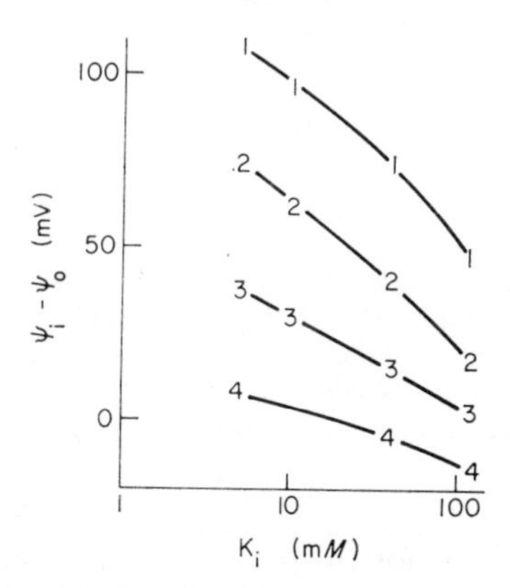

Fig. 6.15. The variation of the electrical potential (ordinate) developed across frog skin as a function of the potassium ion concentration (abscissa) in the medium bathing the inner surface of the frog skin, at various values of sodium ion concentration in the outer bathing fluid. Experiment 1: Na outside = 115 m*M;* Experiment 2: Na outside = 30 m*M;* Experiment 3: Na outside = 6 m*M;* Experiment 4: Na outside = less than 1 m*M.* (Taken with kind permission from Snell and Chowdhury, 1965.)

Finally, the simple picture of Fig. 6.13 which leads, as we have seen, to the prediction that the total potential developed across the membrane be composed of two step potentials at the inner and outer cell membranes, has been directly tested by microelectrode experiments but, unfortunately, the results of such experiments are as yet controversial. Some workers (references in Cereijido and Curran, 1965) have found the predicted two steps, others have found three steps, still others (Snell and Chowdhury, 1965) a continuous change of potential with distance as the microelectrode is inserted through the cell layer. In a very careful study Cereijido and Curran (1965) have not only measured directly by this microelectrode technique the potential across the inner membrane and the potential across the outer membrane but have also studied how these potentials vary with the ionic composition of the medium bathing the two surfaces of the skin. Cereijido and Curran are careful to point out that the interpretation of such studies is extremely difficult since the frog skin is a complex multicellular tissue and the exact location of the tips of the inserted microelectrodes is often uncertain. Nevertheless, their

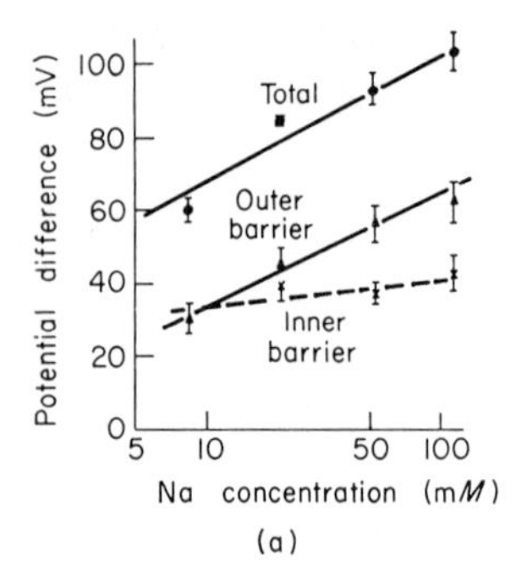

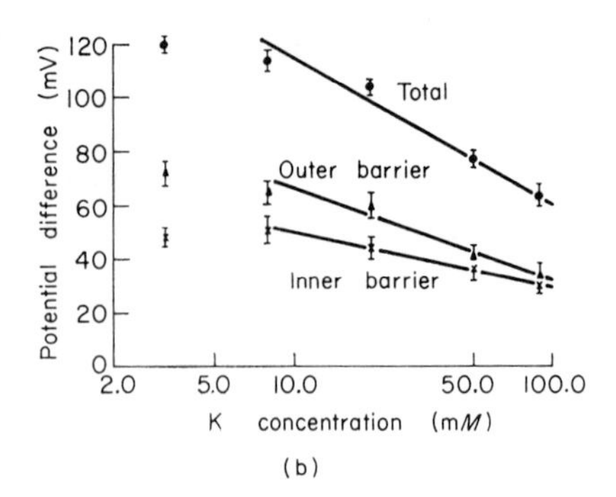

Fig. 6.16. The total potential, ●; the potential between the epithelial interior and the outside solution, ▲; and X, that between the epithelial interior and the inner solution, measured with intracellular electrodes in frog skin, as a function of (a) the sodium concentration in the outside solution and (b) the potassium concentration in the inside solution. Compare Fig. 6.14. (Taken with kind permission from Cereijido and Curran, 1965.)

results (Fig. 6.16) are of much interest. If the concentration of sodium bathing the outer surface of the skin is varied (Fig. 6.16a), the total potential and the potential across the outer cell membrane vary together, the potential across the inner membrane remaining almost constant. This is in excellent agreement with the predictions of the model of Fig. 6.13, although the rate of change of potential with the logarithm of the Na concentration is only two-thirds of that predicted by Eq. (6.13). However, if the potassium ion concentration of the inner medium is varied (Fig. 6.16b), both the potential across the inner membrane and that across the outer membrane vary, a result quite unlike that expected from Fig. 6.13. It is clear that experiments along these lines will, in time, allow a more satisfactory model to be put forward for frog skin but meanwhile no such model is available and we must accept that of Fig. 6.13 as incorrect in detail but probably suggestive of the true nature of the problem.

The solution of this problem will be of more than usual interest since it is apparent that a very similar system of ion pumping is operative in many other important epithelial cell layers, for example, in toad bladder (Leaf, 1960) and perhaps in kidney (Solomon, 1963). As we shall see in Section 7.3, the sodium pumping in these tissues is primarily responsible for their ability to transport water and hence for the role of these tissues in the regulation of the water balance of the animal. An understanding of the problem raised by these epithelial cell layers should lead to an appreciation of the mechanism of action of the antidiuretic hormone, since this hormone (see the review by Orloff and Handler, 1964) controls water balance by affecting sodium transfer. In Chapter 7 we shall briefly consider these studies. Finally, it is becoming clear that a number of toxic substances, for example, the toxin of cholera (Fuhrman and Fuhrman, 1960), the cathartic agents (Phillips *et al.*, 1965), and a large number of substances interfering with kidney action (see Pitts, 1963) act by interference with the salt-transporting activity of epithelial cell layers. The field of such studies may be expected to grow rapidly in the future as the underlying mechanisms of the transport processes involved are clarified.

CHAPTER 7

The Movement of Water

Water is, of course, the major constituent of living matter, and the movement of water across cell membranes is quantitatively the major phenomenon in biological transport. We have considered in Chapter 3 the molecular basis of water movement and especially the view that water movement occurs by diffusion through pores in the cell membrane (Section 3.7). In the present chapter we shall emphasize the physiological, as opposed to the molecular, aspect of water movement. We shall consider the factors which determine the rate of movement of water across cell membranes and into tissues, and also the factors which determine the steady state distribution of water in both cells and tissues.

7.1 The Volume of a Cell at the Steady State

If the tonicity, that is, the concentration of osmotically active material of the medium surrounding an animal cell, is maintained constant and if the cell remains metabolically active, the cell volume will remain constant over a long period of time. Yet if red cells are placed in media containing different concentrations of impermeable substances, the cells respond to the change of external osmotic pressure as if they were, to a very good first approximation, perfect osmometers (LeFevre, 1964). The cell volume is, therefore, capable of varying. Thus the cell is not enclosed by a rigid framework. Yet its volume is maintained constant under physiological conditions, and we can show that this constancy requires the continued performance of metabolic activity. If a tissue is removed from the body into an incubation medium and prevented from metabolizing, it will soon swell through absorbing water (and salts) from the medium (Table 7.1) (Heckman and Parsons, 1959). If metabolism recommences, the accumulated water and salts can be returned to the medium and the initial cell volume regained. The maintenance of the steady state volume is clearly an active, energy-consuming process and we can see at once by the following argument that this must of necessity be the case. Figure 7.1 is a model of a cell, oversimplified, but one which will make the point clearly. We picture the cell as surrounded by a semipermeable membrane M. Within the cell is a set of constituents P_i which are able to diffuse across the membrane and a class A_i to which the membrane is impermeable. In the medium bathing the cells are again the permeable constituents P_e and a set of impermeable constituents Z_e. If, on this simple model, there is no possibility of active transport, then the concentrations of the components P are everywhere the same so that $P_i = P_e$. Only A_i and Z_e contribute to the osmotic pressure difference across the cell, and there will be no such difference, that is, the cell volume will be steady only if $A_i = Z_e$.

TABLE 7.1

EFFECT OF INCUBATION AT 0.5°C ON WATER AND ELECTROLYTE CONTENT OF 0.25 MM THICK RAT LIVER SLICES[a]

Incubation time (min)	Water (kg/kg dry wt.)	Sodium (mequiv per kg dry wt./ kg water)		Potassium (mequiv per kg dry wt./ kg water)		Chloride (mequiv per kg dry wt./ kg water)	
0	2.15	95.2	44.3	305.2	142	97.9	45.4
5	2.26	308	137	179	79.5	230	102
10	2.62	391	150	183	70	303	115
20	2.66	370	139	147	55	324	117
31	3.09	545	176	112	36	434	141
60	3.40	581	171	112	33	424	125
120	3.69	678	184	80	22	560	152
240	3.85	775	201	67	17	591	154
360	4.30	699	162	69	16	651	152
Composition of incubation medium			161		5		166

[a] Data taken from Heckmann and Parsons, 1959.

That this model is untenable is borne out by the data presented in the previous four chapters. We have seen that all the major constituents of conventional bathing media of Fig. 7.1 can penetrate most cell membranes. In particular, the chloride, sodium, and potassium ions move freely, although not necessarily rapidly, into the cell. (For cells suspended in plasma containing a significant proportion of impermeable proteins, part at least of the osmotic pressure gradient due to the intracellular components is neutralized by the plasma proteins, but for cells

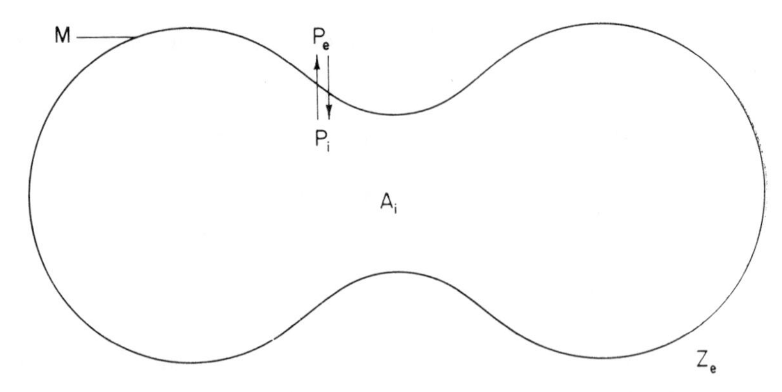

Fig. 7.1. A model of the permeability behavior of a single cell, where M is the cell membrane, Z a nonpenetrating molecule in the external medium, A one within the cell, and P a penetrating molecule. The subscript e refers to the external medium, i to the internal.

in "physiological" saline media, no such contribution can be expected.) In these circumstances, the term Z_e, giving the concentration of the external impermeable component, is zero and by the formulation above, if $P_i = P_e$, a net osmotic difference of A_i, due to the internal constituents A_i, will be present across the membrane. Thus there will be at all times a continuing entry of water into the cell down this osmotic gradient. Unless some restraints were present this influx of water would lead to cell swelling and to the eventual bursting of the cell. Cells exist in a medium containing, as the major constituents, substances to which their membranes are permeable. In addition they must of necessity contain a measure of impermeable intracellular matter. This situation leads to a permanent threat of cell flooding due to water movement down the osmotic gradient, and this threat must be parried if the cell is to exist in a stable state. Plant cells and bacterial cells are enclosed in rigid cell walls which are strong enough to contain the cell volume at a defined level in the face of the enormous pressures (tens of atmospheres) that osmotic water flow imposes at the cell surface. Animal cells, in contrast, have overcome the necessity for rigid cell walls by modification of the membrane itself, leading to the possibility of flexibility and in particular motility in multicellular animal organisms. The evolution of these membrane specializations has been a prerequisite for the evolution of animal life (Leaf, 1959; Tosteson, 1964).

One method of ensuring that excess water would not accumulate within the cell would be to pump it out again just as fast as it enters—by an active transport system. There has, however, been no substantiated evidence for the active transport of water in any cell system, other than for transport across insect cuticle (for which, see Beament, 1964). We must presume that for cells having the usual high water permeability (Section 3.7), such an active transport, even if molecularly feasible, would be energetically impossible. Instead, the method used universally by animal cells is to ensure by the operation of an outwardly driven ion pump, that the intracellular concentration of the permeable components are kept at a low level. The necessary and sufficient condition for osmotic balance is that the total concentration of osmotic material within the cell (permeable and impermeable) is maintained equal to that outside the cell. The amount of impermeable matter within the cell being given, the concentration of permeable matter is reduced by the operation of, generally, the sodium pump until osmotic balance is achieved. The concept is simple, and the following formal description of this procedure in mathematical terms reveals some interesting inter-relationships between the forces determining the steady state cell volume. We shall, at first, follow Post and Jolly (1957) and assume that only undissociated molecules are present in our simple model. The rate of outward pumping of P is determined by the pump rate constant p and the internal concentration P_i. There is also a leak of P determined by the leak rate constant l and the respective concentrations P_i and P_e. We have the following conditions:

At osmotic equilibrium,

$$A_i + P_i = P_e \tag{7.1}$$

Also, the total efflux of P (equal to the efflux by leak plus efflux by pump) is equal to the influx by leak at the steady state, or

$$lP_i + pP_i = lP_e \tag{7.2}$$

Equations (7.1) and (7.2) give us two equations in the two unknowns A_i and P_i, the solution of which is

$$A_i = \frac{p}{l+p} P_e = \frac{1}{(l/p)+1} P_e \tag{7.3}$$

Finally, A_i, the concentration of fixed intracellular matter, is determined by the amount of intracellular matter X_i, a constant, and the volume of the cell V, a variable, as

$$A_i = X_i/V \tag{7.4}$$

Substituting in Eq. (7.3), then,

$$X_i = \frac{V}{1 + l/p} P_e \tag{7.5}$$

or

$$V = (X_i/P_e)(1 + l/p) \tag{7.6}$$

This very interesting result, first obtained by Post and Jolly (1957), shows us that in cells without a rigid cell wall, the volume of the cell is determined absolutely by three factors: the *amount* of fixed intracellular material X_i, the *concentration* of external permeable matter P_e, and the *ratio* of the rate constants of leak to pump. It is easy to see from Eq. (7.6) that if the cell suddenly becomes or is rendered excessively leaky to the components of the surrounding medium (that is, if l increases), the volume of the cell will rise proportionately. (This explains the hemolysis that occurs if red cells are made excessively cation-permeable by treatment with butanol (Ponder, 1948). Similarly, if pumping stops or is reduced (that is, if p decreases) the volume will rise. (This accounts for the swelling that occurs if, as mentioned above, the metabolic processes in tissues are interfered with.) As P_e is increased or decreased, the cell volume alters—the cell behaves as an osmometer. In essence then, Eq. (7.6) accounts (in a very oversimplified manner) for the water balance in the cell. The ratio of the rate constants of the leak to that of the pump will determine the value of P_i, the concentration of permeable material inside the cell. The cell volume then increases or diminishes until the osmotic concentration of fixed intracellular material added to the osmotic concentration of permeable material (determined by the leak to pump ratio) is exactly equal to the external osmotic concentration.

The real cell is, of course, considerably more complex than the above model assumes. The first complexity we should introduce is that the major solutes of the cell and media are ions, and the presence of a fixed ion within the cell (generally an anion) imposes a Donnan distribution (Section 2.7) on the permeable ions of the cell (Fig. 7.2). Also arising

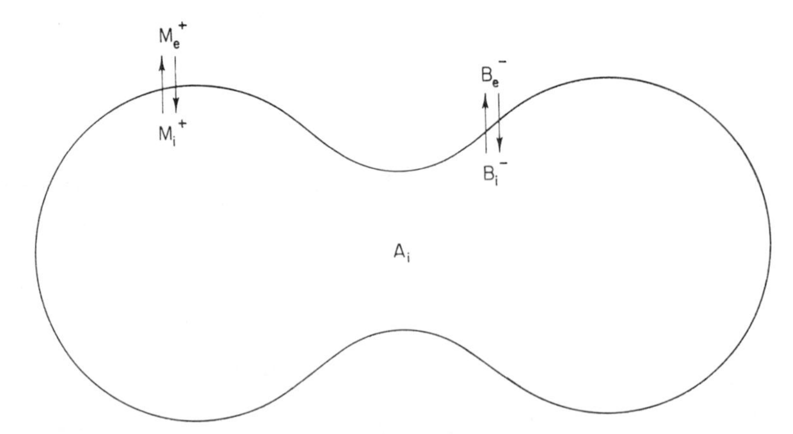

Fig. 7.2. A more complex cell model. The cation M^+ and the anion B^- can penetrate the cell membrane; A is an impermeable anion. Subscripts e and i refer to the external and internal media, respectively.

from the ionic nature of the permeable solutes, we must take into account the fact that to preserve electroneutrality, a penetrating cation must either exchange for a cation going in the opposite direction or must be accompanied by an anion going in the same direction. If no current flow occurs (the usual situation, unless a mechanism for current flow is experimentally imposed), these conditions must apply to ion pumping as well as to ion leakage—an extruded cation must be accompanied by an anion or exchanged for another cation. As we have seen in Section 2.7, these conditions require that the rate of transfer of cation M by a leak is given by the combined product of the leak rate constant l, the concentration of permeable cation M, and the concentration of permeable anion B. If the pump is, as we shall consider at this first stage, able to extrude sodium in company with chloride, the same combined product $M \times B$, but with the *pump* rate constant p, describes the pump flux. These considerations give us, in place of Eq. (7.2), an equation for the steady state transfer of the pumped constituent M across the membrane as

$$lM_iB_i + pM_iB_i = lM_eB_e \tag{7.7}$$

(Note that this is also the equation of balance for the nonpumped constituent B_e.) We have also the condition that there shall be no net charge on either side of the membrane, the electroneutrality condition. This gives us the relationships

$$M_e = B_e \tag{7.8}$$

and

$$M_i = B_i + A_i \tag{7.9}$$

and the osmotic equilibrium condition is now

$$B_i + A_i + M_i = B_e + M_e \tag{7.10}$$

We have then three unknowns B_i, A_i, and M_i, and the three equations [Eqs. (7.7), (7.9), and (7.10)] containing these unknowns, which equations we can solve as before to give, exactly as Eq. (7.6),

$$V = (X_i/M_e)[1 + (l/p)] \tag{7.11}$$

Incidentally, from Eqs. (7.8) through (7.10), it is easy to show that $M_i = M_e$. The concentration of cation is the same on both sides of the membrane. It is the permeable anion that is effectively excluded from the cell. This is because, although it is M that is being pumped out of the cell, B must travel out in company with M, and yet the Donnan condition ensures that M is held within the cell.

More complicated, but closer still to the real situation, is the case where we have two cations M and N (for example, sodium and potassium) present in the medium and in the cell, only one component M being actively pumped by the transport system. If we assume, as in the example just discussed, that the pump here moves a sodium ion M together with an anion B (that is, the pump is *not* a sodium/potassium-linked pump), all the equations (7.7) through (7.10) apply, with the addition that N is present to contribute to the osmotic and electrical balances. We assume that since N is not operated on by the pump, its flow is entirely passive, being determined by an equation of the form of

$$nN_iB_i = nN_eB_e \tag{7.12}$$

where n is the rate constant for leakage of N. The final equation obtained on this model can be readily shown to be identical with Eq. (7.11). We note that the real difference between the two is that in the simple model, with only one cation present, the denominator of Eq. (7.11) refers of course to the total extracellular cation. In the case where two cations are present, only one of these being pumped, only the extracellular concentration of the pumped cation comes into the equation. From Eq. (7.12) we see that the distribution ratios of the passively transferred species are reciprocal to one another, that is, that

$$B_i/B_e = N_e/N_i = \mathbf{r} \tag{7.13}$$

where $\mathbf{r}$ is the Donnan distribution ratio. Also, we no longer have the condition that $M_i = M_e$. Rather we have that

$$M_i = M_e + N_e(1 - 1/\mathbf{r}) \tag{7.14}$$

and consideration of Eq. (7.13) will reveal that M_i is always less than M_e, if it is an anion that is fixed intracellularly. This accords more with our intuitive feeling that the species being pumped out of the cell should be at a lower concentration inside than out.

Finally, we shall consider the case directly applicable to, say, the red blood cell, where the active efflux of sodium is linked to the active influx of potassium. Tosteson and Hoffman (1960) have presented a full treatment of this situation, which we follow here in part. As we have seen in Section 6.4, experiments show that the linkage between active Na^+ efflux and active K^+ influx is, in the red cell, relatively strict with a stoichiometry which a number of workers find to be unity. We shall assume a strict linkage of unity here. In particular, active sodium efflux depends not only on the internal sodium concentration, here M_i, but also on the external potassium concentration, here N_e, and at the commonly found low levels of external potassium concentration (that is, below the half-saturation concentration of the pump with external K^+), we can write with little risk of error:

$$\text{Rate of active Na}^+ \text{ efflux} = \text{Rate of active K}^+ \text{ influx} = pM_iN_e \quad (7.15)$$

(using the symbols of the present section), rather than pM_iB_i as we had for the unlinked sodium pump. We have now two equations for the steady state of sodium [Eq. (7.16)] and for potassium [Eq. (7.17)] as follows:

$$lM_iB_i + pM_iN_e = lM_eB_e \quad (7.16)$$

$$nN_iB_i = pM_iN_e + nN_eB_e \quad (7.17)$$

because the term pM_iN_e gives both the active sodium efflux and the active potassium influx. [Adding together Eqs. (7.16) and (7.17) gives the equation describing the balanced chloride movements across the membrane.] We still have Eqs. (7.8) through (7.10) (but containing the terms N_i and N_e for the potassium ions) giving us four equations in the four unknowns M_i, N_i, B_i, and A_i. The solution of these equations, however, requires the solution of a quadratic equation and is no longer quite as simple as that of Eq. (7.11). We find rather that

$$2A_i = B_e + \frac{p}{l}N_e \pm \left[\left(B_e + \frac{p}{l}N_e\right)^2 - 4\frac{p}{l}N_eM_e\,\frac{(n/l)-1}{n/l}\right]^{\frac{1}{2}} \quad (7.18)$$

where it is the negative sign before the square bracket that is meaningful. We will take two simplified solutions of Eq. (7.18). If, as is usually the case, N_e the concentration of potassium is small compared with M_e, the concentration of sodium, M_e/B_e is close to unity. Also, if n is much greater than l (that is, the rate constant for potassium leakage is much greater than that for sodium leakage) n/l is large and $[(n/l) - 1]/(n/l)$ is unity. We obtain in this case

$$A_i = p/l \cdot N_e \quad \text{for } n > l \quad (7.19)$$

while if we assume on the contrary that n and l are equal

$$A_i = 0 \quad \text{for } n = l \quad (7.20)$$

Clearly, in most real cases A_i will be somewhere between the two limiting values determined by Eqs. (7.19) and (7.20). As before, if A_i is determined, then X_i/V and hence the volume V are determined. It is clear from Eq. (7.20) that if we have a pump which exchanges sodium stoichiometrically for potassium, then it cannot maintain the volume of the cell finite ($A_i = 0$; therefore, $V = \infty$), if the rate constants for the leak of sodium and potassium are identical. A few moments' thought will confirm that this is so: If the pump can only exchange sodium and potassium it can have no osmotic effect if sodium and potassium leak back equally rapidly. There is merely an exchange of two nonidentical particles K^+ and Na^+. However, if the rate of potassium leakage is faster than that of sodium and only if this is the case, the high potassium concentration inside the cell will encourage a leak of potassium and chloride (that is, KCl) out of the cell faster than NaCl leaks into the cell. Osmotic equilibrium can now occur. In the case of the sodium and chloride pump, it is effectively sodium chloride that is forced out of the cell. In the case of the linked sodium/potassium exchange pump it is effectively potassium chloride that is forced out of the cell.

A very beautiful example of the relations introduced here has been provided by Tosteson and Hoffman (1960) and Tosteson (1964) in a theoretical and experimental analysis of ion and water balance in two strains of sheep, the high potassium (HK) and low potassium (LK) strains (Evans, 1954). Evans *et al.* (1956) have shown that mutation at a single gene locus differentiates the two strains of sheep. Careful experimental analysis (Tosteson and Hoffman, 1960) reveals that, as expected, the high potassium strain has a much more active (some tenfold) sodium-potassium exchange pump than the low potassium strain. Our analysis on the basis of Eqs. (7.18) through (7.20) would suggest that A_i, and hence the cell volumes, would be expected to be very different in the two cases. The measured differences in these parameters are, however, only slight. The reason for this, as a consideration of Eq. (7.18) corroborates, is that the low efficiency of pumping of the LK sheep is compensated for by a high specificity for potassium in the leakage system. In fact the potassium/sodium selectivity in the LK strain is some threefold that of the HK strain. If we substitute the membrane parameters obtained by Tosteson and Hoffman (Table 7.2) into Eq. (7.18), we find that A_i for the HK sheep is 60 m*M* (predicted); for the LK strain, 30 m*M*. Table 7.2 records the predicted and experimental values of these various cell parameters. It must be pointed out that the mutation at a single gene locus differentiating the HK and LK strains is associated with two (compensating) changes in membrane parameters—pump rate and K^+/Na^+ selectivity. This is an unexpected result genetically.

There is a range of cell volumes and of potassium levels in the mammalian red blood cells of various species, and it would be of much interest to compare the parameters of pump rate and potassium/sodium selectivity in these cases to see if the accord found by Tosteson and Hoffman is indeed general.

A further test of the validity of Tosteson and Hoffman's treatment is the corollary that if the pump is blocked the cell should swell, and in particular that the rate of cell swelling should depend on (1) the ratio of the potassium to sodium leakage and (2) the ionic compositions of the cell and of the exterior. We shall consider this phenomenon in the following section.

TABLE 7.2
APPLICABILITY OF EQ. (7.18) TO THE DATA ON HK AND LK SHEEP[a,b]

Parameter	Symbol in Eq. (7.18)	Value for sheep of strains HK	Value for sheep of strains LK
Total external cation	B_e	170 mM	170 mM
Ratio of pump to leak rate constant for sodium ion	p/l	121	8.2
External potassium ion concentration	N_e	5 mM	5 mM
External sodium ion concentration	M_e	165 mM	165 mM
Rate constant for passive sodium influx	l	0.00343 hr^{-1}	0.00233 hr^{-1}
Rate constant for passive potassium influx	n	0.0065 hr^{-1}	0.0110 hr^{-1}
Ratio of leak rate constants K^+/Na^+	n/l	1.76	4.76
Concentration of internal non-penetrating anion computed by Eq. (7.18)	A_i	60 mM	31 mM
Concentration of nonpenetrating anion by experiment	A_i	51 mM	42 mM

[a] Data of Tosteson and Hoffman, 1960.

[b] *Note*: For the determination of n and l the cells were poisoned with strophanthidin and placed in an environment designed to create large electrochemical gradients of Na^+ and K^+ across the membrane, hence eliminating much of the exchange diffusion. Thus HK cells were placed in a medium containing little K^+ and much Na^+, while LK cells were placed in a high K^+, low Na^+ medium.

7.2 Why a Linked Sodium/Potassium Pump?

We have seen that both types of sodium pump—the electrogenic in which only sodium is pumped (so that chloride must accompany the transferred sodium ion to maintain electroneutrality), and the neutral pump in which sodium is compulsorily exchanged for potassium—can account for the volume regulatory aspects of ion pumping. [Note that a second type of neutral pump actively transferring both sodium and chloride has been demonstrated (Diamond, 1962).] For the linked Na^+/K^+ pump, a necessary requirement is that the potassium/sodium sensitivity of leakage across the membrane must be greater than unity if efficient volume regulation is to be achieved. It is legitimate to ask why one cell type should have evolved a linked Na/K pump rather than the apparently simpler unlinked pump. Is there a functional significance for linkage? (Of course, these questions of function, to an even greater degree than questions of mechanism, can be asked but may not be answerable.)

Some data obtained by Tosteson and Hoffman (1960) on the cell swelling of pump-inhibited cells provide a clue. Figure 7.3a shows the volume change of LK sheep red cells when the Na/K pump is blocked, the cells being in either a high or a low potassium medium. Figure 7.3b shows similarly the data for HK sheep. There are two striking features of these data. First the volume changes are remarkably slow—in the normal (LK) medium there is less than 4% increase in cell volume in

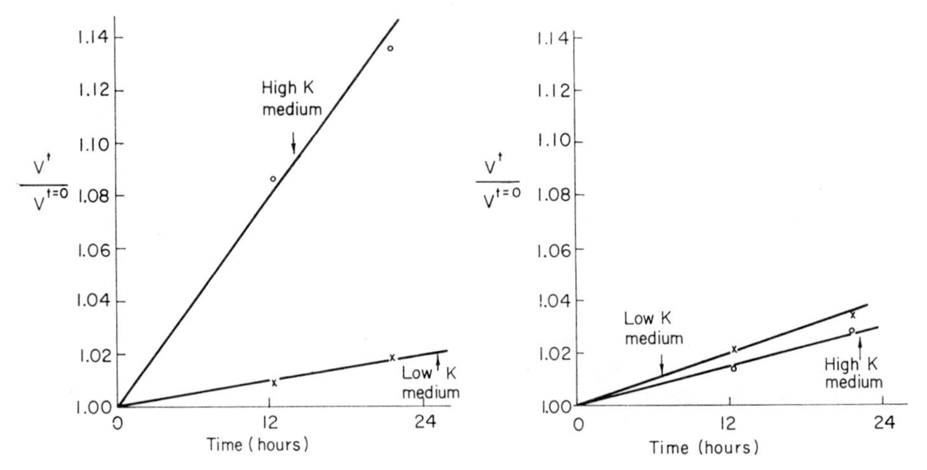

Fig. 7.3. The effect on the rate of cell swelling of varying the cation composition in the incubation medium, when erythrocytes of (a) LK sheep or (b) HK sheep are incubated in the presence of strophanthidin. The ordinate gives the ratio of the volume of the cells, at the time indicated on the abscissa, to the volume at zero time. (Taken with kind permission from Tosteson and Hoffman, 1960.)

24 hr. Second, the LK cells in the high potassium medium are the most unstable. The gradient here of potassium is directed inward. Thus potassium will enter and the cells will swell. For LK cells in the low potassium medium all gradients are small and the volume changes slow. For HK sheep in high potassium medium, the same situation arises. Finally, for the most usual physiological situation, for HK cells in low potassium medium, the gradient of potassium is outward; we might expect the cell to shrink were the potassium movements not compensated for by the sodium movements under an oppositely directed gradient. This last case reveals the possible significance of Na/K linkage. If linkage occurs, the internal potassium is high relative to that outside and the internal sodium low relative to that outside. In such a case, on cessation of pumping (due,

we might imagine, to the momentary depletion of cellular energy reserves) the movement of the ions down their electrochemical gradients is reciprocal and the concomitant water movements are small. Only when a very large interchange of sodium and potassium has occurred will the volume of the cell begin to rise.

In this way the cell can cope with a short-term diminution of its energy supplies for pumping, without the untoward effects of volume increase. In contrast, in unlinked systems, the potassium ion is always present at its steady state distribution ratio and from the moment of cessation of pumping, a net inward flow of sodium occurs with a concomitant and rapid volume increase. It will be easy to see that the most stable cells will be those in which the gradients of potassium and sodium ion are equal and opposite, and the rate constants for leakage of the two are equal. An imbalance in the ion gradients can be compensated for by a corresponding inequality in the rate constants. The full equations for this process are easily derived (Tosteson and Hoffman, 1960) but need not be reproduced here. Considerations of the principles discussed above and of Table 7.2 will enable the changes in Fig. 7.3 to be appreciated.

7.3 Transport across Epithelial Cell Layers

We have seen that the water balance of a cell is maintained by the action of the sodium pump mechanism. It has become clear over the past few years that water balance in the whole animal is likewise dependent on the sodium pump system. We might consider, as a very clear example of the phenomena concerned in whole animal water balance, the profound analysis that Diamond (1964a,b) has made of water movements in rabbit and guinea pig gall bladder.

Isolated bladders can extrude large quantities of fluid from their interior, thereby concentrating the internal fluid, if a continuing metabolism is permitted. A guinea pig bladder will secrete 10% of the internal fluid per hour for long periods, a rabbit bladder some 40%. The addition of cyanide and iodoacetate stops fluid extrusion within a very short time. Careful analysis showed that the extruded fluid was a sodium chloride solution, isotonic with the medium on the inside of the bladder (Fig. 7.7) (Diamond, 1964b). If the sodium chloride on the inside of the bladder was replaced by an osmotically equivalent amount of sucrose, active extrusion of fluid ceased; indeed, the bladder took up fluid from the exterior. The bladder has a significant water permeability when water moves under an osmotic gradient. Gall bladder is also definitely permeable to sodium chloride, but is impermeable to sucrose, raffinose, and various dyes such as Evans blue. These facts can be understood if we consider the model of Fig. 7.4 which is a diagram of a generalized epithelial cell layer—a concentric shell of cells, surrounding an inner region I. In this figure O is the outer fluid and C the cell interior. There are three series of membranes, an inner $M_{I,C}$, an outer $M_{C,O}$, and the intercellular membranes $M_{C,C}$. We can assume, for the moment, that transport across the intercellular region will make no change in the water and solute relations so that we can ignore such transport. A very primitive situation would be where both $M_{I,C}$ and $M_{C,O}$ behaved as does the red

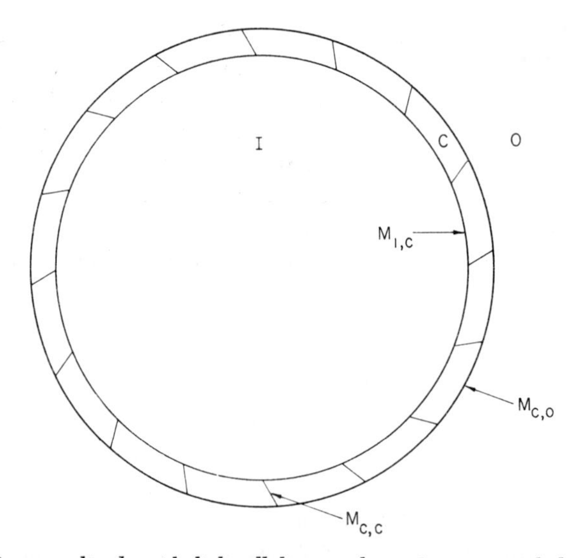

Fig. 7.4. A generalized epithelial cell layer, where C is an epithelial cell, O the external medium, and I the lumen within the cell layer; M, with the relevant subscripts, depicts the various types of membranes possible on this model.

blood cell membrane, possessing a sodium extrusion pump and the associated ion leaks. It will be obvious that at the steady state the composition of the phases I and O will in this case be identical—high Na and low K—and phase C will have the composition of a typical cell (high K, low Na). Let us consider now the consequences of a specialization whereby the sodium pump on the inner $M_{I,C}$ surface is inhibited or absent.

Sodium will now move unidirectionally from C to O, and will enter by the leak from I to C, ensuring a continual one-way transfer of salt from the lumen I to the exterior O. As the region I thereby becomes hypotonic, water will flow from I, through C, to O and a mechanism for the extrusion of the fluid from I (or rather the permeable constituents of this fluid) is available by this simple process. That this discussion is oversimplified and that Fig. 7.4 is not a valid model for the gall bladder is, however, clear from Diamond's (1964b) demonstration that the gall bladder can extrude water against a significant osmotic pressure (up to two atmospheres). A simple scheme such as Fig. 7.4 could not do this. But, as Diamond points out, if the local region into which sodium is extruded across $M_{C,O}$ is shielded from the external fluid at A (Fig. 7.5), fluid will accumulate in this small region (B) (this is the phenomenon known as local osmosis) and will build up a definite hydrostatic pressure. This hydrostatic pressure will force fluid up the tubule C against the prevailing osmotic pressure at A. This is clearly a tenable hypothesis.

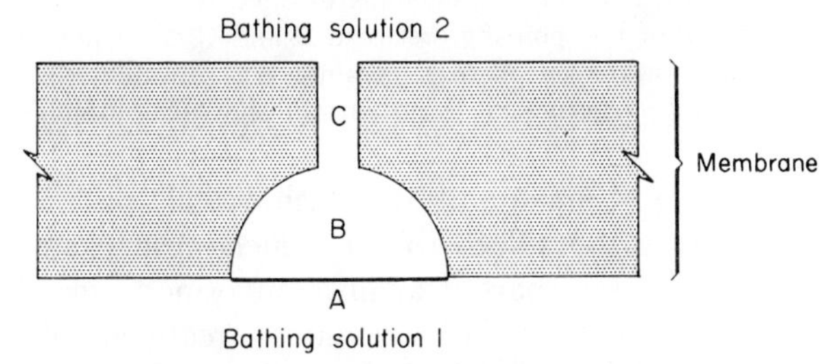

Fig. 7.5. A diagram explaining how local osmosis brought about by active salt transport might move water across a membrane separating two identical solutions. Salt transport across barrier A raises the solute concentration in B, a restricted space within the membrane. Water follows osmotically, and the resulting raised hydrostatic pressure in B forces water and solute out through C. (Taken with kind permission from Diamond, 1964b.)

An ingenious alternative model to account for water extrusion against an osmotic pressure has been put forward by Curran and Macintosh (1962) and by Ogilvie *et al.* (1963) and analyzed in detail theoretically (Patlak *et al.*, 1963). This model, depicted in Fig. 7.6, depends on the fact that the rate of water flow across such a membrane depends on the cross-sectional area of the pores available for flow—Poiseuille's law (Section 3.6). The cellophane membrane between A and B in Fig. 7.6 is permeable to the solute sucrose but has a low permeability to water under a hydrostatic pressure gradient. In contrast, the membrane between B and C—a very large-pored sintered glass disk—has a high water permeability. If there is an osmotic gradient between A and B, water will flow into B under this osmotic gradient. If now the volume of B is kept constant, a hydrostatic pressure will arise in B. Since the membrane between B and C is more porous than that between A and B, water will flow preferentially from B into C rather than from B into A. There will, therefore, be an over-all flow of water from A to C. The osmotic gradient between B and C will not affect the rate of fluid flow. Table 7.3 presents

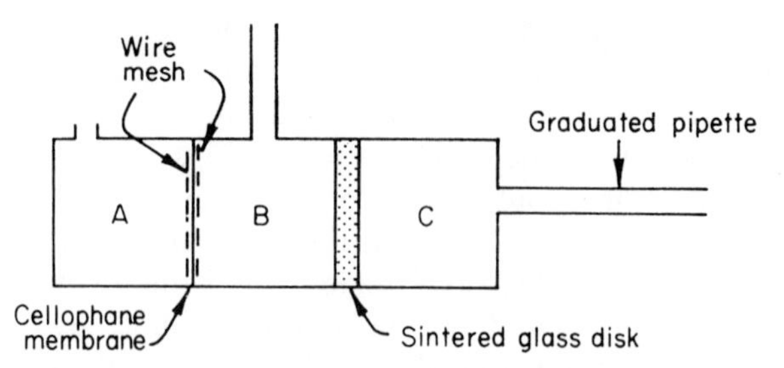

Fig. 7.6. A model system for biological water transport. The solution in compartment *A* contains various concentrations of sucrose in water. Compartment *B* contains 0.5*M* sucrose, compartment *C* initially contains distilled water. Between *A* and *B* is a cellophane membrane of low porosity, between *B* and *C* is a sintered glass disk of high porosity. Water flows from *A* to *C* against the prevailing osmotic pressure gradient. (Taken with kind permission from Curran and Macintosh, 1962.)

some data obtained by Curran and McIntosh (1962) on a model system in which sucrose was the solute that produced the osmotic gradient. Clearly, water can be transported against an osmotic gradient by this "double-membrane" device. A full theoretical treatment of the double-membrane model (Patlak *et al.*, 1963) shows that such a system can allow

TABLE 7.3

VOLUME FLOW IN A SERIES-MEMBRANE SYSTEM[a,b]

Sucrose conc. (mole/liter) in compartment *A*	*B*	*C*	Vol. change in *C* (μl/min)
(1) Compartment *A* Variable			
0.1	0.5	0.02	7.6
0.2	0.5	0.02	5.3
0.3	0.5	0.02	3.6
0.5	0.5	0.03	0.0
0.7	0.5	0.02	−3.5
(2) Compartment *C* Variable			
0.1	0.5	0.0	8.0
0.1	0.5	0.1	8.4
0.1	0.5	0.5	7.8

[a] The apparatus depicted in Fig. 7.6 was used, with sucrose solutions of the molarities indicated in the table being added to compartments *A*, *B*, and *C*. In the first set of experiments below, compartment *C* contained initially distilled water and the value recorded for *C* is the concentration at the end of the experiment. In the second set, the values recorded are the initial values. A positive value for the volume change in column 4 indicates net water movement from *A* to *C*.

[b] Data from Curran and Macintosh, 1962.

an isosmotic flow of fluid to emerge from C only if the parameters describing the permeability of the A, B and B, C membranes are correctly chosen (or have so evolved). If, however, the tonicity of the medium in A is varied, the tonicity of the fluid emerging from C will not follow exactly that of A. Only at one particular tonicity will the emerging fluid be isotonic. Diamond's data on gall bladder (Fig. 7.7) show that the extruded fluid is, over a wide concentration range, strictly isotonic with the fluid bathing the interior of the bladder. The double-membrane model does not, therefore, apply to gall bladder but might conceivably be valid in other circumstances. It is suggestive, for instance, that the degree of

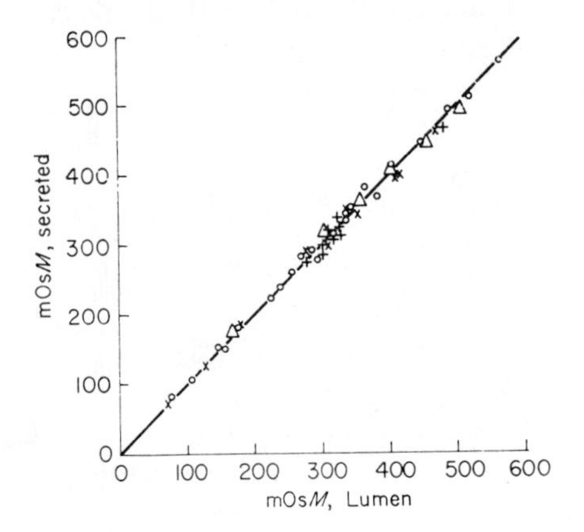

Fig. 7.7. Osmolarity of the fluid secreted out of gall bladder as a function of the osmolarity of the luminal bathing solution. △, No HCO_3^- or glucose; ○, HCO_3^- glucose; X, HCO_3^-; +, sucrose or raffinose. (Taken with kind permission from Diamond, 1964b.)

hydration of the cells of the intestine (Parsons and Wingate, 1961) and of toad bladder (Leaf, 1961) increases when water flow occurs, a result to be expected if the double-membrane model holds. Likewise, an appreciable hydrostatic pressure has been reported to occur in rat intestine when water absorption is taking place (Lee, 1960). We should note that Diamond's "local osmosis" model (Fig. 7.5) is a limiting case of Curran's model where the porous membrane between compartments B and C of Fig. 7.6 is simply the opening of the tube between the regions B and C in Fig. 7.5.

The anatomical location of the structures depicted in either Fig. 7.5 or 7.6 is, of course, obscure. We do not know whether, if indeed such structures exist at all, they exist at a light microscopic level, at the electron microscopic level, or if they are, rather, molecular structures. We might note that water flow due to an imposed osmotic gradient is accompanied by a streaming potential, while water flow due to active salt transport is not. Diamond (1964b) suggests that there must exist a discrete pathway for movement of water by "local osmosis" and that in this case water moves across the cells rather than between the cells of the epithelial cell layer.

Very recently, Diamond and Tormey (1966a) have identified the region in which the local osmosis occurs as the spaces *between* the cells lining the epithelium, that is, between the membranes $M_{C,C}$ in Fig. 7.4. Adjacent cells of the gall bladder epithelium are separated by long, narrow highly tortuous extracellular channels, closed at the luminal end by a terminal bar. If, as in Figs. 7.8 and 7.9, the transport of salt can occur across the

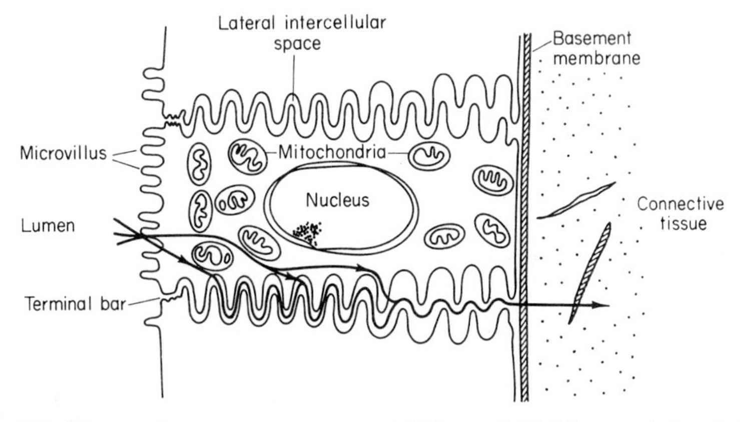

Fig. 7.8. "Route of water transport across rabbit gall bladder, as deduced from combined physiological-anatomical experiments. Salt is pumped from the cells into the lateral intercellular spaces, making them hypertonic. Water enters osmotically along the length of the spaces, distending them, and an isotonic NaCl solution emerges from the ends of the spaces facing the connective tissue." (Taken with kind permission from Diamond and Tormey, 1966a.)

intercellular membranes, into the space between each pair of membranes, water will follow this movement of salt. In this region a local hydrostatic pressure can be built up and water can be forced out of the end of the lumen, against a prevailing pressure. Electron micrographs of the gall bladder epithelium offer convincing evidence of the correctness of this model.

We have previously had occasion (Section 6.7) to consider aspects of ion transport in frog skin. In certain circumstances, under conditions where the animal is dehydrated, frog (or toad) skin can act also as an organ of water transport. In a similar fashion to the gall bladder discussed above, the active accumulation of sodium chloride by frog or toad skin leads to a local osmotic gradient and water can follow this entry of salt. In contrast to the situation in gall bladder, the rate of water transport across anuran skin is under hormonal control. In the absence of antidiuretic hormone, the osmotic water permeability of the tissue is low; in the presence of hormone a marked stimulation in the rate of water permeability occurs (Koefoed-Johnsen and Ussing, 1953) (Table 7.4). Precisely the same phenomenon is found in the toad bladder (Leaf, 1961). These epithelial cell layers have been studied by the methods described in Section 3.6 and measurements made of the rate of penetration of D_2O and of the rate of penetration of water by an osmotic gradient. The data accord with a model in which aqueous channels of some 40 Å in diameter are opened up in the epithelial cell layers when the antidiuretic hormone is present (Leaf, 1961). In toad bladder the hor-

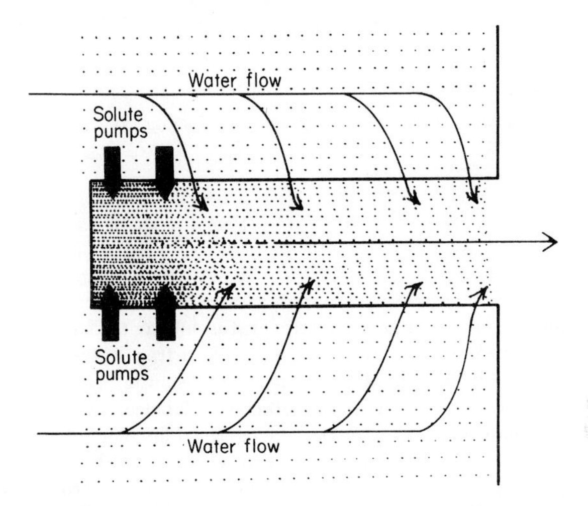

Fig. 7.9. "Standing-gradient osmotic flow: a model for fluid transport across epithelia. Solute is pumped into the closed end of a long and narrow channel, making it hypertonic and pulling in water osmotically. The solute moves toward the open end of the channel as a result of the water flow and of diffusion down its concentration gradient. Osmotic equilibrium is progressively approached as water enters along the length of the channel, until the emergent solution is isotonic. The standing gradient in this flow system is continually maintained by active solute transport. For diagrammatic purposes the solute pump is indicated only at the base of the channel, but it may operate over a greater fraction of the channel length. Given appropriate values for the channel length and water permeability, the model can also yield a hypertonic solution of fixed osmolarity. Epithelial structures to which the model may find application in understanding solute-linked water transport are lateral intercellular spaces (intestine and gall bladder), basal membrane infoldings (kidney, salt gland, ciliary body, etc.), and intracellular canaliculi (stomach)." (Taken with kind permission from Diamond and Tormey, 1966b.)

TABLE 7.4

INFLUENCE OF ANTIDIURETIC HORMONE ON THE DIFFUSIONAL AND OSMOTIC PERMEABILITY COEFFICIENTS FOR WATER[a,b]

Experiment A—Koefoed-Johnsen and Ussing (1953)

No hormone		ADH added	
M_{in}	Δw	M_{in}	Δw
441	12	532	30
460	13	551	36
305	6.7	292	10.8
319	5.0	310	7.7
343	9.7	334	16
370	7.4	404	17
326	11.7	344	21
287	8.0	369	25

Experiment B—Andersen and Ussing (1957)

k_{in}	k_{out}	Δw	k_{in}	k_{out}	Δw
3.73	3.51	22	5.55	4.51	104
4.09	3.89	20	4.25	3.75	50
3.63	3.42	21	4.27	3.73	54
4.04	3.76	28	4.56	3.85	71

[a] Strips of toad skin were studied in an apparatus similar to that of Fig. 6.1 at room temperature. The outside of the skin was exposed to a frog Ringer solution, the inside to the same solution diluted tenfold, producing an osmotic gradient across the skin. Water permeability was measured using isotopically labeled water, in both directions across the skin. Net movement was calculated from the difference between these measurements. In the first set of experiments the inward water permeability is reported as M_{in} in $\mu l\ cm^{-2}\ hr^{-1}$, and Δw the net water flow, as $\mu l\ cm^{-2}\ hr^{-1}$. In the second set, water permeability is expressed as 10^4 times the rate constant, k_{in} or k_{out}, in cm hr^{-1}, while Δw is again in $\mu l\ cm^{-2}\ hr^{-1}$. Note that the unidirectional fluxes are not nearly as greatly affected by the presence of hormone as is the difference Δw between these fluxes.

[b] The permeability of thiourea also increased greatly in the presence of hormone

mone also increases threefold the rate of active sodium transport (Leaf, 1961) and increases tenfold the rate of transfer of urea and a number of other amides, but it does not increase the transfer rate for a number of other substances (Leaf, 1960). We might note that the presence of antidiuretic hormone raises the water permeability of toad bladder to a value which is of the same order of magnitude as that of other epithelial cell layers (Table 3.14), so that it is the permeability in the absence of hormone that is atypical.

It appears (Leaf, 1960) that it is the outer membrane of the epithelial cell layer that is the major barrier to the movement of salts, water, and urea, and it is this barrier that is lowered when hormone is present. How this occurs and why the effect is confined to only some of the molecules capable of penetrating this outer barrier is obscure. The experiments that established these changes in permeability (Leaf, 1960) deserve further mention.

Leaf argued that if only one of these two permeability barriers ($M_{C,O}$ or $M_{I,C}$ Fig. 7.4) were lowered, the rate of entry of isotopically labeled permeant from that side (*O* or *I*, respectively) should be increased. Table 7.5 shows that the addition of antidiuretic hormone to the serosal medium greatly increased the rate of entry of ^{14}C-urea from the mucosal medium, while—although the data are not shown here—the rate of entry

TABLE 7.5

EFFECT OF NEUROHYPOPHYSEAL HORMONE ON THE UPTAKE OF ^{14}C-UREA INTO EPITHELIAL CELLS OF TOAD BLADDER[a,b]

Time after addition of ^{14}C-urea (min)	Distribution ratio: tissue/medium	
	Without hormone	With hormone
8	0.06	0.10
9	0.06	0.17
60	0.05	0.46
123	0.15	0.30
153	0.13	0.45

[a] Data taken from Leaf, 1960.
[b] Paired bladder halves incubated with ^{14}C-urea added to mucosal medium.

from the serosal medium was unaffected. Similarly, the rate of entry of tritiated water across the mucosal membrane, rather than across the serosal membrane, was enhanced by the addition of hormone. Table 7.6 records the data of Curran *et al.* (1963) on the sodium fluxes across these two membranes of frog skin. Again it is specifically the mucosal membrane, the permeability of which is increased by antidiuretic hormone. Interestingly, the effect of added calcium ions, as also recorded in Table 7.6, is in opposition to that of the antidiuretic hormone. The effects of these two agents are (algebraically) additive. They appear to act at quite different sites in the tissue (Herrera and Curran, 1963). As in so many situations (Heilbrunn, 1952), the effect of calcium is to decrease the permeability of cells or cell layers. The most striking aspect of these results is perhaps that the antidiuretic hormone has to be presented to the serosal surface of the skin for its action at the mucosal surface to be demonstrated. Although this is certainly associated with the general impermeability of the mucosal layer, yet the effect is a fascinating one. [In contrast, calcium exerted its effect when applied to the outside (mucosal) surface of the skin.] It is easier to understand why a cardiac glycoside (strophanthin K) inhibits sodium transport across frog skin only if applied to the serosal surface (Bower, 1964) since it is at this surface that the sodium pump is thought to be located (Section 6.7). We might summarize these findings in the schematic picture of Fig. 7.10.

TABLE 7.6

INFLUENCE OF ANTIDIURETIC HORMONE AND CALCIUM IONS ON THE SODIUM ION PERMEABILITY OF FROG SKIN[a,b]

	$F^{i}_{o,c}$	$F^{e}_{o,c}$	$F^{i}_{c,i}$	$F^{e}_{c,i}$	Net flux$_{o,i}$
Control	2.49	1.09	0.42	1.82	1.40
+ ADH[c] (0.2 unit/ml)	4.16	2.32	0.38	2.22	1.84
Control	2.95	1.60	0.45	1.80	1.35
+ Ca^{2+} [c] (11.3 m*M*)	1.31	0.55	0.42	1.18	0.76

[a] Data of Curran *et al.* 1963.

[b] A kinetic analysis of the rate of uptake of ^{24}Na by skin soaked in isotopically labeled NaCl yields values for the influx and efflux of sodium ions (superscripts *i* and *e*, respectively) from the outside of skin into the epithelial cells (subscript *o,c*) and from the epithelial cells to the inner surface of the skin (subscript *c,i*) separately. These four fluxes are recorded in the table as $F^{i}_{o,c}$, $F^{e}_{o,c}$, $F^{i}_{c,i}$, and $F^{e}_{c,i}$ in μequiv. Na cm^{-2} hr^{-1}. The average net fluxes, determined by the short-circuit current, are also included.

[c] The action of both ADH and Ca^{2+} appears to be largely confined to the outer barrier (*o,c*).

A second way in which water transport in frog skin and gall bladder might appear to differ is in the direction of water movement. We saw that Fig. 7.4 could be a fair description of the behavior of gall bladder, if the outer membrane $M_{C,O}$ contained the sodium pump, directing sodium into the external medium. For frog skin, of course, the reverse is the case. The direction of salt and water movement is from the outer medium into the serosal surface, which would be the lumen of the epithelial bag in Fig. 7.4. This apparent paradox is in reality a problem of topology and is resolved by a consideration of Fig. 7.11. This is a diagram of a "generalized multicellular organism" surrounded by the epithelial cell layer *E*. In this "organism," the salt uptake and water movement are everywhere directed from the outside to the interior across the epithelial cell layer. But within the invagination which comprises the region *I*—the intestine—the "outside" is the contents of the gut, which

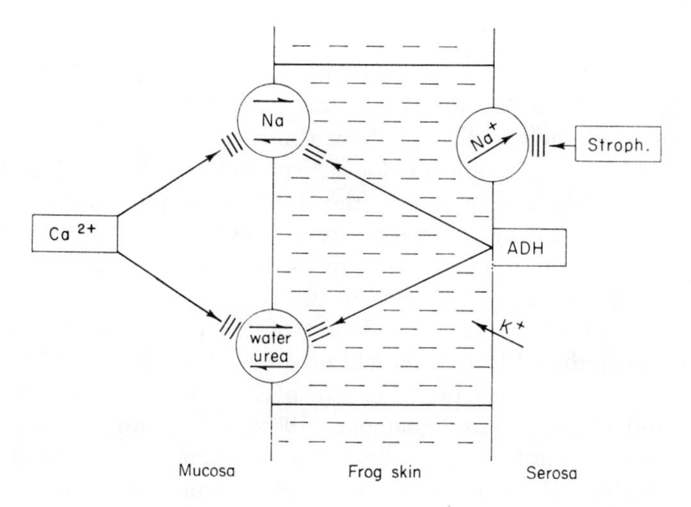

Fig. 7.10. Sites of action of activators (three radial bars on figure) and inhibitors (three tangential bars) of solute movements across frog skin epithelium. Stroph. is the cardiac glycoside strophanthidin; ADH, the antidiuretic hormone.

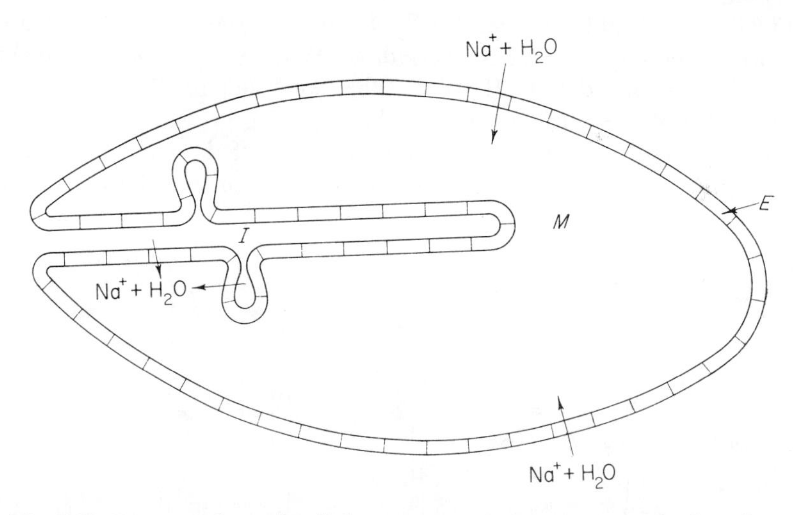

Fig. 7.11. Diagram of a multicellular organism. Active transport of salt and water, which takes place everywhere from the external medium into the interior of the organism, has the effect of accumulating these molecules in that space which is the interior at *M*, but of removing these molecules from the spaces that are the "bladders," invaginations of the intestine *I*.

by the action of the salt pump are brought into the region *M*, the true interior of the organism. Invaginations of *I* will result in the formation of "bladders" which will drain into the interior *M*. Figure 7.4, being a cross section of such an invagination, could be a section of a spherical shell (bladder) or a tubule (kidney). The point to be made is that in skin, intestine, bladder, or kidney the direction of pumping of salt and water is always the same, from the external environment to the interior of the organism. In intestine, bladder, and kidney the external environment is, however, itself internalized within the organism.

We must not expect that the details of the pump and leak systems across these epithelial cell layers will be everywhere the same. It is clear, for instance, that the salt pump in gall bladder actively transports both sodium and chloride ions (Diamond, 1964a), the rate of extrusion of sodium ion and water being unaffected if potassium is removed from the medium bathing the outer face of the membrane. Flux ratio measurements (Section 2.8) suggest that also in intestine, chloride is actively transported (Curran and Solomon, 1957). In frog skin, however, the movement of chloride ion is passive, but here, too (Section 6.7), recent evidence suggests that the sodium pump may not be linked to potassium pumping.

Finally, we might consider briefly the main organ regulating water balance in the higher animals, the kidney. While we cannot go into this vast subject in any detail (H. W. Smith, 1956), our previous discussion might perhaps give a framework for the following grossly oversimplified picture. Figure 7.12 is a very generalized sketch of the main features of the mammalian kidney. The glomerulus delivers an ultrafiltrate of the

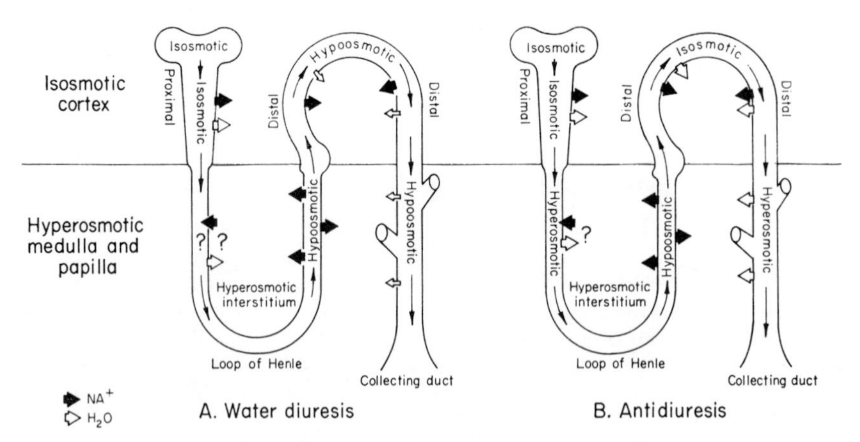

Fig. 7.12. Schematic diagram of nephron, embedded in kidney tissue. (Taken with kind permission from Orloff and Handler, 1964.)

blood, isosmotic with plasma, into the proximal tubule. The proximal tubule secretes salt into the lumen, the movement of salt being followed by water movement, so that an isotonic fluid is extruded into the kidney cortex, the region surrounding the proximal tubule. The proximal tubule thus behaves as does the intestine or gall bladder. The tubular fluid, still isosmotic but greatly reduced in volume, enters the descending loop of Henle. Here, the surrounding region of kidney tissue, the medulla, is itself hyperosmotic. Water therefore flows out of the tubule (and salt enters) so that the fluid at the bottom of the loop of Henle is hyperosmotic to plasma (but isosmotic with the immediately surrounding tissue). The ascending loop is apparently a region in which active extrusion of sodium occurs without appreciable movement of water. (This region behaves, therefore, as does frog skin in the absence of antidiuretic hormone.) The active extrusion of salt in this region, unaccompanied by water, is partly responsible for the local high tonicity. The distal convoluted tubule is a region where (in normal circumstances) an active sodium extrusion occurs accompanied by a limited osmotic movement of water. In the presence of antidiuretic hormone, however, this epithelial cell layer behaves just as does toad bladder or frog skin and demonstrates a high water permeability. Large amounts of water follow the extruded salt back into the circulation if antidiuretic hormone is present, and a retention of water by the body can be achieved. Finally, the region of the collecting tubule is again surrounded by the hyperosmotic medullary contents. Water will leave the tubule in small amounts in the absence of antidiuretic hormone but in large amounts if this hormone is present. There appears to be no significant salt accumulation in the collecting tubule. The major reabsorption of fluid by the kidney (80 to 85%) thus takes place in the proximal tubule and is an isotonic reabsorption. The hyperosmotic condition found in the medulla, brought about by a combination of the high salt extrusion and the low diffusion rates in this region, enables a hyperosmotic urine to be produced in conditions of water shortage when the antidiuretic hormone is in action. The kidney thus exemplifies a combination of many of the properties of the simpler epithelial cell layers and is the organ *par excellence* of active transport.

CHAPTER 8

Molecular Properties of the Transport Systems

In this chapter we intend to draw together as much as possible of the information that has been collected on the molecular, as opposed to the kinetic, properties of the transport systems. Since there has, as yet, been no confirmed isolation of any "carrier" (but see Section 8.4), our knowledge of the molecular properties of these entities has of necessity been derived by indirect means. Many researchers have studied the specificity of transport systems, others the effect of inhibitors and drugs on transport, the effects of hormones and toxins, the temperature dependence, and the stoichiometry of transport. From these studies we can make certain broad generalizations as to the properties of transporting systems. In particular cases, it is possible to make more detailed inferences as to the nature of particular transport systems. We shall once again choose certain studies for a detailed examination rather than deal cursorily with the totality of these investigations.

8.1 Information Derived from Studies of the Specificity of Transport Systems

By "specificity" investigations we shall mean the systematic study of how the rate or extent of transport varies for a *particular* cell type or organ when the molecular architecture of the transported substance is varied. As an indication of the power and limitations of this method, we might consider first how the analysis of specificity has been used in similar studies of enzymes and their mode of action.

We shall base our discussion on the excellent chapter "Enzyme Specificity" in the treatise of Dixon and Webb (1964). These authors point out that the enzymes range from those which are absolutely specific for one particular substrate, allowing no variation in structure, to those in which a fairly small number of closely related molecular species can be acted upon. For the absolutely specific systems little more can be said, but for the less specific systems an extensive comparative study (in which small chemical modifications are made in every part of the molecule separately, and the effect of these modifications are assessed on affinity and reactivity) can be valuable. From such results it is possible to formulate the minimal structures necessary for combination and for reaction and to determine the quantitative effects of additional groups. It is possible to hazard a guess as to the manner in which the substrate unites with the enzyme, as to the groups involved, and (in the best cases) as to the mechanism of the reaction.

We might take as our first example the enzyme hexokinase (2.7.1.1) from yeast (Sols *et al.*, 1958). This enzyme phosphorylates D-glucose and in addition various other sugars. It acts on both the α and β forms of D-glucose, on D-mannose (which differs from glucose in the configuration of the carbon atom at the 2 position), and on the ketose, D-fructose. In addition, it acts on 2-deoxy-D-glucose and on 2-acetylamino-2-deoxy-D-glucose. The enzyme does not act on 3-methyl glucose, on sorbose (which differs from fructose at the 3 and 4 positions), on D-galactose (differing from glucose at the 4 position), nor on D-allose (the 3-epimer of glucose). These results can be understood if the specificity requirement of the hexokinase is for a molecule having the structure of D-glucose in the 3, 4, 5, and 6 position, but tolerating variations of this structure in the 1 and 2 position, even to the extent here of a furanose rather than a pyranose ring.

We might conclude that the enzyme-substrate interaction takes place about the 3, 4, 5, and 6 position of the sugar, but statements on mechanism beyond this are hazardous.

For the lipases (carboxylesterases, 3.1.1.1) an extended study by Webb (see Dixon and Webb, 1964) has led to a detailed, if speculative, model of the substrate-binding site of this enzyme. As might perhaps have been suspected it appears that hydrophobic interaction between the nonpolar groups on the substrate and similar groups on the enzyme dominate the binding of enzyme and substrate.

For the cholinesterases, specificity studies (see Dixon and Webb, 1964) have allowed a distinction to be made between the acetylcholinesterases (3.1.1.7), which hydrolyze acetylcholine some fifty times faster than they hydrolyze butyrylcholine, and the cholinesterases (3.1.1.8), which hydrolyze butyrylcholine twice as fast as acetylcholine. Again, some valuable indications as to the nature of the interaction of enzyme and substrate can be gleaned from the extensive further studies which have been carried out on these systems.

Finally, we might consider the enzymes involved in the digestion of proteins and peptides. The specificity of these peptidases has been intensively investigated over the last fifty years, the results of these investigations being summarized in a convenient table by Dixon and Webb (1964). These studies have enabled a clear classification to be made between the various peptidases; the acceptance of, for example, trypsin (3.4.4.4) and chymotrypsin (3.4.4.5) as separate enzymes was greatly forwarded by such studies. In addition, from the detailed study of such specificity differences, interesting speculations on the mechanism of action of these enzymes can be advanced. Thus, trypsin can only hydrolyze peptide bonds of which the carbonyl group carries as a substituent a basic nitrogen group, the ε-amino of lysine, or the guanidino group of arginine. In contrast, *this* specific requirement is lacking for chymotrypsin, an enzyme which requires an aromatic side chain to be attached to the peptide-bonded carboxyl. Erlanger *et al.* (1965) have gathered evidence suggesting that the basic group of the trypsin substrates itself takes part in the catalytic step. The enzyme is activated by the substrate or by the addition of various amines. No such activation is found for chymotrypsin. These authors infer that in the chymotrypsin molecule a basic amino group which takes part in the catalytic step is present at the active center of this enzyme. In trypsin, on the other hand, no such basic group is present in the enzyme, but the substrate provides the necessary component of activity. We must stress that a good deal of supporting evidence for this postulate is provided in the paper of Erlanger *et al.* (1965) and that it was as much a consideration of this other evidence that led to the understanding of the role of specificity here, as that considerations of specificity led to this view of mechanism.

We must not expect therefore that from the systematic but more limited studies which have been made of the specificity of biological transport systems, we shall be able to make detailed statements as to the molecular basis of transport. Rather we shall find that specificity studies have a classificatory value, helping to distinguish one transport system from another. With such a classification, it will be easier to avoid drawing false deductions as to mechanism from the erroneous collation of data on what may be different systems with different mechanisms.

As a first example, let us take the careful and extended studies that LeFevre and his co-workers (summarized in LeFevre, 1961a) have made on the specificity of the facilitated diffusion system for glucose in the human erythrocyte. We have (in Chapter 4) already discussed the kinetic analyses that have been made of this transport system and have seen that there is broad agreement between various researchers on the experimentally derived values of the parameters K_m and V_{max} for the transport of a number of sugars. To the extent that K_m may be identified with the dissociation constant K_s of the permeant-transport site complex (Section 4.6,G) it is possible to consider that the variation of K_m with the structure of the transported molecule gives a measure of the free energy of the interaction between permeant and transport site. We shall not, however, attempt an absolute computation of the free energy of interaction between substrate and transport system, so that the distinction made in Section 4.6,G between the two possibilities of K_m as K_s, a dissociation constant, on the one hand, or (a less likely situation) as reflecting the ratio of k_2 to k_1, on the other, is not of great significance to us here.

The conclusion that emerges from LeFevre's studies (LeFevre, 1961a) is that K_m, and hence the interaction of permeant and transport site, appears to depend on the precise conformation of the transported sugar as follows: The majority of hexoses exist in solution as pyranose rings and these rings themselves exist in a number of mutually interconvertible forms of differing stability. These forms—the several puckered states of the pyranose ring—are not, of course, true isomers but may be termed "conformers." Two of these forms, those most stable for the common hexoses, are the "chair" conformers depicted in Fig. 8.1. As can be seen from Fig. 8.1 there are two possible chair forms, differing in the orientation of the "back" or the "feet" of the chairs with respect to the plane of the ring. If a sugar is most stable in one of these forms (for example, as D-glucose is most stable in the C1 form), its optical enantiomorph or mirror image is most stable in the alternative form. Thus L-glucose exists in the main as 1C. Whether a particular hexose of the D-series is more stable in the 1C or C1 form depends on the steric distribution of the heavy —OH or other substituents. These can be either in the plane of the ring—"equatorial"—or at right angles to the ring—"axial"—(broken and solid lines, respectively, in Fig. 8.1). That form will be most stable for which the maximum number of heavy (rather than the light hydrogen atom) substituents are able to attain an equatorial position on the ring. In D-glucose, the C1 conformation allows the hydroxyls at the 2, 3, 4, and 5 position to be equatorial. In addition, in the β-anomer of D-glucose, the hydroxyl at carbon 1 is also in the equatorial position. In D-galactose which differs from D-glucose in the configuration at carbon atom number 4, the hydroxyl group at the fourth carbon must be axial in the C1 conformation but the other hydroxyls can be equatorial. Similarly, for D-mannose, the hydroxyl at carbon number 2 must be axial in the C1 conformation. In Fig. 8.2, LeFevre has plotted on a logarithmic scale the experimentally derived values of K_m for a series of 16 sugars against the relative stability of the sugars in the 1C or C1 forms, the stability being assessed by a set of rules enunciated by Reeves (1951). In this figure the blocks are the contributions to the instability in the form listed made by the hydroxyls numbered at the foot of the diagram. Thus comparing D-glucose and L-glucose, the D-isomer is very unstable in the 1C form, all the hydroxyls being axial. Similarly, D-galactose is less unstable in the 1C form than D-glucose. The correlation of increasing K_m with decreasing instability in the 1C form is striking. Apparently, the transport site requires that the permeant exist preferentially in the C1 form (Fig. 8.1) if a strong (or rapid) interaction is to take place.

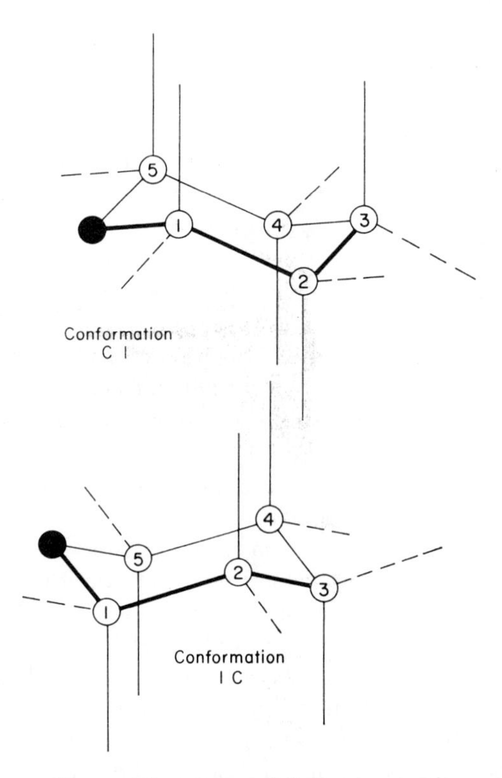

Fig. 8.1. The two chair conformations of the pyranose ring: ●, oxygen atoms; ○, carbon atoms. "Equatorial" bonds are shown as broken lines, "axial" bonds by the vertical solid lines. (Taken with kind permission from LeFevre and Marshall, 1958.)

This specificity requirement is to a marked degree less stringent than that of the many carbohydrate-splitting enzymes that have been studied (see Dixon and Webb, 1964) although such a general preference for the C1 over the 1C conformation is commonly shown. It is clear that the specificity of the transport system is twofold. There is first a preference for those sugars in which the hydroxyl groups are equatorial, that is, do not extend out of the plane of the pyranose ring. Superimposed on this is the requirement that the particular C1 conformation be adopted, that is, that a precise orientation of the 3 carbon or the oxygen of the ring with respect to the binding site is required. Consideration of Fig. 8.1 will suggest that at least a three-point interaction of substrate and binding site is necessary to fix this orientation. This could be interpreted to mean either that the transport site binds to the sugar by three hydroxyl groups and hence that the binding sites are situated on a plane ring surrounding a central crevice or, possibly, that the interaction between the transport site and its substrate is by hydrophobic bonding, the pyranose ring binding to a plane surface of hydrophobic groups on the transport site. In

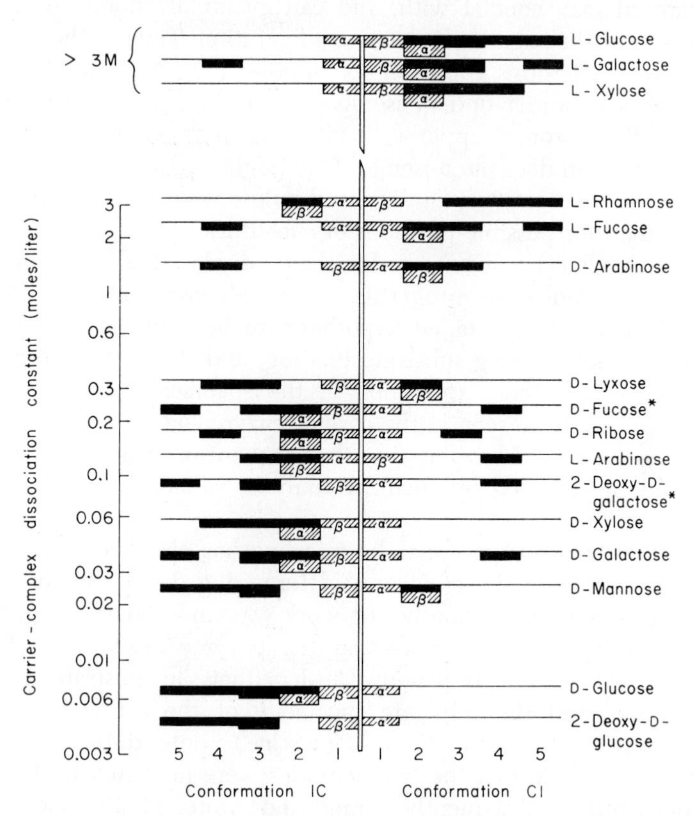

Fig. 8.2. Conformational stabilities of aldoses in relation to K_m (affinity?) for human red blood cell sugar transport systems. Blocks represent instability factors as weighted by Reeves (see Section 8.1) at each ring carbon, as numbered at the bottom of the figure, in the two alternative chair conformations. (Taken with kind permission from LeFevre, 1961a.)

this latter case, the presence of axial hydroxyl groups would interfere with the close apposition of the ring and the hypothetical hydrophobic surface. That all the hydroxyl groups are not involved in binding to the transport site is suggested by the fact that (Fig. 8.2) 2-deoxy-D-glucose (which is also 2-deoxy-D-mannose) interacts with the transport site at least as effectively as does D-glucose or D-mannose. The lower degree of interaction demonstrated by the pentoses (as compared with the hexoses) can as well be ascribed to their possessing one less hydroxyl group if hydroxyls are concerned in the interaction with the transport site, as to their possessing one less hydrophobic binding point, if hydrophobic bonds are concerned.

The requirement that the hydroxyl substituents in the pyranose ring be equatorial may conflict with, and can eventually dominate, any requirement that the sugar be in the D-configuration. Thus as the stabilities of the 1C and C1 forms approach one another (Fig. 8.2) the discrimination between D- and L-optical isomers diminishes and for the pentose arabinose, the L-isomer apparently interacts more vigorously with the transport site than does the D-isomer. One might predict a similar state of affairs for the hexoses altrose, talose, and gulose which differ from glucose at two of the four possible positions of substitution.

It is clear that LeFevre's studies have disclosed a very important feature of the molecular properties of the glucose transport site. The studies allow a set of detailed hypotheses to be made as to the nature of the interactions during substrate binding, and these hypotheses could be tested by a further exploration of the specificity requirements for transport. As we have seen in the case of enzymological studies, it is not necessarily the case, however, that the mechanism of action of the system would be revealed by such an analysis, valuable as such a study might be.

The active transport of sugar by intestine has also been carefully investigated from the point of view of determining the specificity requirements of this system. Originally, the work was undertaken (Sols, 1956a) to test a particular hypothesis—that the enzyme hexokinase (2.7.1.1) was concerned in sugar transport. The fact that the substrate specificity of hexokinase and the substrate specificity of the active transport of glucose were in the same tissue (intestine) quite different, led Sols (1956a) to conclude that the two activities were mediated by two quite different entities. Subsequently, Crane and Krane (1956) showed that 1-deoxyglucose and 6-deoxyglucose could be actively transported, ruling out the possibility of phosphorylation by known enzymes and ruling out also the possibility that transport could occur by mutarotation of sugars [as has been suggested by Keston (1964a,b)]. These specificity studies have since been greatly extended by Crane and his collaborators (reviews in Crane, 1960; Crane *et al.*, 1961), and Table 8.1 records the results of these studies. Sugars are divided into those actively transported and those not actively transported, according to whether a detectable increase in concentration occurs within an everted sac of intestine when this is incubated with the appropriate sugar. Whether specificity is affecting the rate of transport or the affinity of the permeant with the transport site is not considered here. These results can be summarized by Fig. 8.3, which is

TABLE 8.1
SUBSTRATE SPECIFICITY OF THE ACTIVE TRANSPORT SYSTEM FOR SUGARS ACROSS HAMSTER INTESTINE[a]

Compounds Actively Transported	
D-glucose	4-*O*-methyl-D-galactose
1,5-anhydro-D-glucitol	D-allose (the 3-epimer of D-glucose)
2-C-hydroxymethyl-D-glucose	6-deoxy-D-glucose
D-glucoheptulose	6-deoxy-D-galactose
3-*O*-methyl-D-glucose	6-deoxy-6-fluoro-D-glucose
D-galactose (the 4-epimer of D-glucose)	7-deoxy-D-glucoheptose
3-deoxy-D-glucose	α-methyl-D-glucoside

Compounds Not Actively Transported	
D-mannoheptulose	D-gulose (3,4-epimer of D-glucose)
D-mannose (2-epimer of D-glucose)	6-deoxy-6-iodo-D-galactose
D-talose (2-epimer of D-galactose)	6-*O*-methyl-D-glucose
1,5-anhydro-D-mannitol	L-galactose (Optical specificity)
2-deoxy-D-glucose	L-glucose (Optical specificity)
2-deoxy-D-galactose	L-sorbose (Optical specificity)
2-*O*-methyl-D-glucose	6-deoxy-L-galactose
D-glucosamine	6-deoxy-L-mannose
N-acetyl-D-glucosamine	mannitol
2,4-di-*O*-methyl-D-galactose (ketose)	sorbitol
D-fructose	glycerol
3-*O*-methyl-D-fructose	
3-*O*-ethyl-D-glucose	D-xylose (Pentoses)
3-*O*-propyl-D-glucose	L-xylose (Pentoses)
3-*O*-butyl-D-glucose	D-ribose (Pentoses)
3-*O*-hydroxyethyl-D-glucose	L-arabinose (Pentoses)
1,4-anhydro-D-glucitol	D-arabinose (Pentoses)
gold-thioglucose	D-lyxose (Pentoses)

[a] Data taken from Crane, 1960.

a diagram of the minimum requirements for active transport. They are as follows:

(1) A pyranose ring (pentoses are not transported)
(2) The D-configuration (L-sugars are not transported)
(3) A free hydroxyl in the 2 position (the 2-deoxy sugars are not transported)
(4) The 2-hydroxyl having the stereochemical configuration of D-glucose (D-mannose, the 2-epimer of D-glucose, is not transported)
(5) The absence of bulky or charged substituents at the "unspecified" positions (Fig. 8.3) (These lead to loss of potentiality for transport.)

Since D-gulose (which differs from D-glucose in the configuration at both carbon atoms 3 and 4) is not actively transported while D-galactose and D-allose (which differ from D-glucose at only one carbon atom) are actively transported, there is a suggestion here that the C1 conformation is preferred. We can again suggest that an excess of axial hydroxyls might interfere with the interaction between permeant and transport site.

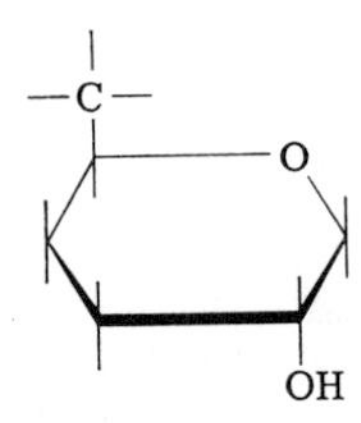

Fig. 8.3. The minimal structural requirements for sugar transport by the active transport system of hamster intestine. Only the substituent at the 2 position and at the methyl group at the 5 position of the hexose ring need be specified. (Taken with kind permission from Crane, 1960.)

In comparison with the specificity requirements of the glucose facilitated diffusion system, those for the active transport system are clearly far more stringent. We have seen in Chapter 5 that there may be less difference in the *mechanism* of action of the two systems than was previously thought to be the case—both are facilitated diffusion systems, but the active transport of glucose requires the presence and co-transport of sodium ions. We cannot yet see clearly how this requirement for sodium ions relates to the more stringent specificity requirements of the intestinal system, if indeed there is any such relation.

A very valuable corollary of mechanistic significance has been derived by Crane and Krane (1959) from a consideration of their diagram which we have reproduced in Fig. 8.3. Since the presence and position of the hydroxyl group at carbon 2 appears to be critically necessary for active transport, the possibility existed that this hydroxyl was involved in a chemical reaction during absorption. This hypothesis was tested by carrying out tracer exchange experiments with D-glucose-2-^{18}O in unlabeled water, and with 1,5-anhydro-D-glucitol in $H_2{}^{18}O$. No exchange of the carbon-bound oxygen with water occurred, ruling out the possibility of a chemical change involving, for instance, a cycle of dehydration and hydration, or of ether formation at the carbon in the 2 position. Similarly, the active transport of the compound 2-C-hydroxymethyl-D-glucose eliminated the possibility that an oxidation-reduction cycle at this carbon atom took place during transport.

Of the other mammalian sugar transport systems that have been studied in any detail, the facilitated diffusion of sugar into rat diaphragm (Kipnis and Crane, 1960), rat heart muscle (H. E. Morgan *et al.*, 1959), the goat placenta (Walker, 1960), isolated lymph node cells (Helmreich and Eisen, 1959), a mouse sarcoma (Luzzatto and Leoncini, 1961), mouse ascites tumor cells (Crane *et al.*, 1957), and isolated mouse fibroblasts in tissue culture (Rickenberg and Maio, 1961) all appear to share the broader specificity requirements of the erythrocyte, rather than those of the intestinal system (see Table 8.2). In yeast, a somewhat unspecific facilitated diffusion system for sugars is present (Burger *et al.*, 1959; Cirillo, 1961). On the other hand, the sugar permeases of *Escherichia coli* (reviewed by Kepes and Cohen, 1962) which, as we have discussed in Chapter 6, are apparently primary active transport systems, have an exceedingly high specificity for their substrate. Thus, in the most extreme case, no sugar or sugar derivative has been found to compete effectively with galactose for uptake by the inducible galactose permease, even when added to a concentration 100-fold greater

TABLE 8.2

SPECIFICITY REQUIREMENTS OF THE SUGAR AND AMINO ACID TRANSPORT SYSTEMS[a]

Cell or tissue	Specificity requirement
Sugars:	
Erythrocyte, human	Broad—D-preferring but not absolute. Cl conformation most affine. Accepts hexoses, ketohexoses, or pentoses (LeFevre, 1961a); accepts erythritol (Bowyer and Widdas, 1955); inhibited by disaccharides (Lacko and Burger, 1962) but must be 4,1-linked (Chen and LeFevre, 1965)
Erythrocyte, rabbit	Broad—carbon atoms 1, 2, and 3 need not be specified (Hillman *et al.*, 1959)

(*continued*)

TABLE 8.2—*Continued*

Cell or tissue	Specificity requirement
Ascites tumor cells, mouse	Broad—accepts pentoses, hexoses, or ketoses (Crane *et al.*, 1957)
180 sarcoma, mouse	Broad—accepts pentoses or hexoses, including 2-deoxyglucose, and mannose (Luzzatto and Leoncini, 1961)
Lymph node cells	Broad—inhibited by mannose, fructose, and 2-deoxyglucose (Helmreich and Eisen, 1959)
Fibroblasts, mouse tissue culture	Broad—2-deoxyglucose an effective inhibitor (Rickenberg and Maio, 1961)
Muscle, diaphragm, rat	Broad—2-deoxyglucose transported (Crane, 1960)
Muscle, heart, rat	Broad—hexoses, pentoses, or 2-deoxyglucose transported (H. E. Morgan *et al.*, 1959)
Placenta, goat	Broad—mannose and xylose transported (Walker, 1960)
Placenta, sheep	Broad—fructose accepted (Nixon, 1963)
Intestine, hamster Sodium co-transport	Less broad—neither pentoses, mannose, nor 2-deoxyglucose accepted (see Table 8.1 and Crane, 1960)
Intestine, dog (fructose system)	Less broad—galactose does not inhibit fructose uptake (Annegers, 1964)
Blood-cerebrospinal fluid barrier (human chorion)	Broad—glucose as well as arabinose transported (Battaglia *et al.*, 1962)
Yeast cells	Broad—pentoses, hexoses, ketohexoses accepted (Cirillo, 1961)
Escherichia coli	
(galactose active transport)	Very narrow—absolutely specific for galactose (see Kepes and Cohen, 1962)
(glucuronide, active transport)	Narrow—requires β-glucuronide residue (see Kepes and Cohen, 1962)
(β-galactosides, active transport)	Narrow—separate systems for methyl-β-galactoside (MG) and methyl-β-thiogalactoside (TMG) (Rotman, 1959); indeed, two inducible systems for TMG induced by specific inducers, one system (TMG permease I) accepting lactose and *o*-nitrophenyl galactoside and TMG, and the other induced by galactinol (TMG permease II), not accepting the two former substrates (Prestidge and Pardee, 1965)
Amino acids:	
Erythrocyte, human and rabbit	Fairly narrow—accepts Leu, Val, Met, but not Gly, Ala (Winter and Christensen, 1965)
Reticulocytes, rabbit	
Sodium-dependent	Narrow—accepts Gly, Ala, α-Me-ala (Winter and Christensen, 1965)
Sodium-independent	Same as rabbit erythrocyte

TABLE 8.2—*Continued*

Cell or tissue	Specificity requirement
Ascites tumor cell, mouse	
Sodium-independent, *L*-site	Fairly narrow—accepts Val, Leu, Pro, Met (Oxender and Christensen, 1963b)
Sodium-dependent, *A*-site	Fairly narrow—Gly, Ala, α-Me-ala, Met accepted (Oxender and Christensen, 1963b)
Sodium-dependent, β-site	Narrow—β-alanine and taurine accepted (Christensen, 1964b)
Dibasic site	Broad—but requires dibasic amino acids, probably a number of sites of overlapping specificity (Christensen, 1966)
Dicarboxylic amino acids	Broad—but requires dicarboxylic amino acids (Christensen, 1966)
Intestine, rat Sodium-dependent	Fairly narrow—probably at least 4 separate systems (somewhat as in ascites cells, above), one accepting many neutral amino acids, and an additional system (insulin-sensitive—Akedo and Christensen, 1962b) for Gly and Pro (Newey and Smyth, 1964); also, a dibasic amino acid system and one accepting *N*-substituted amino acids (Larsen *et al.*, 1964)
Intestine, hamster Sodium-dependent	Fairly narrow—at least 4 systems, viz., one for neutral amino acids (Matthews and Laster, 1965b), one for dibasic, and another for *N*-substituted amino acids, the dibasic and neutral systems competing with one another (Hagihira *et al.*, 1962; Lin *et al.*, 1962), while a possibly distinct system transporting cystine requires an intact —S—S— linkage in its substrates (Spencer *et al.*, 1965)
Kidney, rat	Fairly narrow—at least three systems: one for certain neutral amino acids (Gly, Ala, Phe) (L. E. Rosenberg, *et al.*, 1961), another for dibasic amino acids, and a third for cystine (L. E. Rosenberg *et al.*, 1962)
Escherichia coli	Very narrow—distinct systems for each of the amino acids Gly, Arg, Pro, His, Lys, Ala, Tyr, and Try (this last inducible); a system which accepts either Thr or Ser, another accepting Met or nor-Leu and, finally, one which accepts Val, Leu or iso-Leu (see Kepes, 1964)
Salmonella typhimurium	Narrow—a histidine site, absolutely specific for this amino acid; less broad—an aromatic site accepting His, Try, Phe, or Tyr (see Kepes, 1964)

[a] All sugars are the D-enantiomorph, all optically active amino acids the L-enantiomorph, unless otherwise designated. Amino acids abbreviated conventionally. Specificity classified, somewhat arbitrarily, as broad, less broad, fairly narrow, narrow, or very narrow.

than that of galactose. All β-galactosides and β-thiogalactosides that have been tested have an affinity for the somewhat less specific β-galactoside permease (although the affinities for different substrates, measured as K_m, range over at least 50-fold). Interestingly, in this latter system the specificity requirements for the permease differ from that for the induction of the permease and this differs again from the specificity requirement of the genetically linked enzyme β-galactosidase (3.2.1.23).

The data collected in Table 8.2 do suggest a tentative generalization: The specificity of the transport systems increases with their ability to accumulate substrate actively. It is not immediately clear, however, what significance, if any, this generalization may have for our understanding of the molecular basis of transport.

We shall conclude this section with a review of the specificity requirements of the amino acid transport systems, where the major advances have come from the work of Christensen and his school (Christensen *et al.*, 1962).

We discussed in Section 5.5 two of the systems which act to transfer amino acids into the Ehrlich ascites tumor cells. These were the *A*-system and the *L*-system (Oxender and Christensen, 1963a,b). The *A*-system has a particularly high affinity for glycine, alanine, and α-aminoisobutyric acid. It appears to discriminate especially against branched-chain amino acids containing 5 to 6 carbon atoms, but has a marked affinity also for methionine. As we saw in Section 5.5, this system can accumulate its substrate to high levels. In contrast, the affinity of the *L*-system toward its substrates increases as the length and bulk of the amino acid side chains increase. Thus valine and leucine are transported largely by the *L*-system. Methionine is effectively transported by both the *L*- and the *A*-systems. The *L*-system does not accumulate its substrates to the same high level as does the *A*-system. Both systems show a high, but not absolute, specificity for the L- as opposed to the D-amino acids. It appears, however, that in the series glycine, D- or L-alanine (α-methylglycine), α-aminoisobutyric acid (α,α'-dimethylglycine), the addition in the case of D-alanine of a single methyl group in an apparently unfavorable position is overcome with the change to α-aminoisobutyric acid by the addition of the second methyl group. Both systems require a free amino and a free carboxyl group. The amino group can be secondary (proline is transported by, apparently, the *L*-site) but cannot be tertiary. If the pK of these groups is displaced too far, the ability to be concentrated is lost. Thus D,L-trifluoromethylalanine with pK's of 0.5 and 5.94 for the carboxyl and amino groups is not transported.

A distinct system (the β-system) is involved in the major part of the penetration of β-alanine (Christensen, 1964b) and for this system, the carboxyl group of the amino acid can apparently be replaced by a sulfuric acid group—taurine (the sulfuric acid analog of β-alanine) and β-alanine mutually competing for transport by this system.

All of these three systems have overlapping affinities, and it has been a difficult problem to dissociate the several systems from one another. In contrast, the neutral amino acids have no affinity for the dibasic amino acid system, a distinct and very effective entity which may itself be subdivisible into an *L*- and an *A*-site. The degree of separation of the two amino groups in a dibasic amino acid determines its ability to be transported by the dibasic site, a maximal degree of accumulation being reached with α,γ-diaminobutyric acid. Likewise, the dicarboxylic amino acids are transported by a separate system. Christensen (1966) has formally codified the procedures for analyzing systems presenting such overlapping affinities.

A consideration of such specificity requirements for amino acid transfer, together with some evidence on the possible role of pyridoxal (vitamin B_6) in this transport, led Christensen (Christensen *et al.*, 1956) to propose the imaginative hypothesis that transport occurred as a result of complex formation between the amino acid, pyridoxal, and a metal. The two former compounds were supposed to combine to yield a Schiff's base, the metal thereafter combining by chelation. The requirement for Schiff's base formation would eliminate tertiary amino acids as candidates for transport, while metal chelation would be possible for α- and β-amino acids and would be highly favored by the well-accumulated dibasic amino acids. The more recent findings that pyridoxal appears to act by retarding the efflux of amino acids rather than by enhancing their uptake (Oxender, 1962), and the demonstration of the involvement of sodium ions (sodium is not effectively chelated) in amino acid transport, have led to abandonment of this hypothesis (Christensen, 1960), but it remains of much interest as a demonstration of the way in which considerations of specificity may lead to plausible models as to the mechanism of transport.

As far as other tissues are concerned, in rat kidney (L. E. Rosenberg *et al.*, 1962) a system which concentrates the dibasic amino acids is present and a separate system accumulates cysteine. In the intestine (Akedo and Christensen, 1962b) the neutral amino acids are apparently actively transported by two systems of similar specificity to those of the Ehrlich ascites cells, while the diamino and dicarboxylic amino acids do not seem to be actively transported (Smyth, 1961). In human erythrocytes, it is largely the *L*-site that survives into maturity, although the reticulocyte possesses systems similar to the *L*- and *A*-sites (Winter and Christensen, 1965). The amino acid permeases of bacteria, like their sugar permeases, appear to be very narrowly selective (reviewed in Kepes and Cohen, 1962), many being absolutely specific for the amino acid that they transport. There may be as many as twelve different permeases for the twenty amino acids (Table 8.2), each permease possessing a very sharp specificity. It has recently been shown (reviewed in Kepes, 1964) that for a number of amino acids, two transport systems are present, one of high affinity and narrow specificity, the second of lower affinity and broader specificity. Thus for histidine uptake by *Salmonella typhimurium,* the high affinity site has a K_m of around $2 \times 10^{-7}M$ and the rate of uptake by this system is quite unaffected by the addition of other amino acids, added as competitors. Another site for which histidine exhibits a K_m of $1 \times 10^{-4}M$ has a fivefold higher capacity for histidine uptake, and this uptake is competitively inhibited by tryptophan, phenylalanine, or tyrosine. It is thus an aromatic-specific site. All of these amino acid permeases, other than one for tryptophan (Boezi and de Moss, 1961), appear to be constitutive, in contrast to the sugar permeases which are in general inducible. Kepes points out a possible physiological significance of this fact. For optimal bacterial growth, any carbon source—glucose, galactose, lactose, etc.—is as good as any other in being able to support growth (all calories are equivalent). Thus only one source at a time need be transported into the bacterium and the maintenance of a high steady state accumulation of the other sugars, which would be wasteful of energy, is eliminated. The transport system for the commonest carbon source, glucose, is constitutive; the others, inducible. For efficient protein synthesis, however, all twenty amino acids have to be present simultaneously (the amino acid building blocks are by no means equivalent). It is clearly of value to have all the amino acid permeases functioning all the time. In conditions under which one particular amino acid is in excess, the internal *synthesis* of this amino acid is suppressed, and a balance is thereby achieved. We should note, however, that even if some of the amino acid permeases were inducible, we would be most unlikely to notice this fact; the presence of free amino acids in the amino acid pool of a bacterium engaged in protein synthesis would probably ensure that any inducible system is *always* active.

We have seen that the different transport systems demonstrate a range of specific requirements for their substrates. Many, especially those of the bacteria, mimic the commoner enzymes in the narrowness of these requirements; others, especially the sugar facilitated diffusion systems, have a much broader specificity than that of most enzymes. These studies allow of a detailed classification of the transport systems by the class of substrate transported. As we have seen in the similar study of enzyme mechanisms, the amount of information that such studies have produced as to the detailed mechanism of action of the transport systems is not particularly impressive, save in one or two cases. It is possible, however, that a more intensive study, *with this aim in view,* might be productive.

8.2 Molecular Significance of the Action of Drugs on Transport

The study of the action of drugs on enzyme systems has greatly extended our knowledge of the mechanism of action of enzymes. A noteworthy example of this approach has been the elucidation of the mechanism of toxicity of the fluorophosphate nerve gases such as diisopropylfluorophosphate (DFP) (reviewed in Holmstedt, 1959). The toxic action of these agents was early shown to be due to an inhibition of acetylcholinesterase (3.1.1.7). The irreversible inactivation of acetylcholinesterase could be shown to be accompanied by the transfer of a diisopropylphosphoryl residue from DFP into the enzyme (see Davies and Green, 1958) and this analysis led to the identification of the diisopropylphosphoryl acceptor as one particular serine residue at the "active center" of the enzyme. Many esterases and proteases have since been found to incorporate a diisopropylphosphoryl residue when inactivated by DFP, and in many cases the amino acid sequence around the phosphoryl-accepting serine residue, presumably at the active center of the enzyme, has been determined (a review in Oosterbaan and Cohen, 1964). Although there is no general agreement at the time of writing as to the mechanism of action of these hydrolases, most of the current models include the participation of a serine residue in the catalytic process (see Bender and Kézdy, 1965). Quite apart from its pharmacological importance, then, the analysis of the mode of action of DFP has opened up the attack on enzyme active centers.

We might hope that a similar investigation of the action of drugs on transport systems would likewise lead to a significant advance in our understanding of the mechanism of transport. Let us consider one of the most extensively studied cases—the action of the drugs phloretin and phlorizin on sugar transport—to assess just how far these hopes are, at present, justified.

The drug phlorizin (Fig. 8.4) has been known for eighty years (review in H. W. Smith, 1951) to produce renal glycosuria. Although phlorizin has been found to inhibit many enzyme activities (review in Crane, 1960)—in particular those enzymes requiring adenosine nucleotides as participants or cofactors and also a number of phosphatases and the enzyme mutarotase (5.1.3.3)—the levels of drug required to produce these effects are, in general, at least an order of magnitude higher than those required to inhibit sugar transport (Smyth, 1961). It is therefore accepted that the toxic effect of phlorizin results from its action on the active sugar transport by the kidney and intestine. (For our present purposes—the analysis of the mechanism of action of transport rather than the pharmacology of the drug—any residual controversy on this point is irrelevant, and we shall concentrate entirely on the effect of phlorizin on membrane transport.)

Phlorizin inhibits the active transport of sugar by hamster intestine in, apparently, a strictly competitive fashion (Alvarado and Crane, 1962) the K_i for inhibition (the concentration for 50% inhibition) being of the

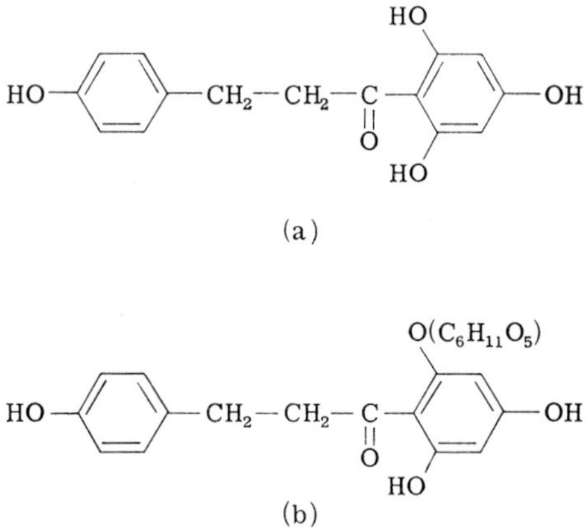

HO

HO—⟨benzene ring⟩—CH_2—CH_2—C(=O)—⟨benzene ring⟩—OH

HO

(a)

$O(C_6H_{11}O_5)$

HO—⟨benzene ring⟩—CH_2—CH_2—C(=O)—⟨benzene ring⟩—OH

HO

(b)

Fig. 8.4. Chemical structures of (a) phloretin and (b) its glucoside phlorizin.

order of $5 \times 10^{-7} M$. The aglucone of phlorizin, phloretin, has in this system only 1% of the effect of phlorizin itself. In contrast, in its action on the facilitated diffusion of glucose into human erythrocytes, the aglucone is some 100 times as effective as phlorizin (T. Rosenberg and Wilbrandt, 1957a). We see here again, as in Section 8.1, a specificity difference between the facilitated diffusion and the active transport (that is, sodium plus glucose co-transport) systems. Extending this correlation (Table 8.3), the active transport system of kidney is far more sensitive to the action of phlorizin than to the aglucone, while in contrast, the facilitated diffusion system in mouse fibroblasts (L-cells) is 30 times more sensitive to phloretin than to the glycoside. Of the other systems collected in Table 8.2, the facilitated diffusion systems transporting sugar into Ehrlich ascites cells, lymph node cells, and muscle are all inhibited by phlorizin, but data enabling a comparison to be made with their sensitivity to phloretin are not available. The amino acid systems are insensitive to phlorizin. The analysis of the differing actions of phloretin and phlorizin on the sugar transports is, once again, of significance in the classification of these transport systems. The significance for studies of *mechanism* of the action of these drugs derives mainly from the work of LeFevre (1961a) and T. Rosenberg and Wilbrandt (1957a) on the erythrocyte system. These studies consist of a systematic analysis of the effects of changes in the molecular structure of the drugs on their inhibitory potency. In phloretin (Fig. 8.4) there are four free hydroxyl

TABLE 8.3

SENSITIVITY OF FACILITATED DIFFUSION SYSTEMS TO INHIBITION BY PHLORETIN OR PHLORIZIN

Substrate	Cell or tissue	Sodium dependence	Sensitivity to Phloretin	Sensitivity to Phlorizin	References
Glycerol	Human erythrocyte	0		+	Bowyer (1954)
Glucose	Human erythrocyte	0	+++	+	T. Rosenberg and Wilbrandt (1957a) LeFevre (1959)
Glucose	Mouse fibroblasts	0	+++	+	Rickenberg and Maio (1961)
Glucose	Mouse ascites cells	0		++	Crane *et al.* (1957)
Glucose	Lymph node cells guinea pig	0		++	Helmreich and Eisen (1959)
Glucose	Muscle, rat heart	0		++	H. E. Morgan and Park (1957)
Glucose	Rat intestine	+	+	+++	Jervis *et al.* (1956)
Glucose	Kidney, cat	+	+	+++	Chan and Lotspeich (1962)
Amino acids	Intestine, rat	+	0	0	Smyth (1961)
Nucleosides	Mouse ascites cells	0		++	Jacquez (1962)
Glucose	Rat adipose cells	0	+++		Blecher (1966)

groups in the 2, 4, 6, and 4′ position capable of being substituted by organic functions. Since the 2 and 6 positions are equivalent, there are 12 possible derivatives in which from 0 to 4 groups are substituted in various combinations. Rosenberg and Wilbrandt, who studied 10 of these possibilities, found that methylation or glucosidation at the 2 or 4 position reduced the inhibitory potency of phloretin about 100-fold, while substitution at the 4′ position resulted in a 1000-fold loss of potency. The 2-methyl or 2-glucosidyl (phlorizin) derivatives were almost equally effective suggesting that the glucose residue plays no role here in the inhibiting action of phlorizin. [In contrast, in an active transport system in dog kidney, Diedrich (1961) finds that the substitution of galactose for glucose in the 2 position of phlorizin decreases the potency of this drug by an order of magnitude, a result which emphasizes once more the specificity difference between these facilitated diffusion and active transport systems.] Of the di- and tri-substituted derivatives of phloretin only the 2,4-dimethyl derivative was effective on the erythrocyte system (and that surprisingly so, with a potency as much as one-tenth that of the unsubstituted phloretin). Any derivative in which both the 2- and 4′-hydroxyls were blocked was quite ineffective. These considerations suggested to Rosenberg and Wilbrandt that the 2- or 6-hydroxyls might be involved in a hydrogen bond (or a metal chelate) with the keto group on the bridge between the two aromatic rings (Fig. 8.4) forming a structure resembling the steroids. In line with this suggestion, it was shown (T. Rosenberg and Wilbrandt, 1957a) that deoxycorticosterone did indeed somewhat inhibit glucose transport in the red cell, but only at levels some 1000-fold greater than that of phloretin. Although it seems unlikely that the transformation of phloretin into a form which is so much *less* active than phloretin itself can account for the mechanism of action of this drug, a tenable explanation of this sort would throw much light on the mechanism of transport. If, for instance, a steroid structure were involved in the inhibition by phloretin, this might suggest that membrane sterols were concerned in the sugar transport.

LeFevre's (1961a) studies have been concerned with evaluating the separate contributions of the two halves of the phloretin molecule to its inhibitory potency and with evaluating the role of the orientation and spacing of the terminal groups. [The molecular fragments of phloretin have only a very small inhibiting effect (Table 8.4), either when acting alone or in combination; a similar finding has been reported for the inhibition of the intestinal sugar transport (Larralde *et al.*, 1961).] A series of some forty compounds, more or less closely related to phloretin, was studied and for each the variation of the degree of inhibition with the concentration of inhibitor was investigated. This latter investigation proved to be necessary since for only a few compounds, including phloretin itself, did the degree of inhibition of sugar transport depend on the concentration of inhibitor in the manner to be expected if a 1 : 1 combination of drug and its receptor was concerned. Rather did it appear that in many cases, the degree of inhibition increased with concentration faster than would be predicted from a simple 1 : 1 interaction, somewhere between 1 and 2 molecules of inhibitor apparently taking part in preventing access of glucose to each transport site. We shall see in a later section (Section 8.5) and discuss in Chapter 9 the significance of the findings that there are now a number of similar examples of an apparent bivalency in the interaction between transport sites and their inhibitors. Here we shall consider the role that structural modifications of the inhibiting phlorizin derivatives play in respect to the potency of inhibition.

The many compounds studied by LeFevre as depicted in Fig. 8.6 fell in the main into the following four classes:

(a) Those in which a single carbon atom formed a bridge between two *p*-hydroxyphenyl groups, and the degree and type of substituents at-

TABLE 8.4

EFFECTIVENESS OF FRAGMENTS OF THE PHLORETIN MOLECULE AS INHIBITORS OF GLUCOSE TRANSPORT ACROSS THE HUMAN ERYTHROCYTE MEMBRANE[a,b]

Phenolic fraction[c]		Phloroglucinol fraction[c]	
$HOC_6H_4(CH_2)_3CH_3$		$CH_3CH_2COC_6H_2(OH)_3$	
p-butyl phenol	(1.5)	phlorpropiophenone	(7)
$HOC_6H_4(CH_2)_2COOH$		$HCOC_6H_2(OH)_3$	
phloretic acid	(0.7)	phloroglucinaldehyde	(0.2)
$HOC_6H_4CH_2CH_3$		$CH_3C_6H_2(OH)_3$	
p-ethyl phenol	(0.1)	methyl phloroglucinol	(0.1)
$HOC_6H_4CH_3$		$C_6H_3(OH)_3$	
p-cresol	($\ll$0.1)	phloroglucinol	($\ll$0.1)
HOC_6H_5 phenol	($\ll$0.1)		

[a] Data taken from LeFevre, 1959.

[b] Phloretin (see Fig. 8.4) $HOC_6H_4CH_2CH_2COC_6H_2(OH)_3$ (100).

[c] The numbers in parentheses after each compound listed represent the potency ratings, the reciprocals of the approximate millimolar concentration of the drug required to produce a 50% inhibition of glucose transport.

tached to the bridging carbon were varied; the crosses in Fig. 8.6 below record the series *p*-$HOC_6H_5C(R_1,R_2)C_6H_5OH$.

(b) A similar series where a pair of *m*-methyl-*p*-hydroxyphenyl groups were attached to the bridging carbon atom, that is, the series [*m*-CH_3-*p*-$HOC_6H_5]_2CR_1R_2$, the triangles of Fig. 8.6.

(c) A small series in which the number of bridging methyl groups between two *p*-hydroxyphenyl groups was varied, crosses within circles on Fig. 8.6.

(d) A series of derivatives based on stilbestrol, hexestrol, and dienestrol (Fig. 8.5) with allyl, chloroallyl, and propyl residues attached in the *meta* position on both of the phenol groups of these stilbestrols, solid circles on Fig. 8.6.

LeFevre noted that as the degree of substitution on the bis(hydroxyphenyl) skeleton was increased, the potency of the drug—expressed as the reciprocal of the millimolar concentration of drug required to give

50% inhibition of glucose transport—likewise increased. We may express this more quantitatively as in Fig. 8.6 where we have plotted the logarithm of the potency of inhibition (data taken from LeFevre, 1961a) against the total number of carbon atoms present in the drug in addition to that of the simple bis(*p*-hydroxyphenyl) skeleton. We include both substitution in the bridge or on the phenyl groups themselves in this

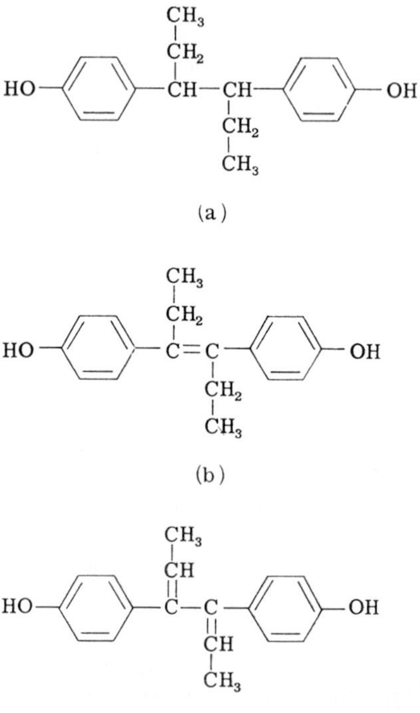

(a)

(b)

(c)

Fig. 8.5. Structure of various sterols, active as inhibitors of erythrocyte glucose transport: (a) hexestrol, (b) stilbestrol, and (c) dienestrol. (Compare Fig. 8.4.)

total. The correlation between bulk and potency is striking. The only drugs that do not obey a strictly linear relation are those where the bridging group is a cyclo structure (numbers 1 and 2 in Fig. 8.6) or a phenyl ring (number 3), where the deviation is small, and the series where the number of methylene bridging groups is increased (numbers 4, 5 and 6 in Fig. 8.6). The major effect of substitution in the basic bis(*p*-hydroxyphenyl) skeleton is presumably to increase the lipophilicity of the drug and it appears to matter very little, if at all, whether the lipophilicity is added in the bridge or to the bridged phenol groups.

These facts suggest strongly that the drug interacts with a hydrophobic region of the transport site. Supporting this view is the evidence presented by LeFevre and Marshall (1959) that the degree of absorption of a series of phloretin drugs by red cells parallels the potency of the drug. Both these properties would be enhanced by an increase in lipophilicity of the drug. The striking effect of increasing the length of the bridge between the two *p*-hydroxyphenyl groups emphasizes, however, that the interaction between drug and transport site is by no means wholly hydrophobic. The hydroxyl groups of the drug must be free (T. Rosenberg

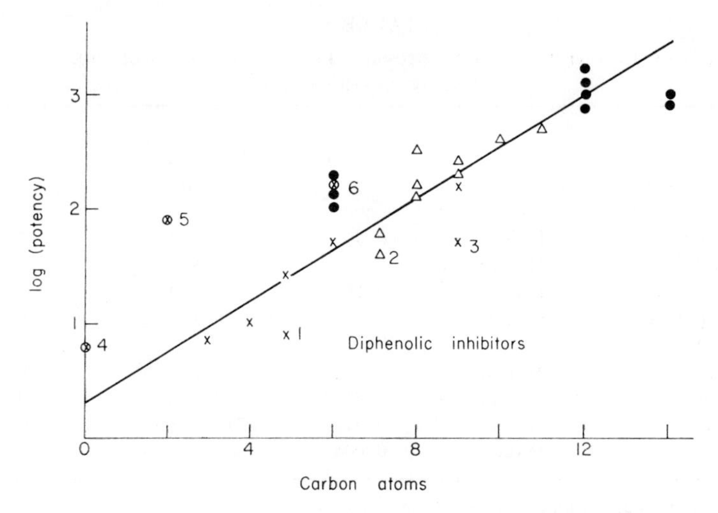

Fig. 8.6. The relation between hydrophobicity and inhibitory potency in a series of diphenolic inhibitors of the erythrocyte glucose transport system. Abscissa: number of carbon atoms in the substituents attached to the phenol rings or to the carbon atom joining the phenol rings. Ordinate: the logarithm of the potency of the inhibitor, where potency is defined as the reciprocal of the millimolar concentration required to half-inhibit glucose transfer. The symbols and numbers used are described in the text (Section 8.2).

and Wilbrandt, 1957a; LeFevre, 1961a), both must be present in a single molecule, and must apparently be spaced apart by at least 3 carbon groups, for optimal binding to the transport site. The active center of the glucose transport site behaves as if it possesses a pair of hydroxyl binding groups spaced perhaps some 8 to 10 Å apart (LeFevre, 1961a) and perhaps surrounded by a hydrophobic region (Fig. 8.6).

These findings recall those of Stein and Danielli (1956) on the competitive inhibition of the facilitated diffusion of glycerol across red blood cell membranes by various hydroxy-containing compounds and by narcotics. Here again (Table 8.5) there was a close correlation between inhibitory action and the lipophilicity (measured as the oil-water partition coefficient) of the inhibitor. But those agents (for example, the glycols) possessing a pair of hydroxyl groups at the extremes of the molecule, were up to forty times as effective in inhibition as their lipophilicity would predict. In Section 8.3 we shall discuss the evidence that this glycerol facilitated diffusion system is bivalent toward the action of certain inhibitors. The glycerol and glucose systems share the properties of bivalency, of being inhibited by molecules possessing pairs of hydroxyl

TABLE 8.5

COMPETITIVE INHIBITORS OF THE GLYCEROL TRANSPORT SYSTEM OF THE HUMAN AND RABBIT ERYTHROCYTES[a,b]

Compound	Cell species	K_m or K_i A	Olive oil-water partition coefficient B	Product of A and B
Glycerol	Rabbit	0.5	1×10^{-4}	5×10^{-5}
Glycerol	Human	0.7	1×10^{-4}	7×10^{-5}
Ethylene glycol	Human	0.1*	5×10^{-4}	5×10^{-5}
1,2-Propylene glycol	Rabbit	0.1*	6×10^{-3}	6×10^{-4}
Ethanol	Human	0.1*	2×10^{-2}	20×10^{-4}
Butanol	Human	0.015*	13×10^{-2}	20×10^{-4}
Urethane	Human	0.015*	13×10^{-2}	20×10^{-4}
Phenol	Human	0.003*	10	300×10^{-4}

[a] Data from Stein and Danielli, 1956.

[b] Glycerol uptake was measured by the photometric method, K_m for the substrate and K_i (indicated by an asterisk) for the inhibitors being determined from the variation of the half-time of glycerol uptake with the concentration of the reagent studied.

groups on an extended structure (but the specificity of inhibition is very different since the glycols do not affect glucose transport, although phloretin inhibits glycerol transfer), and of possessing a hydrophobic region at the active center for substrate binding.

There is little apparent correlation between the structure of these variants on the phloretin pattern and the parameter m, the number of drug molecules required to inhibit a single transport site. Those drugs in which there is little structure other than an unbranched bridge and the bis(p-hydroxyphenyl) group, have values of m equal or close to unity. If the site is bivalent, these drugs are apparently able to bind across the two halves of the site's active center and simultaneously inactivate both hydroxyl receptors. In the other drugs, an interaction between a pair of molecules might have to occur to form the bridging species.

Although all of these inferences from LeFevre's and from Rosenberg and Wilbrandt's studies are very much in the nature of preliminary hypotheses, these are the first molecular statements that we have yet been able to make in our treatment so far, and the beginnings of an approach toward the analysis of mechanism are clearly apparent here. Yet the conclusions that can be drawn on the basis of such studies must still be very tentative.

8.3 Inhibition by Chemical Reagents

Much of the information which has disclosed the chemical components of enzyme active centers and which has thereafter been used as a basis for models for the mechanism of enzyme action has come from the detailed study of the inhibiting action of chemical reagents possessing defined specificities. Thus the reaction between the enzyme ribonuclease (2.7.7.16) and the alkylating agents bromoacetic acid (Barnard and Stein, 1959) and iodoacetic acid (Crestfield *et al.*, 1963) has revealed the presence of a pair of histidine groups at the active center of this enzyme. The study of the inhibition of ribonuclease by protons has confirmed this analysis (Mathias *et al.*, 1964) and this information, together with knowledge of the specificity of enzyme action, has led to the proposal of plausible models for ribonuclease action (Mathias *et al.*, 1964; Witzel and Barnard, 1962). In addition, studies on the inactivation of ribonuclease by dinitrofluorbenzene have identified a specific lysine residue near the active center of this enzyme (Hirs, 1962) and roles for certain other amino acid side chains have also been suggested (Anfinsen and White, 1961). All these studies and similar investigations of other enzymes, have been made meaningful by the choice of reagents which have a defined specificity of attack toward only certain amino acid side chains, or which react at defined rates with a somewhat wider spectrum of side chains (Ray and Koshland, 1961). In the subsequent analysis of the inhibition process it has proved possible by careful investigation to eliminate all but one side chain type from consideration and then to identify the particular amino acid in the protein chain which is concerned in the inhibition sequence. It will be useful in this section to review the more limited steps which have been taken along these lines in the analysis of transport systems. Most of the work has been carried out on the glucose transport system of the human erythrocyte. Other work that we shall discuss concerns the glycerol transport system of that cell. At the present stage of these investigations, the aim of such studies is to identify the groups present at the transport active center. Conclusions as to mechanism will be found to be as yet premature.

The erythrocyte facilitated diffusion systems for glucose and glycerol are inhibited by the chemical reagents listed in Table 8.6. For each inhibitor the range of possible amino acid side chains that may be attacked by this reagent is given. None of the reagents listed in Table 8.6 is by itself absolutely specific for any amino acid residue, but if the different results are taken together a rather more meaningful picture emerges. For instance, the glucose system is inhibited by a number of reagents, which

TABLE 8.6

CHEMICAL REAGENTS AS INHIBITORS OF THE FACILITATED DIFFUSION OF GLUCOSE AND GLYCEROL ACROSS THE HUMAN ERYTHROCYTE MEMBRANE

Reagent	Amino acid side chains attacked	Inhibition[a]	
		Of glucose system	Of glycerol system
Protons pH range 5–9	α-Amino, His, Cys(—SH)	Slight (1)	Intense (2, 3)
Mercuric chloride	Lys, His, Cys(—SH)	Intense, reversed by thiols (4)	Intense, reversed by histidine (5)
p-Chloromercuribenzoate	Lys, His, Cys(—SH)	Intense, reversed by thiols (4)	Intense, reversed by 0.17*M* sodium chloride (5)
p-Chloromercuribenzoate sulfonate	Lys, His, Cys(—SH)	Intense, reversed by thiols (6)	?
Copper chloride	Lys, His, Cys(—SH)	Not inhibited (4)	Intense, reversed by histidine (5)
N-Ethyl maleimide	Lys, Cys(—SH)	Intense (7)	?
2,4-Dinitrofluorbenzene	Lys, His, Cys(—SH), Tyr	Intense (8)	Less intense (8)
Allyl isothiocyanate, Bromoacetophenone, Chloropicrin	Cys(—SH)	Intense (9)	Ineffective (9)
Iodoacetate	His, Cys(—SH), Met	Ineffective (4)	Ineffective (10)
Diazonium hydroxide	Lys, Cys(—SH), Tyr, His	Ineffective (11)	Effective (12)
Mild deamination	Lys	Ineffective (8)	Ineffective (10)
Mild acetylation	Lys	Ineffective (8)	?
Tannic acid	Lys	?	Effective (13)

[a] *References* (numbers in parentheses): (1) Sen and Widdas (1962a). (2) Stein (1962b). (3) Wilbrandt *et al.* (1955). (4) LeFevre (1948). (5) Stein (1958a). (6) van Steveninck *et al.* (1965). (7) Dawson and Widdas (1963). (8) Bowyer and Widdas (1956). (9) Wilbrandt (1954). (10) Stein (1962e). (11) Bowyer (1954). (12) Stein (1958b). (13) Hunter *et al.* (1965).

share in common the property of reacting with the free thiol residue of cysteine. For the reversal of mercurial inhibition, the addition of thiol is necessary, suggesting that the glucose site binds mercury with the affinity characteristic of the thiols. [The dissociation constant for the complex formation between mercury and thiol is very high, some 10^{20} (Stricks and Kolthoff, 1953).] Yet the system is not affected by prolonged treatment with iodoacetate, an effective thiol reagent. It is known, however, that thiol groups in proteins can be more or less "sluggish" in their reactivity toward different thiol reagents (Cecil and Thomas, 1965). The glycerol system, although inhibited by mercurials, is protected against this inhibition by the presence of histidine, alanine, or even 0.17*M* sodium chloride (Stein, 1958a). It is clear that a thiol residue is not involved in mercury binding in this case. Copper inhibits the glycerol system when present at a lower concentration than is needed for mercuric ion inhibition. The affinity of the glycerol transport system for copper [an association constant of $10^{11.65}$ is reported (Stein, 1962c)] is sufficiently high that chelation of the bound copper by at least two nucleophilic centers must be postulated. The two systems differ also in their response to changes in the ambient pH. The glycerol system is over 90% inhibited by a decrease in pH over the range pH 6.0–7.0 (Stein, 1962c) suggesting the presence of an imidazole grouping (from the amino acid histidine) or an α-amino group at this active center. The glucose system is inhibited by only 30% over the range of 6.0–8.0 (Sen and Widdas, 1962a), and no particular conclusion can be drawn from this pH profile. Finally, the two systems differ in their response to the action of dinitrofluorbenzene (DNFB). At 24°C and in the presence of 1.4 m*M* DNFB, half-inhibition of the glucose system occurs in 20 minutes, but for the glycerol system 160 minutes incubation time is necessary to reach the same level of inhibition (Bowyer and Widdas, 1956).

All these results are consistent with the presence of a thiol residue at the active center of the glucose system and at least two groupings (amino or imidazole groups) at the active center of the glycerol system, but this view would seem to be an oversimplification. Thus, for the glycerol system, the dependence of the degree of inhibition on the pM (the negative logarithm of the free metal ion concentration) for copper ions (Fig. 8.7) is that given for a system in which two copper ions are simultaneously required to inhibit transport (theoretical curve *B* of Fig. 8.7) rather than a single copper ion (theoretical curve *A* of this figure). The data for inhibition by protons similarly show a very steep pH profile, suggesting the presence of a pair of proton-binding groups at the active center of the glycerol system. The variation of the copper-binding affinity with pH suggests that each copper-binding site is itself composed of a pair of titratable groups with pK some 9.3 units (Stein, 1962c) (perhaps lysine residues), making four such groups in all present at this active center. Furthermore, rates of inhibition by dinitrofluorbenzene of both the glucose system (Bowyer and Widdas, 1956) and the glycerol system (Stein, 1962e) depend on the square of the dinitrofluorbenzene concentration rather than, as would be expected on a naive view, with the first power of this concentration. These findings recall the analogous results of LeFevre (1961a) on the inhibition of glucose transport by phloretin analogs (Section 8.2). Thus both systems behave toward certain inhibitors as if a pair of binding sites were present at the active center of each transport system.

There is a further complication in the interpretation of the glucose inhibition data: it may be that the sites of action of DNFB and *N*-ethyl maleimide (NEM)—both thiol reagents—are not identical. Thus inhibition by NEM (Dawson and Widdas, 1963) does not depend on the square of the inhibitor concentration, nor is it accelerated by the presence

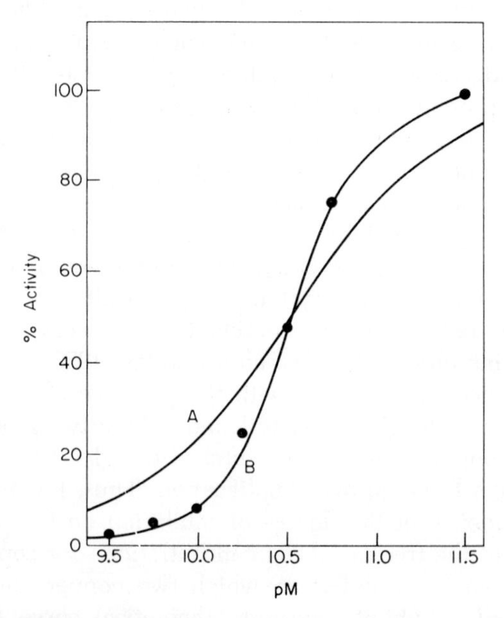

Fig. 8.7. The inhibition by copper ions of the transfer of glycerol across the erythrocyte membrane of man. Ordinate: percentage activity with respect to uninhibited transfer rate. Abscissa: logarithm of the free copper ion concentration. Curve *A*, theoretical on the assumption that a single molecule of copper is required to inhibit glycerol transfer. Curve *B*, theory for the model in which a pair of copper ions must simultaneously be present to inactivate glycerol transfer. ●, Experimental points. (From Stein, 1962c.)

of the substrate (glucose) to the same extent as is the inhibition by DNFB. Nevertheless, the dependence of the rate of inactivation by these two reagents on the ambient pH (Dawson and Widdas, 1963; Stein, 1962e) is very similar. Finally, the inhibition by NEM is blocked if *p*-chloromercuribenzoate is first added to the red cell suspension (Dawson and Widdas, 1963). In contrast, inhibition by DNFB is not blocked if mercuric chloride is present simultaneously (Stein, 1962e).

The author has attempted to delineate more clearly the nature of the DNFB-blocked group in the active center of the glucose transfer system, by making use of the fact that the rate of inhibition is proportional to the square of the DNFB concentration (Stein, 1964b). By a dual-labeling procedure in which ^{14}C-labeled DNFB is used at a high concentration for a short time while ^{3}H-labeled reagent is used for a longer period at a low concentration, and the products of reaction thereafter mixed and analyzed, groups reacting with DNFB according to these unusual kinetics can be identified. The major group in the membrane reacting with DNFB turns out to be cysteine, but substantial amounts of reaction (some 10% that of the thiol) occur at lysine, tyrosine, and histidine groups. At the time of writing, the group reacting with DNFB at a rate dependent on the square of the concentration of inhibitor has not been fully characterized but, in addition to any thiol cysteine that may react in this fashion, at least one other residue is so affected. We shall discuss in Section 8.4 how this double-label procedure might be used to isolate the glucose-transferring system from the cell membrane.

In addition to providing information which will in the future certainly enable the active center of these transport systems to be characterized, these studies can be used to set an upper limit to the number of transport sites present per cell. Thus glucose transport is some 50% inactivated under conditions in which radioactivity corresponding to only 40 mμ moles DNFB/ml of packed red blood cells is bound to the cell membrane (Stein, 1964b). Of this bound DNFB only one-half at most has reacted at a rate proportional to the square of the DNFB concentration. Thus for full inactivation of transport, the square-dependent material is certainly not more than 40 mμ moles/ml packed cells, and if two DNFB molecules bind to each site (see Table 8.8) the number of sites is not more than 20 mμ moles/ml packed cells or 2×10^{-18} mole/cell (since there are approximately 10^{10} cells in 1 ml of packed red cells) or 1.3×10^{6} sites per cell. This number is overestimated to an unknown extent if there are other sites in the red cell membrane which react with DNFB according to the square of the inhibitor concentration, and the glycerol-binding site is one such group.

Another estimate of the number of glucose transport sites per red cell

TABLE 8.7

CHEMICAL REAGENTS AS INHIBITORS OF TRANSPORT SYSTEMS[a]

Reagent, with amino acid side chains possibly affected	System	References
Mercury and mercurials: Cys(—SH) and also Lys and His	Sugars, erythrocytes	See Table 8.6
	Glycerol, erythrocytes	See Table 8.6
	Sugars, yeast cells	See Passow *et al.* (1961)
	Glucose, intestine	See Passow *et al.* (1961)
	Glucose, kidney	See Passow *et al.* (1961)
	Amino acids, kidney	See Passow *et al.* (1961)
	Monocarboxylic acids, rat intestine	Foulkes and Paine (1961)
	Galactosides, *E. coli*	Kepes (1960)
	Sodium ion, frog skin	Linderholm (1952)
	Sodium-potassium-activated ATPase, rat brain	Skou (1964)
	Axonal function, lobster	H. M. Smith (1958)
Copper ions: Cys(—SH), Lys, His	Glycerol, erythrocytes	See Table 8.6
	Glycerol, Schwann cells, squid axon	Villegas and Villegas (1962)
	Chloride ions, frog skin	Ussing and Zerahn (1951)
	Axonal function, lobster	H. M. Smith (1958)
Uranyl ions: polyphosphates	Sugars, yeast	Rothstein (1954)
N-Ethyl maleimide: Cys(—SH) and perhaps Lys	Glucose, erythrocytes	See Table 8.6
	Galactosides, *E. coli*	Koch (1964)
	Axonal function, lobster	H. M. Smith (1958)
	Glucose, adipose cells	Blecher (1966)
2,4-Dinitrofluorbenzene: Cys(—SH), Lys, His, and Tyr	Glucose, erythrocytes	See Table 8.6
	Glycerol, erythrocytes	See Table 8.6
	Leucine, ascites tumor cells	Oxender and Whitmore (1966)
Tannic acid (Lys ?)	Urea, erythrocytes	Hunter *et al.* (1965)
	Glycerol, erythrocytes	Hunter *et al.* (1965)

[a] The results quoted are those in which the chemical reagent appears to inhibit the transport process directly, as opposed to interfering with metabolic energy production by the cell.

comes from studies on the inhibition of this transport by mercurials (van Steveninck *et al.*, 1965).

The thiol reagent *p*-chloromercuribenzoate sulfonate does not appear to penetrate the erythrocyte membrane. It reacts with only 0.6% of the membrane sulfhydryl groups (and there are 180×10^{-18} mole of these per red cell) yet at this stage inhibits glucose transport. This gives a maximum of 1.1×10^{-18} mole/cell as being involved in sugar transport, or 700,000 sites per cell, a figure consistent with that derived from the DNFB studies. Finally, by measuring the amount of glucose bound to erythrocyte ghosts at very low concentrations of glucose, LeFevre (1961c) has estimated that no more than 200,000 glucose transport sites are present per cell. The work on the glucose and glycerol systems has been treated in detail since the available information on other transport systems is often even more fragmentary. Some representative data are collected in Table 8.7. The assembled data suggest a possible general involvement of thiol residues in transport phenomena, but this probably reflects more the ready accessibility of reputedly thiol-specific reagents than any real property of transport systems.

8.4 Attempts at the Isolation of Transport Systems from Cell Membranes

We shall consider in turn the several different types of approach that have been used in attempts to isolate transport systems from cell membranes.

A. Methods Based on Substrate Binding

This method makes use of a property generally found in transport systems, that of the binding of substrate. One such approach emphasizes particularly the view that the binding of substrate to carrier might (or should) produce a lipid-soluble complex which could then cross the lipid cell membrane. Thus in this scheme, membranes are fractionated with a view to finding a component which binds the particular substrate chosen and which renders the "carrier"-substrate complex soluble in hydrophobic solvents. This approach has been used in attempts to isolate the cation transport systems of cell membranes (Solomon *et al.*, 1956; Kirschner, 1957), the sugar transport system of the human erythrocyte (Reinwein, 1961; LeFevre *et al.*, 1964) and the iodide-accumulating component of thyroid gland (Schneider and Wolff, 1965). In all these cases certain lipids have been isolated from the cell membrane which have the sought-for property of solubilizing the substrate in organic solvents or of binding the substrate. Solomon and his colleagues (Solomon *et al.*, 1956) found that a number of purified phospholipids (for example, phosphatidylserine, sphingomyelin, and acetal phosphatide)—but not cholesterol—enabled potassium and sodium to enter the chloroform phase when the lipids were present in an emulsion of chloroform and a cation-containing aqueous buffer solution. Potassium was solubilized in preference to sodium by a factor of 8- to 14-fold. On reextracting the chloroform phase with fresh buffer, the cation-phospholipid complex dissociated and the cations reentered the aqueous phase. Reinwein (1961) found similarly that phospholipid fractions from erythrocyte membranes were able to carry ^{14}C-labeled glucose into the chloroform phase when added to an emulsion of chloroform and water. This property of solubilizing glucose in a lipid solvent disappeared if the sugar transport inhibitor, phlorizin, was present in the system. Unlike the glucose transport system, the phospholipids did not display any specificity toward different sugars. Glucose and sorbitol (a nontransported sugar) were similarly solubilized. LeFevre and his colleagues (LeFevre *et al.*, 1964), who have studied the ability of different membrane fractions to complex sugar when dried down together with sugar at 50°C, found a parallel between the affinity of different sugars for the glucose transport system and the ability of phospholipids to solubilize these sugars. The phospholipids do not, however, discriminate between optical enantiomorphs, evidence raising a very serious objection to the role of these phospholipids in glucose transfer. In contrast to Reinwein (1961), LeFevre's group could not substantiate the claim that these phospholipids could move sugar out of an aqueous phase into a nonaqueous phase. Several further properties of the glucose transport system are, however, paralleled by these phospholipid-sugar interactions. Thus the interactions are, like glucose transport, diminished by stilbestrol. In parallel with the inhibition of glucose transport, on reacting intact red cells with DNFB, there is a decrease in the ability of phospholipid extracted from these cells to take ^{14}C-labeled glucose into hexane. In a similar fashion, phospholipids bind iodide and other anions (Schneider and Wolff, 1965), and these other anions displace bound iodide in a sequence which parallels their ability to displace iodide bound within slices of thyroid.

These properties of phospholipids—their ability to bind sugar and ions—have, it appears (references in Solomon *et al.*, 1956; LeFevre *et al.*, 1964), been known for a good number of years. The suggestion of a definite role for the phospholipids in membrane transport is, on the other hand, new and stimulating. It is quite clear, however, that the interactions described above are not by themselves sufficiently specific to account for the whole behavior of the transport systems. Park (1961) suggests that the specificity might derive from a protein component which catalyzes the interaction between the phospholipid and the permeant, while LeFevre (1965) feels that, in addition, the protein is necessary to bring about an interaction between sugar and phospholipid in the aqueous phase. We shall consider such models more carefully in Chapter 9. Meanwhile, it is necessary only to comment that it is agreed by all that no complete transport system has been isolated by these extraction procedures.

In contrast to the above methods, one approach used in the author's laboratory has been to search for sugar-binding capacities among the *protein* components of the cell membrane. Here the stromal proteins have been solubilized by treatment of the membranes with *n*-butanol, with Triton X-100, or with molar sodium iodide, by the procedures discussed in Section 1.2. To investigate binding, the author has developed a highly sensitive method in which the material to be investigated (whole stroma or stromal extracts) is adsorbed onto some inert, solid, supporting medium (for example, DEAE-cellulose) and this protein-cellulose complex used to form a chromatographic column. The ability of such a column to differentiate between, say, glucose and sorbose can be tested by passing a mixture of the two substrates—one labeled with ^{14}C, the other with ^{3}H—down the column and then analyzing the emerging effluent. If glucose is bound to the column in preference to sorbose, that is, if the glucose transport system is present on the column, the radioactivity due to glucose will emerge from the column after that due to sorbose. Using this technique, Bobinski and the author (Bobinski and Stein, 1966) have shown that such a differential binding of glucose in preference to sorbose is indeed found for cell stroma. This differential binding is eliminated if the column is treated with $10^{-4}M$ phloretin, the inhibitor of glucose transport, and binding disappears in parallel with a loss of sugar transport activity if the cells are treated with DNFB before being made into the columns. Finally, the ability to bind glucose is a saturable function of the glucose concentration and the half-saturation concentration for differential binding has at room temperature a similar value (5–20 mM) to the half-saturation concentration for glucose transport (12 mM). The iodide extract of this stromal material retains the ability to bind glucose, and this activity is associated with a particular one of the five ultraviolet-absorbing peaks found on chromatographic fractionation of the iodide extract. Bonsall and Hunt (1966) have found in addition that the Triton X-100 extract of red cell stroma is able to bind glucose in preference to sorbose. Glucose binding is specific for the D-isomer.

These findings, if confirmed in other laboratories, would suggest that the procedure outlined here has indeed led to the isolation of at least one component of the glucose transport system. Since the technique used, namely, substrate binding, is a general one, it may be possible to study on these lines many more transport systems.

Indeed, Pardee and Prestidge (1966) have used a similar procedure to isolate a sulfate-binding component, apparently involved in sulfate active transport, from *Salmonella typhimurium.* This organism can accumulate sulfate ions and thiosulfate ions to high distribution ratios by an energy-dependent active transport system (Dreyfuss, 1964). If spheroplasts of this organism are prepared, that is, if the cell wall is digested away, sulfate uptake ceases (Dreyfuss and Pardee, 1965). [This is in contrast to the situation for β-galactoside uptake by *E. coli*—here the spheroplasts are fully competent (Sistrom, 1958).] This finding suggested that a sulfate-binding component external to the cell membrane was present in *S. typhimurium.* It is this component that has apparently been isolated by Pardee and Prestidge (1966). In sonicates or osmotic lysates of *S. typhimurium,* a nonsedimentable component can be demonstrated which has the ability to release sulfate into solution, when this sulfate is bound to an ion-exchange resin. The sulfate transport system of this organism is repressed if cysteine is added to the growth medium or de-repressed, if djenkolic acid is added as the sulfur source. The sulfate-binding component could not be demonstrated in repressed cells but was present in the de-repressed cultures. Gel filtration studies suggested a molecular weight of some 70,000 for this sulfate-binding material.

Although the correlation between repression and the appearance of sulfate-binding is excellent and although again both sulfate-binding and sulfate transport disappear when *Salmonella* are converted into spheroplasts, yet the precise role that binding plays in transport is, as yet, unclear. Sulfate transport in these organisms can be shown to be dependent on the activity of at least three linked genes (Mizobuchi *et al.*, 1962) yet sulfate-binding activity is apparently normal in a mutant in which all three genes have suffered deletions (Dreyfuss and Pardee, 1965). We shall see in Section 8.4,C that it is becoming apparent that β-galactoside uptake by *E. coli* is similarly a complex process. One must, however, agree with Pardee and Prestidge (1966) that "these results open the avenue to characterization of such binding material and a study of active transport at the subcellular level."

B. Methods Based on the Use of Chemical Reagents as Inhibitors of Transport

In the previous section we considered the binding of copper ions to the active center of the glycerol facilitated diffusion system. It was originally proposed (Stein, 1958a) that the group binding copper ions was an *N*-terminal histidine residue, and a search was therefore made for *N*-terminal histidine-containing fractions among the components of the erythrocyte membrane. Later analysis (Stein, 1962c) indicated instead that a concatenation of lysine residues is responsible for this copper binding, and the search for *N*-terminal histidine components has therefore been abandoned. More promising are the attempts that have been made (Stein, 1964b) to identify the components concerned in the irreversible binding of dinitrophenyl residues to the glucose transport system of the human erythrocyte. The author's ^{14}C and tritium double-labeling technique discussed in Section 8.3 has been modified so as to reveal, rather than the amino acid side chain, the entire moiety involved in the reaction with DNFB. Samples of erythrocytes have been separately treated with ^{14}C-labeled DNFB at a particular concentration for a particular time and with the tritium-labeled reagent at one-tenth this concentration for ten times the time, the samples thereafter mixed and then extracted to liberate the solubilized proteins. The proteins and lipids have been separately fractionated chromatographically and the ^{3}H to ^{14}C ratio measured in the different fractions. A fraction has been identified which appears to react with DNFB in a kinetic fashion similar to that found for the inhibition of the glucose transport system by this reagent. It is, however, still premature to designate this fraction as being the one concerned in glucose transport. We discuss in Section 8.4,C the powerful use made by C. F. Fox and Kennedy (1965) of the irreversible inhibition of the galactoside permease by NEM.

A procedure that has not yet been applied in the study of transport systems, but which has had some success in studies on pure enzymes, is the deliberate design of specific inhibitors, based on the specific binding properties of the active centers. Thus the enzyme chymotrypsin (3.4.4.5) acts on substrates bearing an aromatic ring one carbon atom distant from the susceptible peptide bond, while the enzyme trypsin (3.4.4.4) acts rather on substrates bearing an amino or guanidine group at a little distance from the susceptible bond. In the brilliant studies of Schoellmann and Shaw (1963) and Mares-Guia and Shaw (1963) the alkylating reagent α-*N*-tosyl-L-phenylalanylchloromethane (Fig. 8.8a), which fulfils the structural requirements of the chymotryptic site, and α-*N*-tosyl-L-lysylchloromethane (Fig. 8.8b), fulfilling the requirements of the tryptic site, were shown to be powerful and specific irreversible inhibitors of chymotrypsin and trypsin, respectively. The chymotrypsin reagents reacted with a histidine residue in a known position in the primary protein chain, presumably at the active center of this enzyme, while the trypsin reagent also appears to react with an active center histidine, although this has not yet been identified in the protein sequence. A number of reagents similar in structure to the chymotrypsin inhibitor, also potent inhibitors of this enzyme, react at neighboring methionine and serine residues, rather than the histidine residue (Hartley, 1964). Clearly, this is a very powerful approach, and the wealth of information on the specificity of the binding sites of the transport systems (Section 8.1) suggests that important advances in the identification of the active centers of these systems and their subsequent isolation would result if such procedures were applied. [Since the above was written, this approach has indeed been used (Hokin *et al.*, 1966) to label irreversibly the steroid-binding site of the ATPase of brain. The reagents used were strophanthidin-3-haloacetates.]

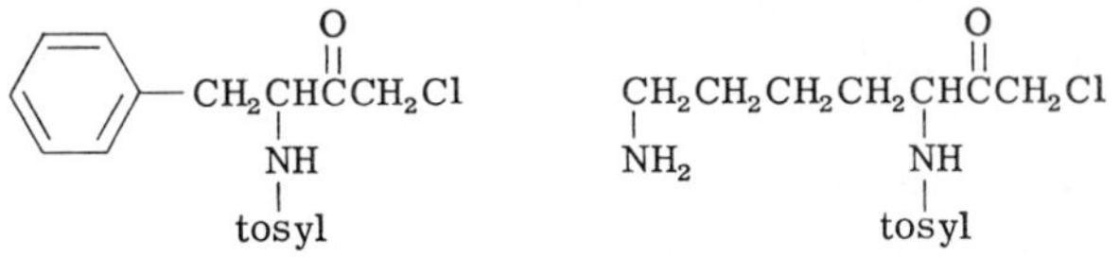

(a) $C_6H_5-CH_2CH(NH\text{-tosyl})C(=O)CH_2Cl$

(b) $CH_2(NH_2)CH_2CH_2CH_2CH(NH\text{-tosyl})C(=O)CH_2Cl$

Fig. 8.8. Structure of designed specific inhibitors of (a) chymotrypsin—the inhibitor is the tosyl derivative of phenylalanylchloromethyl ketone, and (b) trypsin —the inhibitor is the tosyl derivative of lysylchloromethyl ketone.

C. METHODS BASED ON THE INDUCIBILITY OF THE β-GALACTOSIDE PERMEASE OF *Escherichia coli*

The β-galactoside permease of *E. coli* is inducible [that is, in the absence of the specific substrates, no permease is synthesized by the cell, but within minutes of the addition of substrate—the "inducer"—to a growing cell population, the presence of a vigorous permease system can be demonstrated (Rickenberg *et al.*, 1956)]. The gene coding for the β-galactoside permease is part of the *Lac* operon—a set of linked genes, the gene-products of which are coordinately expressed (Jacob and Monod, 1961). The only protein products coded for by the structural genes of the *Lac* operon are apparently the β-galactoside permease, the enzyme β-galactosidase (3.2.1.23), and the enzyme thiogalactoside transacetylase (2.3.1.b) (Zabin *et al.*, 1962). These last two proteins are well-characterized, crystallizable enzymes (Wallenfels and Malhotra, 1960; Zabin, 1963) while, in contrast, the permease has up to now been merely a hypothetical entity introduced to account for the transport behavior of the β-galactosides under various genetic and kinetic conditions. The fact that only these three proteins (if the permease is a protein!) are formed as the expression of the *Lac* operon has enabled Dr. A. R. Kolber and the author (A. R. Kolber and Stein, 1966) to develop a method for the isolation of the permease, using a dual-isotope labeling technique as follows. To one culture of *E. coli* is added ^{14}C-labeled phenylalanine together with a small amount of thiomethyl-β-galactoside, a substance which will induce the formation of the enzymes of the *Lac* operon. To the other culture is added ^{3}H-labeled phenylalanine, but no inducer. All proteins present in the cells of the ^{14}C-labeled culture will be present also in the ^{3}H-labeled culture, except for those proteins coded for by the genes of the *Lac* operon—these proteins will have incorporated only the ^{14}C label. The ratio of ^{14}C to ^{3}H in the proteins extracted from a mixture of the two cultures should be everywhere the same, except for the three proteins from the *Lac* operon. Figure 8.9 shows a radioactivity assay of the effluent from a chromatographic separation of the proteins of such a mixed culture. One large region of ^{14}C enrichment is exactly

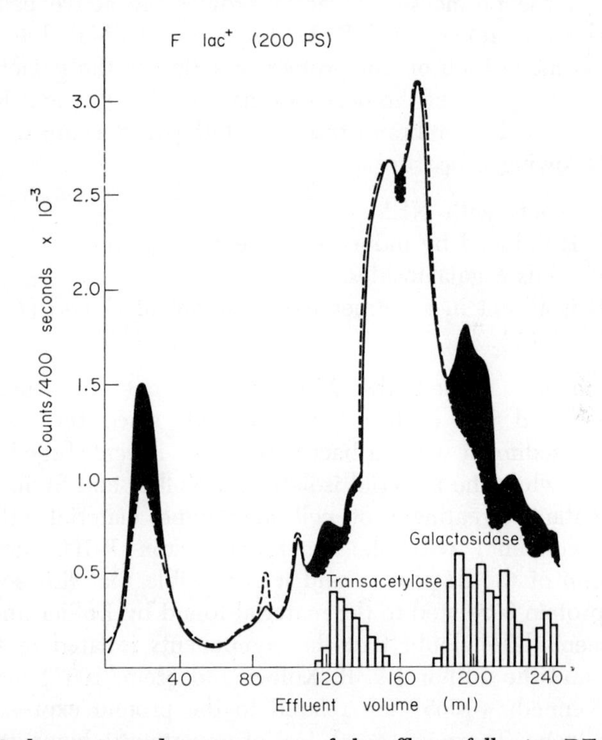

Fig. 8.9. Radioactivity and enzyme assay of the effluent following DEAE-cellulose chromatography of a cell extract from *E. coli.* The solid line indicates the ^{14}C activity, the dashed line, the ^{3}H activity. The solid fills are the regions which are relatively enriched in ^{14}C. The blocks indicate enzymatic activities. ^{3}H-phenylalanine was added to the noninduced culture, ^{14}C-phenylalanine to the induced culture. (Taken from A. R. Kolber and Stein, 1966.)

coincident with the region where β-galactosidase activity is intense, a second small region of enrichment coincides with the thiogalactoside transacetylase region. A third region of enrichment has neither enzyme activity. To confirm the hypothesis that this third region is associated with the permease, a study on similar lines of various mutant strains of *E. coli* was undertaken. In a permease-less mutant, no area of enrichment corresponding to the putative permease could be found, while in a galactosidase-less mutant, only the ^{14}C-enrichment area corresponding to the galactosidase peak was absent.

In a somewhat similar study, C. F. Fox and Kennedy (1965) have isolated from *E. coli* a rather different protein. Their procedure is to treat cultures of *E. coli* with *N*-ethyl maleimide, a reagent which is an irreversible inhibitor of the β-galactoside permease. In the presence of β-galactosides, *N*-ethyl maleimide (NEM) does not inhibit. Thus the substrates of the permease appear to protect the active center of the system against attack by NEM. An ingenious dual-label technique in which use is made both of this protective action of the galactosides and also of the inducibility of the permease has enabled Fox and Kennedy to identify in the cell membrane fraction of the bacterium, a component with the following properties:

(1) It reacts with NEM.
(2) It is induced by inducers of the *Lac* operon.
(3) It binds β-galactosides.
(4) It is absent in a permease-less mutant of *E. coli* (C. F. Fox *et al.*, 1966).

This component, termed the *M*-protein, is not the same material that Kolber and the author have isolated, since the *M*-protein is found in the sediment when a bacterial extract is centrifuged for 30 min at 100,000 *g*, while the material isolated by Kolber and Stein remains in the supernatant. Treatment of cell membrane material (that is, the 100,000-*g* sediment) with the detergent Triton X-100, leads to the solubilization of the *M*-protein and it is possible that this soluble form of the *M*-protein is related to the material found by Kolber and Stein.

There seems little doubt that the components isolated in the studies of Kolber and the author (A. R. Kolber and Stein, 1966) and of C. F. Fox and Kennedy (1965) are related to the protein expression of the "permease" gene. There is a good deal of uncertainty, however, as to the precise role that these materials can play in β-galactoside transport.

We saw in Section 6.5 (Koch, 1964) that when the energy metabolism of *E. coli* is blocked by sodium azide, the active accumulation of galactosides is also blocked, but the downhill transport of galactoside continues unimpaired. It is this facilitated diffusion of galactoside that is inhibited by *N*-ethyl maleimide. Thus it is the protein concerned in the facilitated diffusion of galactoside rather than their active transport that Fox and Kennedy have labeled with NEM. We have seen that the "permease" isolated by Fox and Kennedy (but not that isolated by Kolber and the author) is found in the sediment when a bacterial extract is centrifuged at 100,000 *g*. This fraction contains also the ribosomes, largely located in the membrane of *E. coli.* Thus the permease is presumably also membrane-located. If the isolated permease is indeed the membrane-situated carrier, the relationship of this protein to the energy input (that is, to ATP splitting) must be investigated. We might ask whether the permease itself splits or uses ATP or whether some second enzyme is concerned in the energy input process. Perhaps it is this second enzyme that is the material identified by Kolber and the author.

That more than one protein may be concerned in such active accumulation is suggested by the experiments of Egan and Morse (1965) on the systems involved in sugar transport in *Staphylococcus aureus.* Lactose uptake by these organisms is an active inducible system. A single gene, *lac,* controls the appearance of such active and inducible uptake of the specific sugar lactose. A second gene, *car,* determines the ability of the bacterium to take up a wide variety of sugars. It appears that the defective mutant, lac^-, loses the ability to accumulate lactose against a concentration gradient but can still take up lactose by the residual facilitated diffusion system. This latter system, however, is absent in the defective mutant, car^-. Thus the gene, *car,* determines the capacity for facilitated diffusion, that is, codes for the carrier, while the gene, *lac,* controls the link with the metabolic energy input. Egan and Morse suggest that the same carrier, coded for by the *car* gene, is shared by a large number of transport systems, the active transport link introducing the substrate specificity. It may be that this interesting model may not prove in its entirety applicable to the β-galactoside system of *E. coli.* This study suggests, however, that the material isolated by Fox and Kennedy may be the expression of a *car* gene, while the material of Kolber and Stein may be more closely related to the energy input link and hence a *lac* gene. Clearly, we can hope for much exciting information to come out of the study of these bacterial systems in the near future. It is the details of the molecular events associated with the binding of β-galactoside with permease and with the (subsequent?) linkage to the metabolic energy input that will have to be carefully explored if the role of these components in membrane transport is to be understood (see Chapter 9).

D. The Isolation of the Sodium- and Potassium-Activated Membrane ATPase

We have seen (Section 6.3) that the active extrusion of sodium and the concomitant accumulation of potassium by the erythrocyte and by the nerve cell are accompanied by the hydrolysis of adenosine triphosphate. Both sodium and potassium ions are necessary if this hydrolysis is to occur. These facts suggested to a number of investigators that the presence in the membrane of an ATPase activated by the combination of sodium and potassium was indicated. The successful demonstration of the presence of such an enzyme in crab nerve cells (Skou, 1957) and in human erythrocytes (Post *et al.*, 1960; Dunham and Glynn, 1961) and the establishment of a close link between the properties of this enzyme and those of cation transport are certainly among the most significant advances made over the last ten years in the field of transport.

The crab nerve enzyme (Skou, 1957), which has a mandatory requirement for magnesium ions, seems to be associated with a submicroscopic fraction of cell homogenates, probably containing disintegrated nerve membrane fragments. In this preparation, the ATPase activity in the absence of both sodium and potassium ions is extremely low but as the potassium ion concentration is varied over a range of values, at a series of fixed sodium ion concentrations (Fig. 8.10), a great stimulation of activity results. (Potassium can be replaced, more or less effectively, by other monovalent cations — Fig. 8.11.) Conversely, as the sodium ion

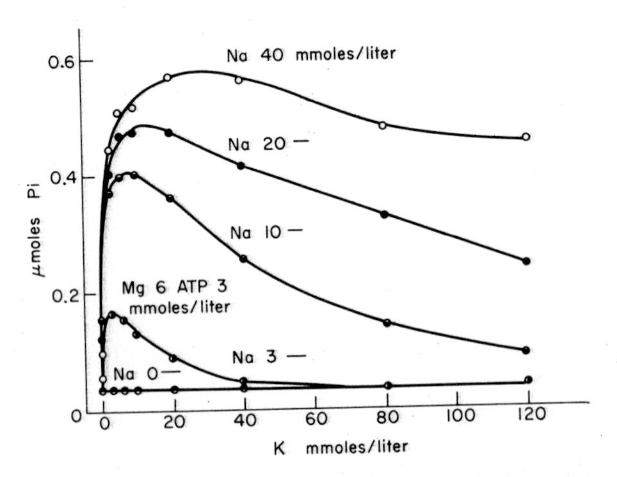

Fig. 8.10. The effect of increasing potassium ion concentrations (abscissa) on the activity of the crab nerve ATPase (ordinate) at a series of different concentrations of sodium ion (designated next to each curve) and in the presence of 6 mmoles/liter of magnesium ion. (Taken with kind permission from Skou, 1964.)

concentration is raised, at a series of potassium ion concentrations, a gradual but striking enhancement in ATPase activity is found. This ATPase activity is greatly inhibited by the addition of the cardiac glycoside *g*-strophanthin, a specific inhibitor of cation transport.

The studies on human erythrocytes (Post *et al.*, 1960; Dunham and Glynn, 1961) have proved more difficult because of the presence in the cells of substantial amounts of an ATPase which is not dependent on the sodium or potassium ions and which is not inhibited by the cardiac

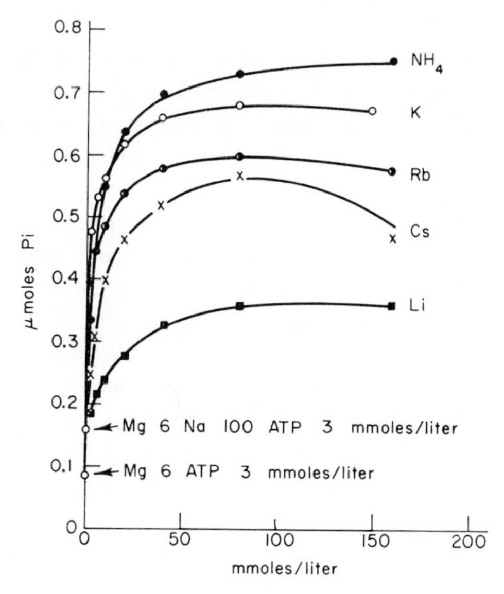

Fig. 8.11. The effect of increasing concentrations of various monovalent cations (abscissa) on the activity of the crab nerve ATPase (ordinate) at a fixed value of the sodium ion concentration (100 mmoles/liter) and magnesium ion concentration (6 mmoles/liter). (Taken with kind permission from Skou, 1964.)

glycosides. The presence of this enzyme leads to a high basal level of ATP splitting which has to be subtracted from the experimentally observed level to obtain a result for the transport-linked activity. On the other hand, the wealth of kinetic information on the cation transport of this cell has enabled a number of significant correlations of enzyme activity and cation transport to be made. Thus the concentration of potassium ion required for half-maximal stimulation of the univalent cation-dependent ATPase activity is 3 m*M*, while that for half-maximal stimulation of sodium efflux is 2.1 m*M*. Similarly, the concentration of sodium half-maximal for enzyme activity is 24 m*M*, for transport some 20 m*M*. Finally, the cardiac glycoside ouabain present at $10^{-7}M$ half-maximally inhibits the enzyme activity while at $3–7 \times 10^{-8}M$, it half-maximally inhibits transport. These data have been taken from Post and Albright (1961). Similar data were obtained by Dunham and Glynn (1961). There appears little doubt from the information assembled by these investigators that cation transport is brought about by the membrane's Na- and K-dependent ATPase. Similar enzymes (ouabain-inhibited and Na- and K-dependent ATPases) have now been found in a wide variety of tissues, in many cases (Bonting *et al.*, 1961) the distribution of enzyme paralleling the intensity of cation transport in the tissue studied (review in Csáky, 1965).

Recent work on these enzymes has consisted of attempts to solubilize and isolate the ATPase from its membranous surroundings. This has so far proved extremely difficult. Skou (1961) found initially that treatment of the particulate enzyme from crab nerve with phospholipase *A* removed the activating effect of sodium and potassium, suggesting that the organization of the particle and especially the specific presence of lipids were essential for the full activity of the system. Similar observations have been made on the erythrocyte system (Schatzmann, 1962). Supporting this, Tosteson and his associates (Tosteson *et al.*, 1964) showed that increasing times of sonication of erythrocytes—which leads to a progressive decrease in the size of the particles released by sonication—were followed by a loss in the sodium and potassium dependency of the membrane ATPase. In contrast to these findings, Wheeler and Whittam (1964) present evidence that the erythrocyte enzyme can be extracted by organic solvents, presumably with the removal of most of the lipid material, while retaining substantial enzyme activity. Further attempts at solubilization, which involved treatment of the extracted stroma with sodium hydroxide at pH 9, led to a complete loss of activity.

Of much interest in this context has been the demonstration by Post *et al.* (1965) that an acceptor group in the erythrocyte stroma is phosphorylated when ATP is split in the presence of Na and K and that the kinetics of phosphorylation parallel those of the membrane ATPase and thus those of cation transport by the erythrocyte. A number of enzymes involved in phosphate transfer similarly incorporate phosphate when incubated with appropriate substrates. For phosphoglucomutase (2.7.5.1) incorporation occurs at a serine residue presumably located at the active center of the enzyme (review in Najjar, 1962). Recently, however, Heinz and Hoffman (1965) and also Skou (1966) have shown that the phosphorylation of membrane proteins is not on the direct pathway of the action of the ATPase in ion translocation. Skou finds that incorporation of phosphate by a highly purified preparation occurs only if K is absent. With potassium present, ATP is indeed split, but without

TABLE 8.8

SITUATIONS IN WHICH SUBSTRATES OR INHIBITORS OF TRANSPORT SITES INTERACT WITH THEM IN A HIGHER-THAN-FIRST-ORDER MANNER

System	Substrate (S) or inhibitor (I)	Comments and references
Erythrocyte, human (Glucose transport)	Dinitrofluorbenzene (I)	Second order (Stein, 1964b)
	Dihydroxyphenols (I)	Some first, some approaching second order (LeFevre, 1961a)
	Phloretin (I)	First order but requires both terminal hydroxyls to be free (Table 8.4) (LeFevre, 1959)
	D-arabinose (S) D-ribose (S) D-fructose (S)	At low substrate levels show a second-order dependence of uptake rate on concentration (Wilbrandt and Kotyk, 1964)
(Glycerol transport)	Dinitrofluorbenzene (I)	Second order (Stein, 1962e)
	Protons (I)	Second order (Stein, 1962c)
	Copper ions (I)	Second order (Stein, 1962c)
Pancreas slices, mouse (Proline uptake)	Pro (S) Gly (I) Met (I) 1-Amino cyclopentane-carboxylic acid (I)	Substrate (and inhibitors) demonstrate a second-order dependence of proline uptake rate on concentration (Bégin and Scholefield, 1965b)
Ascites tumor cells, mouse (Tryptophan)	Met, nor-Leu, nor-Val, His, Leu, Cys (I)	These and some other amino acids stimulate tryptophan uptake with a 2 : 1 stoichiometry (Jacquez, 1963)
Erythrocytes, pigeon (Glycine uptake)	Na^+ (S)	Sodium is co-transported by the system; shows a 2 : 1 stoichiometry with glycine (Vidaver, 1964a)
Rat brain (Na^+-K^+-activated ATPase)	Na^+, K^+, ATP, Mg^{2+} (S)	All show a higher-than-first-order dependence of activity on concentration (Squires, 1965)
Frog muscle (Na^+ efflux)	Na^+ (S)	A third-order dependence of efflux rate on internal Na^+ concentration, H^+ being able to substitute for Na^+ (Keynes, 1965)

the transfer of its terminal phosphoryl group to the enzyme active center. Sen and Post (1966) have found that the incorporation of phosphate into protein is not the first step in the reaction sequence catalyzed by the membrane ATPase, but rather a transient binding of ATP can be demonstrated to occur before the incorporation of phosphate.

Hokin and Hokin (1961), on the basis of ^{32}P incorporation studies, originally suggested that a cyclic phosphorylation and dephosphorylation of the phosphoryl residue of phosphatidic acid were connected with cation transport in the avian salt gland (and in a variety of other tissues) and that phosphatidic acid was indeed the cation carrier. More recent data (Hokin and Hokin, 1963; Glynn *et al.*, 1965) have led the Hokins' to discard the view that phosphatidic acid itself is the membrane carrier but a role for phosphatidic acid turnover during the excitation of secretion in avian salt glands seems likely.

It it clear that it is with the membrane's Na + K-linked ATPase that transport studies are brought closest to the enzyme level. The intense application of scientific energy to the study of the ATPases is most likely in the next few years to lead to an understanding of transport phenomena at the molecular level. In the next and final chapter we shall consider some of the current thoughts on this topic.

8.5 Data Suggesting That a Number of Transport Systems May Be Bivalent toward Substrates or Inhibitors

We have mentioned in a number of places that there are indications that a number of transport systems behave as if they interact with a pair of substrate, inhibitor, or activator molecules. Since current views on the modification of enzymes by substrates and activators (Monod *et al.*, 1965) suggest that such "allosteric" phenomena may be of some mechanistic significance, we collect in Table 8.8 the available data on the bivalency of the transport sites. We might note that Vidaver (1966) has put forward a detailed treatment which accounts for counter-transport on an "allosteric" model, while Wong (1965) and Britton (1966) have provided kinetic treatments of polyvalent carrier transport.

CHAPTER 9

Possible Mechanisms for Mediated Transfer

In this final chapter we shall consider briefly some of the suggestions that have been made over the last ten years as to the detailed molecular mechanisms of mediated transfer, that is, of facilitated diffusion and active transport. These suggestions are, of course, essentially speculative, but to reduce the area available for speculation we intend to preface our catalogue and critique (Section 9.3) by a list of those properties of the transport systems that the models must account for (Section 9.1). We intend also to be as precise as possible in specifying the criteria for a useful model (Section 9.2).

The aim of the next generation of transport studies must be to give a molecular description of the phenomena of transport. This will have to be preceded by an identification and listing of all the molecular species that take part in a given transport and the molecular description will then consist of a detailed analysis of the movements of all these species concerned in the transport event. To further such a study we shall have to be as precise as possible in our statements on mechanism and to phrase these in molecular terms.

9.1 What the Models Have to Account For: A Summary of the Properties of the Mediated Transfer Systems

In the following summary, references are given to the section or sections of this book where the particular point concerned is given detailed treatment.

(a) We shall assume that the major barrier to diffusion is at the cell membrane (Section 3.3).

(b) To a first approximation, the membrane is probably a lipid bilayer, coated by protein (Section 1.2). We cannot exclude the possibility that certain regions of the membrane may consist of globular micelles (Section 1.5). We must accept that in these and other regions, protein pores or plugs may traverse the membrane, forming a continuous nonlipid region from one face of the membrane through to the other face (Sections 3.5 and 3.6).

(c) The permeability of a large number of compounds can be satisfactorily accounted for on the assumption that their penetration occurs across a lipid bilayer (Section 3.3) which is then the major permeability barrier of the cell. It is possible that the penetration of water and of ions occurs at least in part through water-filled pores extending through this lipid bilayer (Section 3.6). A number of other very small permeants may also penetrate through these pores, but most compounds cannot do so by reason of their large size (Section 3.6).

(d) The specialized mediated transfer systems are present in most cell membranes (Chapters 4, 5, and 6). These systems exist as minor components of the membrane and display kinetic properties formally equivalent to saturation phenomena (Sections 4.4, 5.3, and 6.4).

(e) There are a large number of such systems in each cell, each system being more or less specific, many being optically specific (Section 8.1). The systems are separately inhibitible either by substrate analogs or by chemical reagents (Section 8.3).

(f) Many of the inhibitors of transport are reagents attacking the side chains of proteins (Section 8.3).

(g) The membrane alone (appropriately fortified with energy sources) can transport many substrates (Sections 5.2 and 6.3). These transport systems are therefore wholly located in the membrane.

(h) There are two major classes of transport systems: on the one hand, the primary active transports (Chapter 6) which are directly linked to an input of chemical bond energy and are essentially unidirectional (vectorial) and, on the other hand, the class (Chapters 4 and 5) containing the facilitated diffusion systems and the co-transport systems (the secondary active transports).

(i) There is good evidence (Section 4.5) that for these systems, at least a part of the active center for interaction with the substrate must move physically from one membrane interface to the other during transport.

(j) During transport, the substrate is not apparently modified by covalent changes (Section 8.1). This suggests that the interaction between transport system and permeant during transport involves only hydrogen, hydrophobic, and electrostatic bonds—the forces that determine the binding between enzymes and their substrates. The temperature coefficient for mediated transfer is *less* than would be expected for the movements of the substrates across a simple lipid bilayer (Section 4.2).

(k) The phospholipid components of the membrane can bind sugars and cations in a manner reminiscent of the binding of these substances to the transport systems (Section 8.4).

(l) The interactions described in (k) above fall, however, a long way short of accounting for the behavior of transport systems. Although labeled membrane proteins and particles of membranes, having enzymic activity suggestive of their role in cation transport, have been obtained from cell membranes (Section 8.4) and, in addition, the product of the permease gene has been identified, no known "carrier" has been isolated. This situation is, however, changing rapidly (see Sections 8.4,A and 8.4,C).

(m) In a number of cases the kinetics of the interaction of substrates and inhibitors with the transport system suggest that the active centers for transport are bivalent (Section 8.5).

This is a formidable list of properties to be accounted for by a transport model. It must be emphasized that this list brings together information on many different transport systems. Perhaps no single model of transport will or should account for all these properties. In particular, as explained in (h) above, the two general classes comprising mediated transfer might require two quite different types of model for their adequate description. While bearing these precautions in mind, we shall look for a unitary hypothesis at this stage and until it is shown that a more complex scheme is necessary. It must be emphasized that a number of the above statements are made here more emphatically than the evidence in truth permits. In particular, (a), (h), and (j), which should be the strongest, may prove on further analysis to be misleading.

9.2 The Criteria of Acceptability for Model Systems of Transport

As general criteria we can insist that an adequate model for transport must:

(1) be molecular;

(2) list all the molecular species concerned in the transport;

(3) account for the properties listed in Section 9.1 and in particular account for the specific interaction that occurs between the substrate and the transport system; that is, it must account for the binding and for the specificity of this binding. Eventually, although this might require an X-ray crystallographic investigation of the interaction between substrate and transport system, a listing of those groups at the active center of the transport system which are concerned in binding will be required for a complete understanding of specific binding. This is essentially a problem within the sphere of

"mechanism of enzyme action" studies insofar as the binding of enzyme and substrate is a part of such a study;

(4) be physicochemically plausible within the context of the structure of the cell membrane. The model must make reasonable demands as to the structure of this membrane, and the molecular movements of substrate and transport system that are postulated by the model must be such as to be within the known capabilities of such species. It is possible, however, that the physical chemist may have to learn from those studying transport phenomena of certain new properties of molecules or molecular arrays;

(5) be energetically acceptable. In particular this applies to models for the primary transport systems, but it is relatively easy to build unacceptable models for the secondary transports, which contradict the second law of thermodynamics;

(6) be testable. Ideally a model should lead to predictable hypotheses, preferably quantitative, which could be used to test the model and hence lead to its rejection or temporary acceptance.

These criteria are sufficiently general to be hardly disputable. It is, however, their application to the available models that may lead to controversy.

Of the particular problems that a model for transport must solve we should emphasize first the elucidation of the details of the movement of the transported species through the membrane. It has to be established whether this movement occurs through the lipid region of the membrane (does the diffusion of the transported species take place between the hydrocarbon chains of the lipids comprising the membrane?) or whether this movement occurs through nonlipid regions, that is, by diffusion across a hydrophilic surface or through a hydrophilic region, perhaps the interior of a membrane protein. This is an absolutely crucial question which has to be answered before an acceptable description of transport can be provided. Does the transported species move in combination with some other molecule—a "carrier"—or does it move over the surface of a fixed membrane component? Property (i) of Section 9.1 would have that at least at some stage of the transport process there has to be an interface-to-interface movement of the transporting as well as the transported species. Our question then is best asked in two parts: (a) How far does the transporting species move—from membrane surface to membrane surface (some 50 Å) or only over some critical distance (say, a few angstroms) within the membrane? (b) Is the "movement" a change in the position of some atomic groupings, with respect to the membrane as the fixed axes, or is it a movement of electrons—merely a movement of an "affinity site?" Either of these would give an acceptable explanation of property (i) in Section 9.1.

It is essential for us to keep asking these questions of any proponent of a particular model. Although no easy method of experimentally testing the answers to these questions is yet available, by casting our questions in such molecular terms, we shall be able in time to reach the necessary precision that a satisfactory model must achieve. In the process of asking and answering these questions, the role and relevance of the contemporary attack on the problem of the mechanism of enzyme action will become clearer, and many of the techniques found to be useful in the solution of the enzyme problem will be found to be applicable in the context of transport studies.

The second major problem for our transport studies is the elucidation of the details of the coupling of the input of chemical bond energy into the primary active transport systems. How does the splitting of ATP by the membrane ATPase ensure that a coupled interchange of sodium and potassium ion takes place across the membrane? Is this a forced movement of the "carrier" akin perhaps to the relative movements between actin and myosin that occur during muscular contraction? If so, we will have to provide a description of how the splitting of ATP brings about these contractile movements, a problem that has not yet been completely solved for muscle. Again, we are entitled to ask whether a large part of the membrane moves during this process of ion pumping, or whether merely some subtle movement of an affinity center (and a change in that affinity) is sufficient for transport. Is the "movement," once again, electronic or atomic? It will be important to establish whether there is an obligatory phosphorylation of the active center of the transport system during ATP hydrolysis and what significance this phosphorylation may have in transport. It is very probable that the mechanisms for the primary transport of cations and of metabolites (for example, the sugar permeases) will prove to be distinct and that the latter mechanisms will be found to be far closer to the mechanisms for facilitated diffusion and co-transport.

It will probably be important to keep distinct the two questions of the detailed movements of the transporter and of the details of the energy input. Each question has its own peculiar difficulties and is probably best solved within its own context. The mechanism of the energy coupling may prove to be intimately concerned with facilitating or preventing particular movements of the transporter within the membrane, but the detailed analysis of these movements need not accompany the demonstration that the modulation of specific movements is indeed the mechanism of energy input.

If the above paragraphs are read, bearing in mind the material presented in the preceding chapters, it is clear that we are a long way from being able to propose an acceptable model (in the specific sense used here) for transport. A brief review of the models that have been proposed will make clear where lie the deficiencies in our knowledge of transport.

9.3 Models for Mediated Transfer

Our emphasis on the cell membrane as the diffusion barrier in cellular transport precludes us from considering those models for mediated transfer which propose that the cell interior (Troshin, 1961; Ling, 1962; E. J. Harris and Prankerd, 1957; F. H. Shaw *et al.*, 1956) or else a gel-like layer beneath the membrane (Miller, 1960) is responsible for the specific transport phenomenon. A critique of these views has been presented by Glynn (1959), and a debate on this topic is recorded in the report of the Prague Symposium (Kleinzeller and Kotyk, 1961).

The membrane-based models can be classified perhaps arbitrarily into four classes: (a) the mobile carrier models, (b) the pore models, and, finally, two classes of enzyme models (c) those based on superficial enzymes, and (d) those based on enzymes located within the membrane. We shall need to discuss these in turn and in detail and in the context of Sections 9.1 and 9.2.

A. Mobile Carrier Models

By a mobile carrier most people would imply a transport component which forms a complex with the substrate, the whole complex diffusing across the membrane. In their very full and erudite review of membrane transport, Wilbrandt and Rosenberg (1961) marshal much of the evidence in favor of the mobile carrier model. The interaction between carrier and substrate may be catalyzed by an enzyme present at either or both surfaces of the membrane. (We discuss below the difficulty that this assumption raises.) The specificity of interaction between substrate and carrier may reside in the enzyme component, in which case specificity is easily accounted for by invoking the similar properties of well-known enzymes, or else may reside in the carrier itself, a somewhat less plausible hypothesis if the carrier must yet be small enough to diffuse through the membrane. The carrier-substrate complex traverses the membrane through the lipid (that is, between the lipid side chains) as a result of simple diffusion. The complex must on this assumption behave as a simple permeant—it must be small and must not have many free hydrogen-bonding groups to anchor it in the aqueous phase. The mobile carrier model specified here in such detail is probably not seriously accepted by any investigator in the field, for two main reasons. First, although lipid-soluble derivatives or complexes of glucose, amino acids, glycerol, and the cations are known, no serious contender for the role of carrier has been found. Second, if it is held likely that the carrier must be small and held unlikely that a small carrier would possess the necessary specificity for binding, the consequent necessity to postulate a membrane-located enzyme raises a serious difficulty. If it is this enzyme that is inhibited by the protein reagents which are inhibitors of transport then, since a nonpenetrating inhibitor serves to inhibit both entrance and exit (Section 4.5,A), the carrier must interact with the enzyme both during the formation of the complex with the substrate and during the breakdown of this complex. It is most unlikely that by the random process of thermal diffusion, a carrier-substrate complex can find its way directly, or at all, to the "exit" enzyme after it has passed through the membrane—unless its passage through the membrane is constrained within a "pore." If such a "pore" is required, the essential simplicity of the carrier model is lost.

A detailed description of the mobile carrier model raises some serious difficulties that this model has to overcome. A "carrier" may still, however, be ultimately found to provide the most acceptable model for transport. In this context the studies of, for instance, Gaines (1960) on the overturning of stearic acid molecules in spread monolayers may well prove to be of great relevance.

B. Pore Models

This type of model in some of its forms makes fewer demands on novel properties of membrane constituents than does the carrier concept, and it has therefore been supported by a number of investigators. The assumption here is that a hydrophilic region extends through the membrane from interface to interface. Substrates move (diffuse) through this region rather than between the hydrophobic side chains of the lipid molecules bordering the pore. The specificity of transport depends on the specific binding properties of the (enzyme-like) proteins which line or form the pore. Danielli with the author (Stein and Danielli, 1956), and also Zierler (1961), have shown how the formal equations describing saturation can be obtained from a pore model. Adair (1956) has suggested how the substrate itself may induce a pore to open, allowing the specific translocation of a substrate. Burgen (1957) has described oscillating pores that can produce active transport and Patlak

(1956, 1957) a reorienting pore with similar capacities. Skou (1964) has accounted for cation transport by a membrane ATPase which forms such a pore.

The simple pore models of Stein and Danielli and of Zierler are now held in disfavor since they fail to account adequately for the mobility of the transport sites. Only if the pores are unidirectional can the competitive exchange diffusion and counterflow phenomena be accounted for, and such unidirectional pores must be rejected as failing to accommodate the data on the action of nonpenetrating inhibitors and on the phenomenon of accelerative exchange diffusion.

The more complex "pores" such as those postulated by Burgen (Fig. 9.1) make very severe demands on membrane structure and are, for

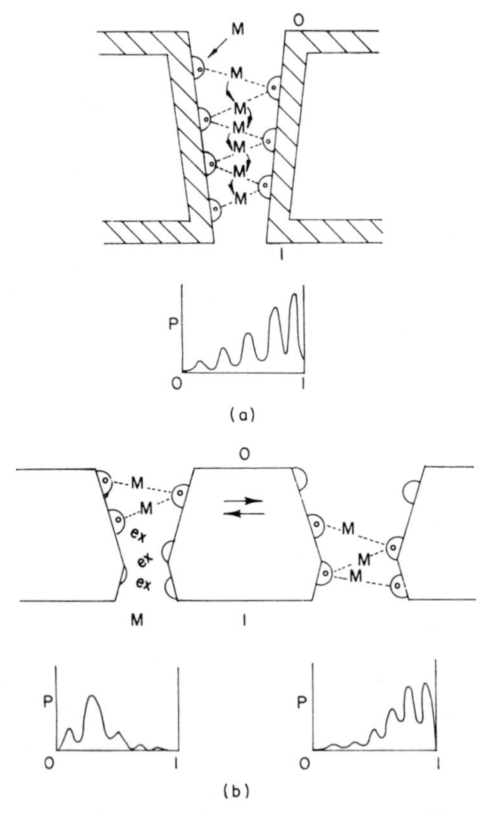

Fig. 9.1. Hypothetical models of pores in the cell membrane: (a) a conical matrix with, below, the probability distribution of molecules within this matrix; (b) an oscillating matrix in the contracted state (left) or expanded state (right) with, below, probability distribution of the molecules. *M* are the permeant molecules, *O* are active center sites on the pore. (Taken with kind permission from Burgen, 1957.)

that reason, less acceptable than some of the enzyme models to be discussed below.

C. Superficial Enzymes

These must be discussed only for the sake of completeness. The author (Stein, 1962c) has proposed that at each face of the membrane is an enzyme-like protein termed a "dimeriser" which has the property of enabling a pair of substrate molecules to interact with one another in such a way that each becomes the carrier for the other, the two molecules mutually satisfying all their hydrogen-bonding capacities. This model, attractive at first and accounting for the property of bivalence possessed by a number of transport systems, must be rejected since it fails to account for the inhibiting action on efflux of the nonpenetrating inhibitors, a phenomenon observed in the glycerol and glucose facilitated diffusion systems for which the model was proposed (Section 4.7). A similar objection must be raised against any model of this type, and also those models which involve an enzyme-catalyzed interaction between permeant and the phospholipid of the membrane, to account for the solubilization of the permeant in the lipid. This general argument was also raised above in the discussion on the "mobile carrier."

A new dimension has been added to the concept of superficial enzymes by Crane's (1966) suggestion that enzymes situated at, say, the outer surface of the membrane may, if the membrane in this region undergoes a localized thinning, be made available at the inner surface. Figure 9.2 depicts the sort of model that Crane is proposing. It must at once be stated that the lipid bilayer which we depicted in Fig. 1.8 is an idealized picture and presumably far too rigid to be an accurate model of the membrane. It is quite conceivable that thermal agitation of the membrane's components may bring about a situation where a local thinning of the membrane, as is required by Fig. 9.2, can occur. Detailed calculations need to be made, however, to test this ingenious hypothesis. The major evidence for Crane's view comes from studies suggesting an association between the various protein components of the membrane. For instance, enzymes hydrolyzing sugars are present at the external surface of the cells lining the intestine (see Crane, 1966). These enzymes appear to be intimately linked to the glucose transport system, since the glucose liberated by sugar hydrolysis is more rapidly transported across the cell membrane than is glucose added to the cell exterior. Another example of possible interaction derives from data suggesting that the maximum velocity of transport is the same for many sugars as it is for amino acids. This would fit the model of Fig. 9.2 in that a localized thinning may bring together at one time a relatively substantial portion of the outer and inner surfaces containing a number of carriers, allowing the simultaneous transport of both amino acids and sugars.

Clearly other interpretations of these findings are possible so that there is as yet no direct indication of the correctness or otherwise of the model of Fig. 9.2. Yet this is a stimulating suggestion, and experiments to test this scheme would be well worth designing. The model of

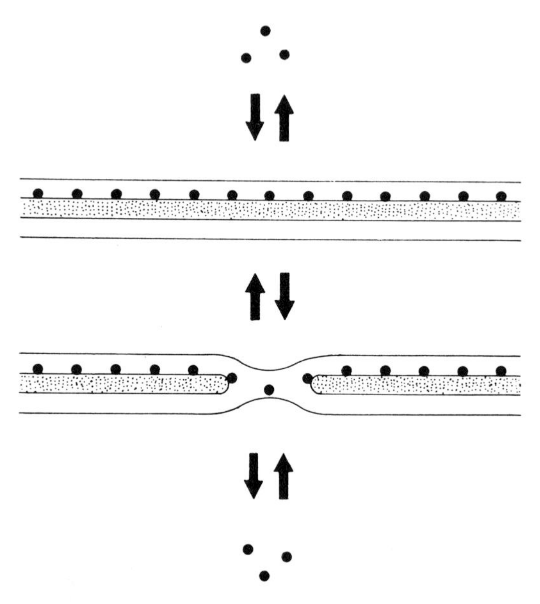

Fig. 9.2. Model of the mobile membrane concept of Crane. ●, Molecules (permeants) adsorbed to the membrane exterior and later released to the interior. The stippled region is the bimolecular lipid layer. (Taken with kind permission from Crane, 1966.)

Fig. 9.2 appears to be compatible with many of the criteria for mediated transfer. Thus it can account for the requirement that the equilibria between "carrier" and substrate at the inner and outer surfaces be shielded from one another, by proposing that these equilibria take place while the lipid bilayer insulates the two protein faces of the membrane. Yet the ability of nonpenetrating inhibitors to inhibit both exit and entrance of permeants can be accounted for by the flow of "carrier" that occurs during a period of local membrane thinning, when both inner and outer "carriers" are available to the external inhibitor.

D. Transport "Enzymes" As Part of the Cell Membrane

In a brilliant series of essays over a number of years, P. Mitchell (1957, 1961) has propounded the view that the phenomena of mediated transport are due to the action of enzymes situated within (in fact, part of) the cell membrane and that the action of these enzymes differs little, if at all, from their action during the processes that occur in enzyme catalysis. This view has slowly gained wider acceptance and must surely come to dominate the future development of transport studies. In this view, it is assumed that a part of the cell membrane consists of plugs of protein. These may be the normal enzymes of cellular metabolism or may be the membrane's sodium- and potassium-dependent ATPase or the ATPases that might be concerned in the primary transport of metabolites, or they may be specialized "enzymes" concerned only with transport. In this latter case the enzymic function is possibly very unlike those of normal enzymes in that, as we have seen, no covalent bond need be formed or split during the mediated transfer of certain substrates. In certain cases, Mitchell has shown how the normal enzymes of metabolism, if located in an appropriate fashion in the membrane (and if these enzymes are assumed to have certain specialized active center properties), may be used for transport of, for instance, phosphate across the membrane (see Fig. 9.3) (P. Mitchell, 1957). Keston (1964a,b) has presented evidence that the enzyme mutarotase (5.1.3.3) may be the active species in intestinal sugar transport.

We shall take rather the view—conceded, but not favored, by Mitchell (see the discussions in Kleinzeller and Kotyk, 1961)—that in many cases specialized enzymes are concerned in transport. One can then make rather detailed suggestions as to the properties of such enzymes. Thus in order that they be anchored in the membrane they must, as do the protein moieties of the lipoproteins, make firm bonds with the membrane lipids. The transport enzymes may well be lipoproteins or even proteolipids (Folch and Lees, 1951). Such enzymes therefore are likely to have an amino acid composition reflecting a high proportion of nonpolar side chains (see Hatch, 1965). This outer shell of nonpolar groupings might then enclose a hydrophilic interior—in direct contrast to the structure of those water-soluble globins that have been analyzed by X-ray crystallography, where a hydrophobic interior is surrounded by hydrophilic groupings (Kendrew, 1962). The hydrophilic interior of the protein plug would then be the pore through which the substrate must pass. Diffusion through the structure of the plug will occur by simple thermal movements until the active center of the enzyme is reached. Here bond formation between the specific substrate and the binding groups of the enzyme can occur, a normal process of chemical equilibration. It is the subsequent thermally induced movement of this active center that is the transport event. What is being postulated here is a substrate-induced change in the conformation of the enzyme—a postulate enunciated by Koshland (1960) in order to explain the specificity of action of certain hydrolytic enzymes. After the internal movement of the (active center plus substrate) complex, the substrate enters into chemical equilibrium with the medium on the opposite face of the

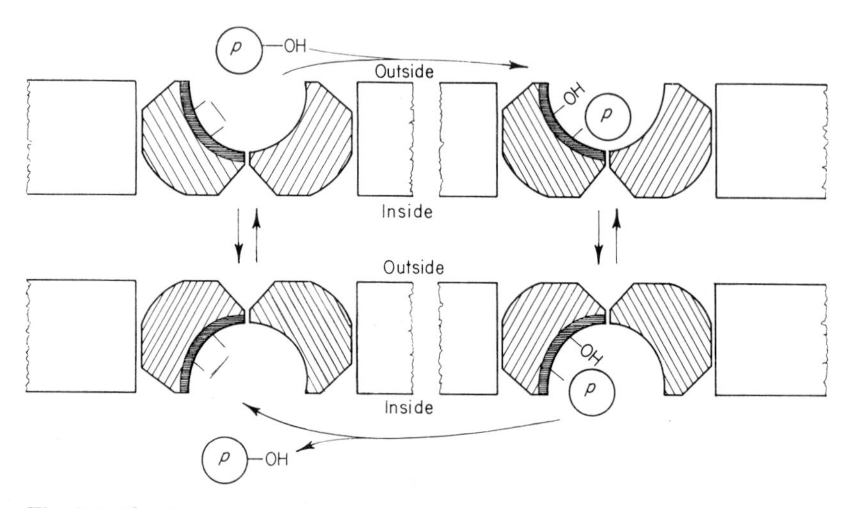

Fig. 9.3. The "translocase" model of Mitchell, shown here as catalyzing the translocation of a phosphoryl anion, the ringed *P*, from the outside to the inside of the cell. The hatched areas represent the "translocase," a membrane-bound "enzyme" which is available alternately to the inner and outer faces of the membrane, whether free or in combination with the permeant. (Redrawn from Mitchell, 1957.)

membrane. It then dissociates from the enzyme active center and diffuses through the plug to the aqueous phase. The protein plug is at once the barrier to the movement of those molecular species that cannot combine with the specific active center grouping, and the vehicle of movement for those species that can activate the substrate-induced conformational change in this enzyme.

Stated thus, such a model would seem to be a very reasonable explanation of the mechanism of action of the facilitated diffusion and co-transport systems. It can be extended to cation co-transport with little difficulty as there are a number of possibilities whereby the energy input can so couple as preferentially to allow ingress rather than egress through the membrane enzyme. Thus Skou (1966) suggests that the binding of ATP to the membrane ATPase changes the affinity of one site of the enzyme for potassium, while changing in addition the affinity at some other site for sodium. On splitting of the ATP, subsequent to the binding of the cations in the correct positions, a rearrangement of the protein molecule enables the sodium to leave and the potassium to enter the cell, after which the ATPase relaxes to its original condition. Of the properties listed in Section 9.1, many seem to be peculiarly favorable to an interpretation on the basis of such models. Thus, for example, the bivalency properties are very reminiscent of those found in allosteric enzymes (Monod *et al.*, 1965) where a substrate- or modifier-induced conformational change in the enzyme has been postulated. The validity of these models depends heavily on the currently fashionable but not, therefore, necessarily correct, view that the active centers of enzymes are not fixed but are motile, their motility playing a direct role in their catalytic function. As to the tests of this model, these will have to take the form of the direct isolation of transport enzymes and the demonstration by physicochemical methods that, on interaction with their specific substrates, conformational changes of sufficient magnitude to account for the phenomenon of transport can occur.

9.4 The Future of Transport Studies

It would be extremely rash—and a source of much amusement for anyone who happened to pick up this book in ten years' time—to attempt to forecast the future development of transport studies. Nevertheless, it may be important to indicate those areas and methods which, in the author's belief, are likely to prove valuable in the future. It has been the aim of this book to summarize the achievements of transport analysis thus far and to show the extent to which the dominant kinetic approach has illuminated those areas to which it has been applied. It has been a source of much satisfaction to the investigators in this field to discover how many transport systems exist and how many of them can be satisfactorily accounted for by the more-or-less simple treatment of the Michaelis–Menten equation. Nevertheless, we must agree with Christensen (1961a) when he writes: "I hesitate to say how dismaying I find a situation where the description of just one more transport [system] meeting the usual conditions may be treated as a major gain. I wonder whether we are not being a little too discouraged in holding that this is not yet the time to investigate whether we are dealing with moving micromolecules, or instead with rotating, relaxing or contracting macromolecules. I believe that we may with advantage at this time apply some good protein chemistry to these exposed sites on cells." We must agree, too, with Danielli (in Stein and Danielli, 1956) that "the urgent requirement in this field is the isolation of the effective membrane components."

It is clear that transport must advance on two fronts. It must become, on the one hand, "applied transport"—applied to the solution of physiological and clinical problems. As has been so successful in the past, the analysis of kidney function, digestion, secretion, hormone action, and the action of relevant drugs and toxins must continue to be pursued on the basis of the current picture of transport. This is Wilbrandt and Rosenberg's (1961) "Carrier physiology and carrier pharmacology." In addition, transport must become molecular. It must use the methods and approach of enzyme mechanism studies in an attempt to provide a molecular basis for transport. Each statement on membrane function or structure—the microanatomy of the membrane, the processes of simple diffusion, facilitated diffusion, co-transport, and primary active transport—must be put in molecular terms, the molecular components identified and listed, and the totality of molecular movements analyzed and described.

General statements, such as "diffusion through pores" or "transport by shuttling carriers" will not prove adequate in future studies unless the components of the pores are specified, their composition determined, and their location in the membrane ascertained; while the carriers, too, must be identified and the notion of "shuttling" given an interpretation in terms of the details of the movement of the "shuttling" components. We must use the methods of classical enzymology to aid the isolation of the membrane's transport systems, realizing that we may isolate only a part of the complete system, but learning from the enzymologist how to reconstruct the whole system from the fragmented residues of the membrane. In this regard the model membrane systems are likely to prove invaluable. We must use the methods of those analyzing the mechanism of enzyme action to study the interaction between substrate and transport site, realizing that a detailed understanding of the binding events will probably have to await an X-ray crystallographic analysis but that the demonstration of substrate binding by isolated membrane components will be the problem that is the particular concern of the transport investigator. The techniques of the physical biochemist will enable us to demonstrate any possible substrate-induced conformational changes in the isolated transport systems. We will surely need to study the movement of transport sites, using optical rotation, depolarization of fluorescence, and light-scattering techniques. We shall be able to identify the chemical groupings at the active centers of the transport systems using methods based on those of the protein chemists. Finally or, far better, simultaneously, we shall have to study transport at the level of the electron microscope, using this instrument to locate any transport components in the cell membrane. Electron cytochemistry will be invaluable here and, indeed, far earlier, to establish in the initial stages whether the continuous lipid bilayer is the true structure of the living cell membrane.

These are exciting fields—the exploration of the ramifications of transport in the physiology and pathology of man and other animals and of plants, and the exploration of the molecular basis of transport. New methods and new approaches will, of course, have to be developed to aid these explorations. In the meantime, there is an abundance of work to do and of problems to solve by available methods which have not yet been applied.

References

Abrams, A. (1960). Metabolically dependent penetration of oligosaccharides into bacterial cells and protoplasts. *J. Biol. Chem.* **235**, 1281.

Adair, G. S. (1956). Discussion. *Discussions Faraday Soc.* **21**, 285–286.

Adamson, A. W. (1960a). "Physical Chemistry of Surfaces," Chapter I. Wiley (Interscience), New York.

Adamson, A. W. (1960b). "Physical Chemistry of Surfaces," Chapter III. Wiley (Interscience), New York.

Adrian, R. H. (1956). The effect of internal and external potassium concentration on the membrane potential of frog muscle. *J. Physiol.* (*London*) **133**, 631–658.

Ahmed, K., and Judah, J. D. (1964). Preparation of lipoproteins containing cation-dependent ATPase. *Biochim. Biophys. Acta* **93**, 603–613.

Akedo, H., and Chistensen, H. N. (1962a). Transfer of amino acids across the intestine: A new model amino acid. *J. Biol. Chem.* **237**, 113–117.

Akedo, H., and Christensen, H. N. (1962b). Nature of insulin action on amino acid uptake by the isolated diaphragm. *J. Biol. Chem.* **237**, 118–122.

Alvarado, F., and Crane, R. K. (1962). Phlorizin as a competitive inhibitor of the active transport of sugars by hamster small intestine *in vitro*. *Biochim. Biophys. Acta* **56**, 170–172.

Alvarado, F., and Crane, R. K. (1964). Studies on the mechanism of intestinal absorption of sugars. VII. Phenylglycoside transport and its possible relationship to phlorizin inhibition of the active transport of sugars by the small intestine. *Biochim. Biophys. Acta* **93**, 116–135.

Andersen, B., and Ussing, H. H. (1957). Solvent drag on non-electrolytes during osmotic flow through isolated toad skin and its response to antidiuretic hormone. *Acta Physiol. Scand.* **39**, 228–239.

Anfinsen, C. B., and White, F. H. (1961). The ribonucleases: Occurrence, structure and properties. *In* "The Enzymes" (P. D. Boyer, H. Lardy, and K. Myrbäck, eds.), 2nd ed., Vol. 5, pp. 95–122. Academic Press, New York.

Annegers, J. H. (1964). Intestinal absorption of hexoses in the dog. *Am. J. Physiol.* **206**, 1095–1098.

Austin, G., Sato, M., and Longuet-Higgins, H. C. (1966). Water permeability in *Aplysia* neuronal membrane. *Abstr. Am. Biophys. Soc. 10th Ann. Meeting, Boston, 1966* p. 130 [*Biophys. J.* 6 (1966)].

Autilio, L. A., Norton, W. T., and Terry, R. D. (1964). The preparation and some properties of purified myelin from the central nervous system. *J. Neurochem.* **11**, 17–27.

Avigad, G. (1960). Accumulation of trehalose and sucrose in relation to the metabolism of α-glucosides in yeasts of defined genotype. *Biochim. Biophys. Acta* **40**, 124–134.

Baker, P. F., Hodgkin, A. L., and Shaw, T. I. (1961). Replacement of the protoplasm of a giant nerve fibre with artificial solutions. *Nature* **190**, 885–887.

Bangham, A. D., Standish, M. M., and Watkins, J. C. (1965). Diffusion of univalent ions across the lamellae of swollen phospholipids. *J. Mol. Biol.* **13**, 238–252.

Barnard, E. A., and Stein, W. D. (1959). The histidine residue in the active centre of ribonuclease. I. A specific reaction with bromoacetic acid. *J. Mol. Biol.* **1**, 339–349.

Barton, T. C., and Brown, D. A. J. (1964). Water permeability of the foetal erythrocyte. *J. Gen. Physiol.* **47**, 839–849.

Battaglia, F. C., Hellegers, A. E., Meschia, G., and Barron, D. H. (1962). *In vitro* investigations of the human chorion as a membrane system. *Nature* **196**, 1061–1063.

Beament, J. W. L. (1964). The active transport and passive movement of water in insects. *Advan. Insect Physiol.* **2**, 67–130.

Bégin, N., and Scholefield, P. G. (1965a). The uptake of amino acids by mouse pancreas *in vitro*. II. The specificity of the carrier systems. *J. Biol. Chem.* **240**, 332–337.

Bégin, N., and Scholefield, P. G. (1965b). The uptake of amino acids by mouse pancreas *in vitro*. III. The kinetic characteristics of the transport of L-proline. *Biochim. Biophys. Acta* **104**, 566–573.

Bender, M. L., and Kézdy, F. J. (1965). Mechanism of action of proteolytic enzymes. *Ann. Rev. Biochem.* **34**, 49–76.

Bihler, I., and Crane, R. K. (1962). Studies on the mechanism of intestinal absorption of sugars. V. The influence of several cations and anions on the active transport of sugars, *in vitro*, by various preparations of hamster small intestine. *Biochim. Biophys. Acta* **59**, 78–93.

Bihler, I., Hawkins, K. A., and Crane, R. K. (1962). Studies on the mechanism of intestinal absorption of sugars. VI. The specificity and other properties of Na^+-dependent entrance of sugars into intestinal tissue under anaerobic conditions, *in vitro*. *Biochim. Biophys. Acta* **59**, 94–102.

Blank, M. (1965). A physical interpretation of the ionic fluxes in excitable membranes. *J. Colloid Sci.* **20**, 933–949.

Blank, M., and Britten, J. S. (1965). Transport properties of condensed monolayers. *J. Colloid Sci.* **20**, 789–800.

Blecher, M. (1966). On the mechanism of action of phospholipase A and insulin on glucose entry into free adipose cells. *Biochem. Biophys. Res. Commun.* **23**, 68–74.

Bobinski, H. (1966). Personal communication.

Bobinski, H., and Stein, W. D. (1966). Isolation of a glucose-binding component from human erythrocyte membranes. *Nature* **211**, 1366–1368.

Bockris, J. O'M. (1949). Ionic solvation. *Quart. Rev. Chem. Soc.* **3**, 173–180.

Boezi, J. A., and de Moss, R. D. (1961). Properties of a tryptophan transport system in *Escherichia coli*. *Biochim. Biophys. Acta* **49**, 471–484.

Bonsall, R. B., and Hunt, S. (1966). Solubilization of a glucose-binding component of the red cell membrane. *Nature* **211**, 1368–1370.

Bonting, S. L., Simon, K. A., and Hawkins, N. M. (1961). Studies on sodium-potassium-activated adenosine triphosphatase. I. Quantitative distribution in several tissues of the cat. *Arch. Biochem. Biophys.* **95**, 416–423.

Bower, B. F. (1964). Site of cardiac glycoside inhibition of cation transport. *Nature* **204**, 786.

Bowyer, F. (1954). The passage of glucose and glycerol across the red cell membrane. *Nature* **174**, 355.

Bowyer, F. (1957). The kinetics of the penetration of non-electrolytes into the mammalian erythrocyte. *Intern. Rev. Cytol.* **6**, 469–511.

Bowyer, F., and Widdas, W. F. (1955). Erythrocyte permeability to erythritol. *J. Physiol. (London)* **128**, 7P–8P.

Bowyer, F., and Widdas, W. F. (1956). The facilitated transfer of glucose and related compounds across the erythrocyte membrane. *Discussions Faraday Soc.* **21**, 251–258.

Bowyer, F., and Widdas, W. F. (1958). The action of inhibitors on the facilitated hexose transfer system in erythrocytes. *J. Physiol. (London)* **141**, 219–232.

Boyd, G. E., and Soldano, B. A. (1954). Self-diffusion of cations in and through sulfonated polystyrene cation-exchange polymers. *J. Am. Chem. Soc.* **75**, 6091–6099.

Bradbury, M. W. B., and Davson, H. (1964). The transport of urea, creatinine and certain monosaccharides between blood and fluid perfusing the cerebral ventricular system of rabbits. *J. Physiol. (London)* **170**, 195–211.

Branton, D. (1966). Fracture faces of frozen membranes. *Proc. Natl. Acad. Sci.* **55**, 1048–1056.

Bricker, N. S., Biber, T., and Ussing, H. H. (1963). Exposure of the isolated frog skin to high potassium concentrations at the internal surface. I. Bioelectric phenomena and sodium transport. *J. Clin. Invest.* **42**, 88–99.

Britton, H. G. (1963). Induced uphill and downhill transport: Relationship to the Ussing criterion. *Nature* **198**, 190–191.

Britton, H. G. (1964). Permeability of the human red cell to labelled glucose. *J. Physiol. (London)* **170**, 1–20.

Britton, H. G. (1966). Fluxes in passive, monovalent and polyvalent carrier systems. *J. Theoret. Biol.* **10**, 28–52.

Burgen, A. S. V. (1957). The physiological ultrastructure of cell membranes. *Can. J. Biochem. Physiol.* **35**, 569–575.

Burger, M., and Hejmová, L. (1961). Uptake of metabolizable sugars by *Saccharomyces cerevisiae*. *Folia Microbiol. (Prague)* **6**, 80–85.

Burger, M., Hejmová, L., and Kleinzeller, A. (1959). Transport of some mono- and di-saccharides into yeast cells. *Biochem. J.* **71**, 233–242.

Butler, J. A. V. (1951). "Chemical Thermodynamics," 4th ed. Macmillan, New York.

Caldwell, P. C., Hodgkin, A. L., Keynes, R. D., and Shaw, T. I. (1960). Effects of injecting energy-rich phosphate compounds on the active transport of ions in the giant axons of *Loligo*. *J. Physiol. (London)* **152**, 561–590.

Cecil, R., and Thomas, M. A. W. (1965). Nature of the unreactive sulphydryl groups in human haemoglobin. *Nature* **206**, 1317–1321.

Cereijido, M., and Curran, P. (1965). Intracellular electrical potentials in frog skin. *J. Gen. Physiol.* **48**, 543–557.

Cereijido, M., Herrera, F. C., Flanigan, W. J., and Curran, P. F. (1964). The influence of Na concentration on Na transport across frogskin. *J. Gen. Physiol.* **47**, 879–893.

Chan, S. S., and Lotspeich, W. D. (1962). Comparative effects of phlorizin and phloretin on glucose transport in the cat kidney. *Am. J. Physiol.* **203**, 1975–1979.

Chance, B. (1952). The identification of enzyme-substrate compounds. *In* "Modern Trends in Physiology and Biochemistry" (E. S. Guzman Barrón, ed.), pp. 25–46. Academic Press, New York.

Chen, L., and LeFevre, P. G. (1965). Disaccharide specificity of human red cell glucose transport system. *Federation Proc.* **24**, 465.

Chinard, F. P. (1952). Derivation of an expression for the rate of formation of glomerular fluid. Applicability of certain physical and physicochemical concepts. *Am. J. Physiol.* **171**, 578–586.

Chirigos, M. A., Greengard, P., and Udenfriend, S. (1960). Uptake of tyrosine by rat brain *in vivo*. *J. Biol. Chem.* **235**, 2075.

Christensen, H. N. (1960). Reactive sites and biological transport. *Advan. Protein Chem.* **15**, 239–314.

Christensen, H. N. (1961a). Transport by membrane-bound sites or by free shuttling carriers? *In* "Membrane Transport and Metabolism" (A. Kleinzeller and A. Kotyk, eds.), pp. 470–478. Academic Press, New York.

Christensen, H. N. (1961b). Discussion. *In* "Membrane Transport and Metabolism" (A. Kleinzeller and A. Kotyk, eds.), p. 505. Academic Press, New York.

Christensen, H. N. (1962). "Biological Transport." Benjamin, New York.

Christensen, H. N. (1964a). Personal communication.

Christensen, H. N. (1964b). Relations in the transport of β-alanine and the α-amino acids in the Ehrlich cell. *J. Biol. Chem.* **239**, 3584–3589.

Christensen, H. N. (1966). Methods for distinguishing amino acid transport systems of a given cell or tissue. *Federation Proc.* **25**, 850–853.

Christensen, H. N., Riggs, T. E., Fischer, H., and Palatine, I. M. (1952a). Amino acid concentration by a free cell neoplasm: Relations among amino acids. *J. Biol. Chem.* **198**, 1–15.

Christensen, H. N., Riggs, T. E., Fischer, H., and Palatine, I. M. (1952b). Intense concentration of α,γ-diaminobutyric acid by cells. *J. Biol. Chem.* **198**, 17–22.

Christensen, H. N., Riggs, T. E., and Ray, N. E. (1952c). Concentrative uptake of amino acids by erythrocytes *in vitro*. *J. Biol. Chem.* **194**, 41–51.

Christensen, H. N., Riggs, T. E., Aspen, A. J., and Mothon, S. (1956). Concentrative uptake of amino acids by free-cell neoplasms. *Ann. N.Y. Acad. Sci.* **63**, 983–987.

Christensen, H. N., Akedo, H., Oxender, D. L., and Winter, C. G. (1962). Permeability and amino acid transport. On the mechanism of amino acid transport into cells. *In* "Amino Acid Pools" (J. T. Holden, ed.), pp. 527–538. Elsevier, Amsterdam.

Christie, G. S., Ahmed, K., McLean, A. E. M., and Judah, J. D. (1965). Active transport of potassium by mitochondria. I. Exchange of K^+ and H^+. *Biochim. Biophys. Acta* **94**, 432–440.

Cirillo, V. P. (1961). The transport of non-fermentable sugars across the yeast cell membrane. *In* "Membrane Transport and Metabolism" (A. Kleinzeller and A. Kotyk, eds.), pp. 343–351. Academic Press, New York.

Cirillo, V. P., Wilkins, P. O., and Anton, J. (1963). Sugar transport in a psychrophilic yeast. *J. Bacteriol.* **86**, 1259–1264

Clarkson, E. M., and Maizels, M. (1956). Sodium transfer in human and chicken erythrocytes. *J. Physiol. (London)* **129**, 476–503.

Cohen, G. N., and Rickenberg, H. V. (1956). Concentration spécifique réversible des aminoacides chez *Escherichia coli*. *Ann. Inst. Pasteur* **91**, 693–720.

Cohen, J. A., and Warringa, M. G. P. J. (1953). Purification of cholinesterase from ox red cells. *Biochim. Biophys. Acta* **10**, 195–196.

Collander, R. (1949). The permeability of plant protoplasts to small molecules. *Physiol. Plantarum* **2**, 300–311.

Collander, R., and Barlund, H. (1933). Permeabilitats-studien an *Chara ceratophylla*. *Acta Botan. Fennica* **11**, 1–14.

Conway, E. J. (1954). Some aspects of ion transport through membranes. *Symp. Soc. Exptl. Biol.* **8**, 297–324.

Cook, G. M. W., Heard, D. H., and Seaman, G. V. F. (1960). A sialomucopeptide liberated by trypsin from the human erythrocyte. *Nature* **188**, 1011–1012.

Crane, R. K. (1960). Intestinal absorption of sugars. *Physiol. Rev.* **40**, 789–825.

Crane, R. K. (1964). Uphill outflow of sugar from intestinal epithelial cells induced by reversal of the Na^+ gradient: Its significance for the mechanism of Na^+-dependent active transport. *Biochem. Biophys. Res. Commun.* **17**, 481–485.

Crane, R. K. (1965). Na^+-dependent transport in the intestine and other animal tissues. *Federation Proc.* **24**, 1000–1006.

Crane, R. K. (1966). Structural and functional organisation of an epithelial cell brush border. International Society for Cell Biology, Frascati, Italy (to be published).

Crane, R. K., and Krane, S. M. (1956). On the mechanism of the intestinal absorption of sugars. *Biochim. Biophys. Acta* **20**, 568–569.

Crane, R. K., and Krane, S. M. (1959). Studies on the mechanism of the intestinal active transport of sugars. *Biochim. Biophys. Acta* **31**, 397–401.

Crane, R. K., Field, R. A., and Cori, C. F. (1957). Studies of tissue permeability. I. The penetration of sugars into the Ehrlich ascites tumour cells. *J. Biol. Chem.* **224**, 649.

Crane, R. K., Miller, D., and Bihler, I. (1961). Restrictions on the possible mechanisms of intestinal active transport of sugars. *In* "Membrane Transport and Metabolism" (A. Kleinzeller and A. Kotyk, eds.), pp. 439–449. Academic Press, New York.

Crane, R. K., Forstner, G., and Eichholz, A. (1965). Studies on the mechanism of the intestinal absorption of sugars. X. An effect of Na^+ concentration on the apparent Michaelis constants for intestinal sugar transport, *in vitro*. *Biochim. Biophys. Acta* **109**, 467–477.

Crestfield, A. M., Stein, W. H., and Moore, S. (1963). Alkylation and identification of the histidine residue at the active site of ribonuclease. *J. Biol. Chem.* **238**, 2413–2420.

Crofford, O. B., and Renold, A. E. (1965). Glucose uptake by incubated rat epididymal adipose tissue. *J. Biol. Chem.* **240**, 14–22.

Csáky, T. Z. (1961). Significance of sodium ions in active intestinal transport of nonelectrolytes. *Am. J. Physiol.* **201**, 999–1001.

Csáky, T. Z. (1965). Transport through biological membranes. *Ann. Rev. Physiol.* **27**, 415–450.

Csáky, T. Z., and Fernald, G. W. (1960). Absorption of 3-methylglucose from the intestine of the frog, *Rana pipiens*. *Am. J. Physiol.* **198**, 445–448.

Csáky, T. Z., and Thale, M. (1960). Effect of ionic environment on intestinal sugar transport. *J. Physiol.* (*London*) **151,** 59–65.

Csáky, T. Z., and Zollicoffer, L. (1960). Ionic effect on intestinal transport of glucose in the rat. *Am. J. Physiol.* **198,** 1056–1058.

Csáky, T. Z., Hartzog, H. G., and Fernald, G. W. (1961). Effect of digitalis on active intestinal sugar transport. *Am. J. Physiol.* **200,** 459–460.

Curie, P. (1908). "Oeuvres," pp. 118–141. Gauthier-Villars, Paris.

Curran, P. F. (1963). Biophysical nature of biological membranes. *In* "The Transfer of Calcium and Strontium Across Biological Membranes" (R. H. Wasserman, ed.), pp. 3–23. Academic Press, New York.

Curran, P. F., and McIntosh, J. R. (1962). A model system for biological water transport. *Nature* **193,** 347–348.

Curran, P. F., and Schwartz, G. F. (1960). Na, Cl and water transport by rat colon. *J. Gen. Physiol.* **43,** 555–572.

Curran, P. F., and Solomon, A. K. (1957). Ion and water fluxes in the ileum of rats. *J. Gen. Physiol.* **41,** 143–168.

Curran, P. F., Herrera, F. C., and Flanigan, W. J. (1963). The effect of Ca and antidiuretic hormone on Na transport across frog skin. II. Sites and mechanisms of action. *J. Gen. Physiol.* **46,** 1011–1027.

Dainty, J., and Ginzburg, B. Z. (1964a). The measurement of hydraulic conductivity (osmotic permeability to water) of internodal Characean cells by means of transcellular osmosis. *Biochim. Biophys. Acta* **79,** 102–111.

Dainty, J., and Ginzburg, B. Z. (1964b). The permeability of the cell membranes of *Nitella translucens* to urea, and the effect of high concentrations of sucrose on this permeability. *Biochim. Biophys. Acta* **79,** 112–121.

Dainty, J., and Ginzburg, B. Z. (1964c). The permeability of the protoplasts of *Chara australis* and *Nitella translucens* to methanol, ethanol and isopropanol. *Biochim. Biophys. Acta* **79,** 122–128.

Dainty, J., and Ginzburg, B. Z. (1964d). The reflection coefficient of plant cell membranes for certain solutes. *Biochim. Biophys. Acta* **79,** 129–137.

Danielli, J. F. (1954). The present position in the field of facilitated diffusion and selective active transport. *Proc. Symp. Colston Res. Soc.* **7,** 1–4.

Daniels, F., and Alberty, R. A. (1961). "Physical Chemistry," 2nd ed., Chapters 6 and 7. Wiley, New York.

Davies, D. R., and Green, A. L. (1958). The mechanism of hydrolysis by cholinesterase and related enzymes. *Advan. Enzymol.* **20,** 283–318.

Davson, H., and Danielli, J. F. (1943). "The Permeability of Natural Membranes." Cambridge Univ. Press, Cambridge and New York.

Davson, H., and Danielli, J. F. (1952). "The Permeability of Natural Membranes," 2nd ed. Cambridge Univ. Press, Cambridge and New York.

Dawson, A. C., and Widdas, W. F. (1963). Inhibition of the glucose permeability of human erythrocytes by N-ethyl maleimide. *J. Physiol.* (*London*) **168,** 644–659.

Dawson, A. C., and Widdas, W. F. (1964). Variations with temperature and pH of the parameters of glucose transfer across the erythrocyte membrane in the foetal guinea pig. *J. Physiol.* (*London*) **172,** 107–122.

De Almeida, D. F., Chain, E. B., and Pocchiari, F. (1965). Effect of ammonium and other ions on the glucose-dependent active transport of L-histidine in slices of rat-brain cortex. *Biochem. J.* **95,** 793–796.

de Gier, J., and van Deenen, L. L. M. (1961). Quoted in van Deenen and de Gier (1964).

De Groot, S. R. (1959). "Thermodynamics of Irreversible Processes." North-Holland Publ., Amsterdam.

Dervichian, D. G. (1955). *In* "Exposes actuels: Problems de structures, d'ultra-structures et de functions cellulaires," Chapter IV. Masson, Paris.

Dervichian, D. G. (1958). The existence and significance of molecular associations in monolayers. *In* "Surface Phenomena in Chemistry and Biology" (J. F. Danielli, K. G. A. Pankhurst, and A. C. Riddiford, eds.), pp. 70–87, Pergamon, London.

Diamond, J. M. (1962). The reabsorptive properties of the gall bladder. *J. Physiol. (London)* **161,** 442–473.

Diamond, J. M. (1964a). Transport of salt and water in rabbit and guinea pig gall bladder. *J. Gen. Physiol.* **48,** 1–14.

Diamond, J. M. (1964b). The mechanism of isotonic water transport. *J. Gen. Physiol.* **48,** 15–42.

Diamond, J. M., and Tormey, J. McD. (1966a). Role of long extracellular channels in fluid transport across epithelia. *Nature* **210,** 817–820.

Diamond, J. M., and Tormey, J. McD. (1966b). *Federation Proc.* **25,** 1458–1463.

Dick, D. A. T. (1959). Osmotic properties of living cells. *Intern. Rev. Cytol.* **8,** 387–448.

Dick, D. A. T. (1964). The permeability coefficient of water in the cell membrane and the diffusion coefficient in the cell interior. *J. Theoret. Biol.* **7,** 504–531.

Diedrich, D. F. (1961). Comparison of effects of phlorizin and phloretin 2′-galactoside on the renal tubular reabsorption of glucose in the dog. *Biochim. Biophys. Acta* **47,** 618–620.

Dixon, M., and Webb, E. C. (1964). "Enzymes," 2nd ed., Chapters 4 and 6. Longmans, Green, New York.

Dodge, J. T., Mitchell, C. D., and Hanahan, D. J. (1963). The preparation and chemical characteristics of hemoglobin-free ghosts of human erythrocytes. *Arch. Biochem. Biophys.* **100,** 119–130.

Drapeau, G. R., and MacLeod, R. A. (1963). Na^+-dependent active transport of α-amino-isobutyric acid into cells of a marine pseudomonad. *Biochem. Biophys. Res. Commun.* **12,** 111–115.

Dreyfuss, J. (1964). Characterisation of a sulfate- and thiosulfate-transporting system in *Salmonella typhimurium. J. Biol. Chem.* **239,** 2292–2297.

Dreyfuss, J., and Pardee, A. B. (1965). Evidence for a sulfate-binding site external to the cell membrane of *Salmonella typhimurium. Biochim. Biophys. Acta* **104,** 308–310.

Dunham, E. T., and Glynn, I. M. (1961). Adenosinetriphosphatase activity and the active movements of alkali metal ions. *J. Physiol. (London)* **156,** 274–293.

Durbin, R. P. (1960). Osmotic flow of water across permeable cellulose membranes. *J. Gen. Physiol.* **44,** 315–326.

Durbin, R. P., Frank, H., and Solomon, A. K. (1956). Water flow through frog gastric mucosa. *J. Gen. Physiol.* **39,** 535–551.

Eddy, A. A., and Mulcahy, M. (1965). Ion gradients as a factor controlling the accumulation of glycine by ascites tumour cells in the presence of sodium cyanide. *Biochem. J.* **96,** 76P.

Edwards, C., and Harris, E. J. (1955). Long pores and water flow across cell membranes. *Nature* **175,** 262.

Egan, J. B., and Morse, M. L. (1965). Carbohydrate transport in *Staphylococcus aureus.* II. Characterisation of the defect of a pleiotropic transport mutant. *Biochim. Biophys. Acta* **109,** 172–183.

Egan, J. B., and Morse, M. L. (1966). Carbohydrate transport in *Staphylococcus aureus*. III. Studies of the transport process. *Biochim. Biophys. Acta* **112**, 63–73.

Einstein, A. (1905). "Investigations on the Theory of the Brownian Movement" (English translation and notes by R. Fürth). Reprint: Dover, New York, 1956.

Emmelot, P., Bos, C. J., Benedetti, E. L., and Rümke, P. L. (1964). Studies on plasma membranes. I. Chemical composition and enzyme content of plasma membranes isolated from rat liver. *Biochim. Biophys. Acta* **90**, 126–145.

Erickson, B. N., Williams, H. H., Bernstein, S. S., Avria, I., Jones, R. L., and Macy, J. G. (1938). The lipid distribution of posthemolytic residue or stroma of erythrocytes. *J. Biol. Chem.* **122**, 515–528.

Erlanger, B. F., Cohen, W., Vratsanos, S. M., Castleman, H., and Cooper, A. G. (1965). Postulated chemical basis for observed differences in the enzymatic behaviour of chymotrypsin and trypsin. *Nature* **205**, 868–871.

Evans, J. V. (1954). Electrolyte concentrations in red blood cells of British breeds of sheep. *Nature* **174**, 931–932.

Evans, J. V., King, J. W. B., Cohen, B. L., Harris, H., and Warren, F. L. (1956). Genetics of haemoglobin and blood potassium differences in sheep. *Nature* **178**, 849–850.

Ewell, R. H., and Eyring, H. (1937). Theory of the viscosity of liquids as a function of temperature and pressure. *J. Chem. Phys.* **5**, 726–736.

Fernández-Morán, H. (1962). New approaches in the study of biological ultrastructure by high-resolution electron microscopy. *In* "Symposia of the International Society for Cell Biology," Vol. I (R. J. C. Harris, ed.), pp. 411–428. Academic Press, New York.

Fick, A. (1855). *Ann. Physik Chem.* **94**, 59.

Finch, L. R., and Hird, F. J. R. (1960). The uptake of amino acids by isolated segments of rat intestine. II. A survey of affinity for uptake from rates of uptake and competition for uptake. *Biochim. Biophys. Acta* **43**, 278–287.

Finean, J. B. (1961). "Chemical Ultrastructure in Living Tissues." Thomas, Springfield, Illinois.

Finean, J. B. (1962). Correlation of electron microscope and X-ray diffraction data in ultrastructure studies of lipoprotein membrane systems. *In* "Symposia of the International Society for Cell Biology," Vol. I (R. J. C. Harris, ed.), pp. 89–99. Academic Press, New York.

Fisher, R. B., and Parsons, D. S. (1953a). Glucose movements across the wall of the rat small intestine. *J. Physiol.* (*London*) **119**, 210–224.

Fisher, R. B., and Parsons, D. S. (1953b). Galactose absorption from the surviving small intestine of the rat. *J. Physiol.* (*London*) **119**, 224–232.

Fisher, R. B., and Zachariah, P. (1961). The mechanism of the uptake of sugars by the rat heart and the action of insulin on this mechanism. *J. Physiol.* (*London*) **158**, 73–85.

Fishman, R. A. (1964). Carrier transport of glucose between blood and cerebrospinal fluid. *Am. J. Physiol.* **206**, 836–844.

Fleming, W. E. (1957). On the role of hydrogen ion and potassium ion in the active transport of sodium across the isolated frog skin. *J. Cellular Comp. Physiol.* **49**, 129–152.

Folch, J., and Lees, M. (1951). Proteolipides, a new type of tissue lipoproteins: Their isolation from brain. *J. Biol. Chem.* **191**, 807–818.

Foulkes, E. C., and Paine, C. M. (1961). The uptake of monocarboxylic acids by rat diaphragm. *J. Biol. Chem.* **236**, 1019.

Fox, C. F., and Kennedy, E. P. (1965). Specific labelling and partial purification of the M protein, a component of the β-galactoside transport system of *Escherichia coli*. *Proc. Natl. Acad. Sci. U.S.* **54,** 891–899.

Fox, C. F., Carter, J. R., and Kennedy, E. P. (1966). Properties of the membrane protein component of the β-galactoside transport system of *E. coli*. *Federation Proc.* **25,** 591.

Fox, M., Thier, S., Rosenberg, L. E., and Segal, S. (1964). Ionic requirements for amino acid transport in the rat kidney cortex slice. I. Influence of extracellular ions. *Biochim. Biophys. Acta* **79,** 167–176.

Frazier, H. S., Dempsey, E. F., and Leaf, A. (1962). Movements of sodium across the mucosal surface of the isolated toad bladder and its modification by vasopressin. *J. Gen. Physiol.* **45,** 529–543.

Frenkel, J. I. (1946). "Kinetic Theory of Liquids." Oxford Univ. Press, London and New York.

Frey-Wyssling, A. (1953). "Submicroscopic Morphology of Protoplasm," 2nd ed. Elsevier, Amsterdam.

Fuhrman, G. J., and Fuhrman, F. A. (1960). Inhibition of active sodium transport by cholera toxin. *Nature* **188,** 71–72.

Gaines, G. L. (1960). Overturning of stearic acid molecules in monolayers. *Nature* **186,** 384–385.

Gale, E. F. (1954). The accumulation of amino acids within Staphylococcal cells. *Symp. Soc. Exptl. Biol.* **8,** 242–253.

Gardos, G. (1954). Accumulation of K ions in human blood cells. *Acta Physiol. Acad. Sci. Hung.* **6,** 191–199.

Garrahan, P. J., and Glynn, I. M. (1965). Uncoupling the sodium pump. *Nature* **207,** 1098–1099.

Geren, B. B., and Schmitt, F. O. (1955). Electron microscope studies of the Schwann cell and its constituents with particular reference to their relation to the axon. *8th Congr. Cell Biol., Leiden, 1954* pp. 251–260. Noordhoff, Groningen.

Giebel, O., and Passow, H. (1960). Die Permeabilität der Erythrocytenmembran für organische Anionen – zur frage der Diffusien durch Poren. *Arch. Ges. Physiol.* **271,** 378–388.

Gilbert, J. C. (1965). Mechanism of sugar transport in brain slices. *Nature* **205,** 87.

Ginzburg, B. Z., and Katchalsky, A. (1963). Frictional coefficients of the flows of non-electrolytes through artificial membranes. *J. Gen. Physiol.* **47,** 403–418.

Glasstone, S., Laidler, K. J., and Eyring, H. (1941). "The Theory of Rate Processes." McGraw-Hill, New York.

Glynn, I. M. (1956). Sodium and potassium movements in human red cells. *J. Physiol. (London)* **134,** 278–310.

Glynn, I. M. (1957). The ionic permeability of the red cell membrane. *Progr. Biophys. Biophys. Chem.* **8,** 241–307.

Glynn, I. M. (1959). Sodium and potassium movements in nerve, muscle, and red cells. *Intern. Rev. Cytol.* **8,** 449–481.

Glynn, I. M., Slayman, C. W., Eichberg, J., and Dawson, R. M. C. (1965). The adenosine-triphosphatase system responsible for cation transport in electric organ. Exclusion of phospholipids as intermediates. *Biochem. J.* **94,** 692–699.

Goldstein, D. A., and Solomon, A. K. (1960). Determination of equivalent pore radius for human red cells by osmotic pressure measurement. *J. Gen. Physiol.* **44,** 11–17.

Goodman, J., and Rothstein, A. (1957). The active transport of phosphate into the yeast cell. *J. Gen. Physiol.* **40,** 915–923.

Goresky, C. A., Watanabe, H., and Johns, D. G. (1963). The renal excretion of folic acid. *J. Clin. Invest.* **42,** 1841–1849.

Green, D. E., and Fleischer, S. (1964). Role of lipid in mitochondrial function. *In* "Metabolism and Physiological Significance of Lipids" (R. M. C. Dawson and D. N. Rhodes, eds.), pp. 581–618. Wiley, New York.

Guroff, G., and Udenfriend, S. (1960). The uptake of tyrosine by isolated rat diaphragm. *J. Biol. Chem.* **235,** 3518–3522.

Hagihira, H., Wilson, T. H., and Lin, E. C. C. (1962). Intestinal transport of certain N-substituted amino acids. *Am. J. Physiol.* **203,** 637–640.

Halvorson, H. O., Okada, H., and Gorman, J. (1964). The role of an α-methylglucoside permease in the inducted synthesis of *iso*maltose in yeast. *In* "The Cellular Functions of Membrane Transport" (J. F. Hoffman, ed.), pp. 171–192. Prentice-Hall, Englewood Cliffs, New Jersey.

Hanai, T., Haydon, D. A., and Taylor, J. (1965). Some further experiments on bimolecular lipid membranes. *J. Gen. Physiol.* **48,** Suppl. 1, 59–63.

Handovsky, H., ed. (1941). "Tabulae Biologicae," Vol. 19, Part 2. Junk, The Hague.

Harned, H. S., and Owen, B. B. (1958). "Physical Chemistry of Electrolytic Solutions," 3rd ed. Reinhold, New York.

Harris, E. J. (1954). Linkage of sodium and potassium active transport in human erythrocytes. *Symp. Soc. Exptl. Biol.* **8,** 228–241.

Harris, E. J. (1956). "Transport and Accumulation in Biological Systems." Butterworth, London and Washington, D.C.

Harris, E. J. (1964). An analytical study of the kinetics of glucose movements in human erythrocytes. *J. Physiol.* (*London*) **173,** 344–353.

Harris, E. J., and Martins-Ferreira, H. (1955). Membrane potentials in the muscles of the South American frog. *J. Exptl. Biol.* **32,** 539–546.

Harris, E. J., and Prankerd, T. A. J. (1957). Diffusion and permeation of cations in human and dog erythrocytes. *J. Gen. Physiol.* **41,** 197–218.

Hartley, B. S. (1964). The structure and activity of chymotrypsin. *In* "The Structure and Activity of Enzymes" (T. W. Goodwin, B. S. Hartley, and J. I. Harris, eds.), pp. 47–60. Academic Press, New York.

Hatch, F. T. (1965). Correlation of amino-acid composition with certain characteristics of proteins. *Nature* **206,** 777–779.

Hauser, G. (1965). Energy and sodium-dependent uptake of inositol by kidney cortex slices. *Biochem. Biophys. Res. Commun.* **19,** 696–701.

Haydon, D. A., and Taylor, F. H. (1960). On adsorption at the oil/water interface and the calculation of electrical potentials in the aqueous surface phase. I. Neutral molecules and a simplified treatment for ions. *Phil. Trans. Roy. Soc.* (*London*) **A252,** 225–248.

Haydon, D. A., and Taylor, J. (1963). The stability and properties of bimolecular lipid leaflets in aqueous solutions. *J. Theoret. Biol.* **4,** 281–296.

Hays, R. M., and Leaf, A. (1962a). Studies on the movement of water through the isolated toad bladder and its modification by vasopressin. *J. Gen. Physiol.* **45,** 905–919.

Hays, R. M., and Leaf, A. (1962b). The state of water in the isolated toad bladder in the presence and absence of vasopressin. *J. Gen. Physiol.* **45,** 933–948.

Heard, D. H., and Seaman, G. V. F. (1960). The influence of pH and ionic strength on the electro-kinetic stability of the human erythrocyte membrane. *J. Gen. Physiol.* **43,** 635–654.

Heckmann, K. D., and Parsons, D. S. (1959). Changes in the water and electrolyte content of rat-liver slices *in vitro.* *Biochim. Biophys. Acta* **36,** 203–213.

Heilbrunn, L. V. (1952). "An Outline of General Physiology," 3rd ed. Saunders, Philadelphia, Pennsylvania.

Heinz, E., and Hoffman, J. F. (1965). Phosphate incorporation and Na, K-ATPase activity in human red blood cell ghosts. *J. Cellular Comp. Physiol.* **65,** 31–44.

Heinz, E., and Walsh, P. M. (1958). Exchange diffusion, transport, and intracellular level of amino acids in Ehrlich carcinoma cells. *J. Biol. Chem.* **233,** 1488–1493.

Heinz, E., Pichler, A. G., and Pfeiffer, B. (1965). Studies on the transport of glutamate in Ehrlich cells–inhibition by other amino acids and stimulation by H-ions. *Biochem. Z.* **342,** 542–552.

Heisey, S. R., Held, D., and Pappenheimer, J. R. (1962). Bulk flow and diffusion in the cerebrospinal fluid system of the goat. *Am. J. Physiol.* **203,** 775–781.

Helmreich, E., and Eisen, H. N. (1959). The distribution and utilisation of glucose in isolated lymph node cells. *J. Biol. Chem.* **234,** 1958.

Hempling, H. G. (1958). Potassium and sodium movements in the Ehrlich mouse ascites tumour cell. *J. Gen. Physiol.* **41,** 565–583.

Hempling, H. G. (1959). Permeability of the leucocyte and the neoplastic cell. Progress Report 1958–1959, Cornell Medical College (privately circulated).

Hempling, H. G. (1960). Permeability of the Ehrlich ascites tumor cell to water. *J. Gen. Physiol.* **44,** 365–380.

Hempling, H. G. (1967). Analog program for the Widdas model of sugar transport. *Biochim. Biophys. Acta* **135,** 355–358.

Henderson, M. J. (1964). Uptake of glucose into cells and the role of insulin in glucose transport. *Can. J. Biochem.* **42,** 933–944.

Herrera, F. C., and Curran, P. F. (1963). The effect of Ca and antidiuretic hormone on Na transport across frog skin. I. Examination of interrelationships between Ca and hormone. *J. Gen. Physiol.* **46,** 999–1010.

Hillman, R. S., Landau, B. R., and Ashmore, J. (1959). Structural specificity of hexose penetration of rabbit erythrocytes. *Am. J. Physiol.* **196,** 1277.

Hirs, C. H. W. (1962). Dinitrophenylribonucleases. *Brookhaven Symp. Biol.* **15,** 154–178.

Höber, R., and Ørskov, S. L. (1933). Untersuchengen über die Permeiergeschwindigkeit von Anelektrolyten bei den roten Blutkörperchen verschiedener Tierarten. *Arch. Ges. Physiol.* **231,** 599–615.

Hodgkin, A. L. (1958). Ionic movements and electrical activity in giant nerve fibres. *Proc. Roy. Soc.* **B148,** 1–37.

Hodgkin, A. L., and Katz, B. (1949). The effect of sodium ions on the electrical activity of the giant axon of the squid. *J. Physiol.* (*London*) **108,** 37–77.

Hodgkin, A. L., and Keynes, R. D. (1955a). Active transport of cations in giant axons from *Sepia* and *Loligo.* *J. Physiol.* (*London*) **128,** 28–60.

Hodgkin, A. L., and Keynes, R. D. (1955b). The potassium permeability of a giant nerve fibre. *J. Physiol.* (*London*) **128,** 61–88.

Hodgkin, A. L., and Keynes, R. D. (1956). Experiments on the injection of substances into squid giant axons by means of a microsyringe. *J. Physiol.* (*London*) **131,** 592–616.

Hodgman, C. D., ed. (1954). "Handbook of Chemistry and Physics," 36th ed. Chem. Rubber Publ. Co., Cleveland, Ohio.

Hoffman, J. F. (1958). Physiological characteristics of human red blood cell ghosts. *J. Gen. Physiol.* **42,** 9–28.

Hoffman, J. F. (1962). Active transport of sodium by ghosts of human red blood cells. *J. Gen. Physiol.* **45**, 837–859.

Hokin, L. E., and Hokin, M. R. (1961). Studies on the enzymatic mechanism of the sodium pump. *In* "Membrane Transport and Metabolism" (A. Kleinzeller and A. Kotyk, eds.), pp. 204–218. Academic Press, New York.

Hokin, L. E., and Hokin, M. R. (1963). Biological transport. *Ann. Rev. Biochem.* **32**, 553–578.

Hokin, L. E., Mokotoff, M., and Kupchan, S. M. (1966). Alkylation of a brain transport adenosine-triphosphatase at the cardiotonic steroid site by strophanthidin-3-haloacetates. *Proc. Natl. Acad. Sci. U.S.* **55**, 797–804.

Holmstedt, B. (1959). Pharmacology of organophosphorus cholinesterase inhibitors. *Pharmacol. Rev.* **11**, 567–688.

Horecker, B. L., Thomas, J., and Monod, J. (1960). Galactose transport in *Escherichia coli*. II. Characteristics of the exit process. *J. Biol. Chem.* **235**, 1586–1590.

Hoskin, F. G. G., and Rosenberg, P. (1965). Penetration of sugars, steroids, amino acids and other organic compounds into the interior of the squid giant axon. *J. Gen. Physiol.* **49**, 47–56.

House, C. R., and Green, K. (1963). Sodium and water transport across isolated intestine of a marine teleost. *Nature* **199**, 1293–1294.

Hunt, S., and Bonsall, R. B. (1966). See Bonsall and Hunt (1966).

Hunter, F. R., George, J., and Ospina, B. (1965). Possible carriers in erythrocytes. *J. Cellular Comp. Physiol.* **65**, 299–312.

Inui, Y., and Christensen, H. N. (1966). Discrimination of single transport systems: the Na^+-sensitive transport of neutral amino acids in the Ehrlich cell. *J. Gen. Physiol.* **50**, 203–224.

Jacob, F., and Monod, J. (1961). Genetic regulatory mechanisms in the synthesis of proteins. *J. Mol. Biol.* **3**, 318–356.

Jacobs, M. H. (1952). The measurement of cell permeability with particular reference to the erythrocyte. *In* "Modern Trends in Physiology and Biochemistry" (E. S. Guzman Barron, ed.), pp. 149–172. Academic Press, New York.

Jacobs, M. H., and Stewart, D. R. (1936). Distribution of penetrating ammonium salts between cells and their surroundings. *J. Cellular Comp. Physiol.* **7**, 351–365.

Jacobs, M. H., Glassman, H. N., and Parpart, A. K. (1935). Osmotic properties of the erythrocyte. VII. The temperature coefficient of certain hemolytic processes. *J. Cellular Comp. Physiol.* **7**, 197–225.

Jacquez, J. A. (1961). The kinetics of carrier-mediated active transport of amino acids. *Proc. Natl. Acad. Sci. U.S.* **47**, 153–163.

Jacquez, J. A. (1962). Transport and enzymic splitting of pyrimidine nucleosides in Ehrlich cells. *Biochim. Biophys. Acta* **61**, 265–277.

Jacquez, J. A. (1963). Carrier-amino acid stoichiometry in amino acid transport in Ehrlich ascites cells. *Biochim. Biophys. Acta* **71**, 15–33.

Jacquez, J. A., and Sherman, J. H. (1965). The effect of metabolic inhibitors on transport and exchange of amino acids in Ehrlich ascites cells. *Biochim. Biophys. Acta* **109**, 128–141.

Jardetzky, O. (1964). The Curie principle and the problem of active transport. *Biochim. Biophys. Acta* **79**, 631–633.

Jenerick, H. P. (1953). Muscle membrane potential, resistance and external KCl. *J. Cellular Comp. Physiol.* **42**, 427–447.

Jervis, E. L., Johnson, F. R., Sheff, M. F., and Smyth, D. H. (1956). The effect of phlorizin on intestinal absorption and intestinal phosphatase. *J. Physiol.* (*London*) **134,** 675–688.

Jones, G., and Talley, S. K. (1933). The viscosity of aqueous solutions as a function of the concentration. *J. Am. Chem. Soc.* **55,** 624–642.

Jost, W. (1960). "Diffusion in Solids, Liquids, Gases," 2nd ed. Academic Press, New York.

Judah, J. D., McLean, A. E. M., Ahmed, K., and Christie, G. S. (1965a). Active transport of potassium by mitochondria. II. Effect of substrates and inhibitors. *Biochim. Biophys. Acta* **94,** 441–451.

Judah, J. D., Ahmed, K., McLean, A. E. M., and Christie, G. S. (1965b). Uptake of magnesium and calcium by mitochondria in exchange for hydrogen ions. *Biochim. Biophys. Acta* **94,** 452–460.

Kaback, H. R., and Stadtman, E. R. (1966). Proline uptake by an isolated cytoplasmic membrane preparation of *Escherichia coli.* *Proc. Natl. Acad. Sci. U.S.* **55,** 920–927.

Kedem, O. (1961). Criteria of active transport. *In* "Membrane Transport and Metabolism" (A. Kleinzeller and A. Kotyk, eds.), pp. 87–93. Academic Press, New York.

Kedem, O., and Katchalsky, A. (1958). Thermodynamic analysis of the permeability of biological membranes to non-electrolytes. *Biochim. Biophys. Acta* **27,** 229–246.

Kedem, O., and Katchalsky, A. (1961). A physical interpretation of the phenomenological coefficients of membrane permeability. *J. Gen. Physiol.* **45,** 143–179.

Kendrew, J. C. (1962). Side-chain interactions in myoglobin. *Brookhaven Symp. Biol.* **15,** 216–227.

Kepes, A. (1960). Études cinetiques sur la galactoside-perméase d'*Escherichia coli.* *Biochim. Biophys. Acta* **40,** 70–84.

Kepes, A. (1961). Discussion. *In* "Membrane Transport and Metabolism" (A. Kleinzeller and A. Kotyk, eds.), pp. 459–460. Academic Press, New York.

Kepes, A. (1964). The place of permeases in cellular organisation. *In* "The Cellular Functions of Membrane Transport" (J. F. Hoffman, ed.), pp. 155–170. Prentice-Hall, Englewood Cliffs, New Jersey.

Kepes, A., and Cohen, G. N. (1962). Permeation. *In* "The Bacteria" (I. C. Gunsalus and R. Y. Stanier, eds.), Vol. 4, Chapter 5, pp. 179–221. Academic Press, New York.

Keston, A. S. (1964a). Kinetics and distribution of mutarotases and their relation to sugar transport. *J. Biol. Chem.* **239,** 3241–3251.

Keston, A. S. (1964b). Mutarotase inhibition by l-deoxyglucose. *Science* **143,** 698–700.

Keynes, R. D. (1951). The ionic movements during nervous activity. *J. Physiol.* (*London*) **114,** 119–150.

Keynes, R. D. (1961a). Discussion. *In* "Membrane Transport and Metabolism" (A. Kleinzeller and A. Kotyk, eds.), p. 104. Academic Press, New York.

Keynes, R. D. (1961b). The energy source for active transport in nerve and muscle. *In* "Membrane Transport and Metabolism" (A. Kleinzeller and A. Kotyk, eds.), pp. 131–139. Academic Press, New York.

Keynes, R. D. (1965). Some further observations on the sodium efflux in frog muscle. *J. Physiol.* (*London*) **178,** 305–325.

Kinter, W. B., and Wilson, T. H. (1965). Autoradiographic study of sugar and amino acid absorption by everted sacs of hamster intestine. *J. Cell Biol.* **25,** 19–39.

Kipnis, D. M., and Crane, R. K. (1960). Quoted in Crane (1960).

Kirschner, L. B. (1957). Phosphatidylserine as a possible participant in active sodium transport in erythrocytes. *Arch. Biochem. Biophys.* **68,** 499–500.

Klahr, S., and Bricker, N. S. (1965). Energetics of anaerobic sodium transport by the fresh water turtle bladder. *J. Gen. Physiol.* **48,** 571–580.

Kleinzeller, A., and Kotyk, A., eds. (1961). "Membrane Transport and Metabolism." Academic Press, New York.

Koch, A. L. (1964). The role of permease in transport. *Biochim. Biophys. Acta* **79,** 117–200.

Koefoed-Johnsen, V., and Ussing, H. H. (1953). The contributions of diffusion and flow to the passage of D_2O through living membranes. Effect of neurohypophyseal hormone on isolated anuran skin. *Acta Physiol. Scand.* **28,** 60–76.

Koefoed-Johnsen, V., and Ussing, H. H. (1958). Nature of the frog skin potential. *Acta Physiol. Scand.* **42,** 298–308.

Kolber, A. R., and LeFevre, P. G. (1967). *J. Gen. Physiol.* (in press).

Kolber, A. R., and Stein, W. D. (1966). Identification of a component of a transport carrier system: Isolation of the permease expression of the *Lac* operon of *Escherichia coli. Nature* **209,** 691–694.

Kolber, J., and Stein, W. D. (1966a). The β-galactoside permease of *Escherichia coli* as a primary active transport system. *Biochem. J.* **98,** 8P.

Kolber, J., and Stein, W. D. (1966b). Unpublished data.

Kono, T., and Colowick, S. P. (1961). Isolation of skeletal muscle cell membrane and some of its properties. *Arch. Biochem. Biophys.* **93,** 520–533.

Koshland, D. E. (1960). The active site and enzyme action. *Advan. Enzymol.* **22,** 46–97.

Kotyk, A., and Höfer, M. (1965). Uphill transport of sugars in the yeast *Rhodotorula gracilis. Biochim. Biophys. Acta* **102,** 410–422.

Krane, S. M., and Crane, R. K. (1959). The accumulation of D-galactose against a concentration gradient by slices of rabbit kidney cortex. *J. Biol. Chem.* **234,** 211–216.

Kromphardt, H., Grobecker, H., Ring, K., and Heinz, E. (1963). Über den Einfluss von Alkali-ionen auf den Glycintransport in Ehrlich-ascites-tumorzellen. *Biochim. Biophys. Acta* **74,** 549–551.

Lacko, L., and Burger, M. (1961). Common carrier system for sugar transport in human red cells. *Nature* **191,** 881–882.

Lacko, L., and Burger, M. (1962). Interaction of some disaccharides with the carrier system for aldoses in erythrocytes. *Biochem. J.* **83,** 622–625.

Larralde, J., Giraldez, A., and Ron-Noya, J. (1961). Transporte activo de glucosa y papel de la florricina, floretina, fosfate de floretina, ácido floretínico, floroglucina y florina sobre el mismo. *Rev. Espan. Fisiol.* **17,** 193–201.

Larsen, P. R., Ross, J. E., and Tapley, D. F. (1964). Transport of neutral, dibasic and N-methyl-substituted amino acids by rat intestine. *Biochim. Biophys. Acta* **88,** 570–577.

Lassen, U. V. (1961). Kinetics of uric acid transport in human erythrocytes. *Biochim. Biophys. Acta* **53,** 557–569.

Lassen, U. V., and Overgaard-Hansen, K. (1962a). Purine derivatives as inhibitors of uric acid transport into human erythrocytes. *Biochim. Biophys. Acta* **57,** 111–115.

Lassen, U. V., and Overgaard-Hansen, K. (1962b). Hypoxanthine as inhibitor of uric acid transport in human erythrocytes. *Biochim. Biophys. Acta* **57,** 115–122.

Leaf, A. (1959). Maintenance of concentration gradients and regulation of cell volume. *Ann. N.Y. Acad. Sci.* **72**, 396–404.

Leaf, A. (1960). Some actions of neurohypophyseal hormones on a living membrane. *J. Gen. Physiol.* **43**, Suppl., 175–189.

Leaf, A. (1961). Some observations on transport across the toad bladder. *In* "Membrane Transport and Metabolism" (A. Kleinzeller and A. Kotyk, eds.), pp. 247–255. Academic Press, New York.

Leaf, A., and Hays, R. M. (1962). Permeability of the isolated toad bladder to solutes and its modification by vasopressin. *J. Gen. Physiol.* **45**, 921–932.

Lee, J S. (1960). On the "absorption pressure" of rat intestine during water absorption *in vitro*. *Federation Proc.* **19**, 182.

LeFevre, P. G. (1948). Evidence of active transfer of certain non-electrolytes across the human red cell membrane. *J. Gen. Physiol.* **31**, 505–527.

LeFevre, P. G. (1954). The evidence for active transport of monosaccharides across the red cell membrane. *Symp. Soc. Exptl. Biol.* **8**, 118–136.

LeFevre, P. G. (1959). Molecular structural factors in competitive inhibition of sugar transport. *Science* **130**, 104–105.

LeFevre, P. G. (1961a). Sugar transport in the red blood cell: Structure-activity relationships in substrates and antagonists. *Pharmacol. Rev.* **13**, 39–70.

LeFevre, P. G. (1961b). Persistence in erythrocyte ghosts of mediated sugar transport. *Nature* **191**, 970–972.

LeFevre, P. G. (1961c). Upper limit for number of sugar transport sites in red cell surface. *Federation Proc.* **20**, 139 (oral communication accompanying the published report).

LeFevre, P. G. (1962). Rate and affinity in human red blood cell sugar transport. *Am. J. Physiol.* **203**, 286–290.

LeFevre, P. G. (1963). Absence of rapid exchange component in a low-affinity carrier transport. *J. Gen. Physiol.* **46**, 721–731.

LeFevre, P. G. (1964). Osmotically functional water content of the human erythrocyte. *J. Gen. Physiol.* **47**, 585–603.

LeFevre, P. G. (1965). Personal communication to the author.

LeFevre, P. G. (1966). The "dimeriser" hypothesis for sugar permeation through red cell membrane: reinvestigation of original evidence. *Biochim. Biophys. Acta* **120**, 395–405.

LeFevre, P. G., and McGinniss, G. F. (1960). Tracer exchange *vs* net uptake of glucose through human red cell surface. *J. Gen. Physiol.* **44**, 87–103.

LeFevre, P. G., and Marshall, J. K. (1958). Conformational specificity in a biological sugar transport system. *Am. J. Physiol.* **194**, 333–337.

LeFevre, P. G., and Marshall, J. K. (1959). The attachment of phloretin and analogues to human erythrocytes in connection with inhibition of sugar transport. *J. Biol. Chem.* **234**, 3022–3026.

LeFevre, P. G., Habich, K. I., Hess, H., and Hudson, M. R. (1964). Phospholipid-sugar complexes in relation to cell membrane monosaccharide transport. *Science* **143**, 955–957.

Lehninger, A. L. (1965). "The Mitochondrion," Chapter 8. Benjamin, New York.

Lenard, J., and Singer, S. J. (1966). Protein conformation in cell membrane preparations as studied by optical rotatory dispersion and circular dichroism. *Proc. Natl. Acad. Sci.* **56**, 1828–1835.

Levine, M. (1965). The transport of glucose across the human erythrocyte membrane. M.Sc. Thesis, University of Manchester.

Levine, M., and Stein, W. D. (1966). The kinetic parameters of the monosaccharide transfer system of the human erythrocyte. *Biochim. Biophys. Acta* **127,** 179–193.

Levine, M., Oxender, D. L., and Stein, W. D. (1965). The substrate-facilitated transport of the glucose carrier across the human erythrocyte membrane. *Biochim. Biophys. Acta* **109,** 151–163.

Lin, E. C. C., Hagihira, H., and Wilson, T. H. (1962). Specificity of the transport system for neutral amino acids in the hamster intestine. *Am. J. Physiol.* **202,** 919–925.

Lindemann, B., and Solomon, A. K. (1962). Permeability of luminal surface of intestinal mucosal cells. *J. Gen. Physiol.* **45,** 801–810.

Linderholm, H. (1952). Active transport of ions through frog skin with special reference to the action of certain diuretics. *Acta Physiol. Scand.* **27,** Suppl. 97.

Ling, G. N. (1962). "A Physical Theory of the Living State." Ginn (Blaisdell), Boston, Massachusetts.

Longsworth, L. G. (1955). Diffusion in liquids and the Stokes-Einstein relation. *In* "Electrochemistry in Biology and Medicine" (T. Shedlovsky, ed.), pp. 225–247. Wiley, New York.

Love, A. H. G., Mitchell, T. G., and Neptune, E. M. (1965). Transport of sodium and water by rabbit ileum, *in vitro* and *in vivo.* *Nature* **206,** 1158.

Love, W. E. (1953). The permeability of the erythrocyte-like cells of *Phascolosoma gouldi.* *Biol. Bull.* **105,** 128–132.

Lovelock, J. E. (1955). The physical instability of human red blood cells. *Biochem. J.* **60,** 692–696.

Lubin, M., Kessel, D. H., Budreau, A., and Gross, J. D. (1960). The isolation of bacterial mutants defective in amino acid transport. *Biochim. Biophys. Acta* **42,** 535–538.

Lucy, J. A., and Glauert, A. M. (1964). Structure and assembly of macromolecular lipid complexes composed of globular micelles. *J. Mol. Biol.* **8,** 727–748.

Lumry, R. (1959). Some aspects of the thermodynamics and mechanism of enzymic catalysis. *In* "The Enzymes" (P. D. Boyer, H. Lardy, and K. Myrbäck, eds.), 2nd ed., Vol. 1, Chapter 4. Academic Press, New York.

Luzzato, L. (1961). Personal communication.

Luzzatto, L., and Leoncini, G. (1961). Transport of monosaccharides in leukocytes and tumour cells. *Ital. J. Biochem.* **10,** 249–257.

MacRobbie, E. A. C. (1965). The nature of the coupling between light energy and active ion transport in *Nitella translucens.* *Biochim. Biophys. Acta* **94,** 64–73.

McCutcheon, M., and Lucké, B. (1932). The effect of temperature on the permeability to water of resting and of activated cells (unfertilised and fertilised eggs of *Arbacia punctulata*). *J. Cellular Comp. Physiol.* **2,** 11–26.

Maddy, A. H. (1964). The solubilisation of the protein of the ox-erythrocyte ghost. *Biochim. Biophys. Acta* **88,** 448–449.

Maddy, A. H. (1966). The properties of the protein of the plasma membrane of ox erythrocytes. *Biochim. Biophys. Acta* **117,** 193–200.

Maffly, R. H., and Coggins, C. H. (1965). Carbon dioxide production and sodium transport by the toad bladder. *Nature* **206,** 197.

Maizels, M. (1954). Active cation transport in erythrocytes. *Symp. Soc. Exptl. Biol.* **8,** 202–227.

Maizels, M. (1961). Cation transfer in human red cells. *In* "Membrane Transport and Metabolism" (A. Kleinzeller and A. Kotyk, eds.), pp. 256–269. Academic Press, New York.

Maizels, M., Remington, M., and Truscoe, R. (1958). The effects of certain physical factors and of the cardiac glycosides on sodium transfer by mouse ascites tumour cells. *J. Physiol.* (*London*) **140,** 61–79.

Mandelstam, J. (1956). Inhibition of bacterial growth by selective interference with the passage of basic amino acids into the cell. *Biochim. Biophys. Acta* **22,** 324–328.

Mares-Guia, M., and Shaw, E. (1963). The irreversible inactivation of trypsin by the chloromethyl ketone derived from N-tosyl-L-lysine. *Federation Proc.* **22,** 528.

Mathias, A. P., Deavin, A., and Rabin, B. R. (1964). Studies on the active site and mechanism of action of bovine pancreatic ribonuclease. *In* "The Structure and Activity of Enzymes" (T. W. Goodwin, B. S. Hartley, and J. I. Harris, eds.), pp. 9–30. Academic Press, New York.

Matthews, D. M., and Laster, L. (1965a). Kinetics of intestinal active transport of five neutral amino acids. *Am. J. Physiol.* **208,** 593–600.

Matthews, D. M., and Laster, L. (1965b). Competition for intestinal transport among five neutral amino acids. *Am. J. Physiol.* **208,** 601–606.

Mauro, A. (1957). Nature of solvent transfer in osmosis. *Science* **126,** 252–253.

Mauro, A. (1960). Some properties of ionic and nonionic semipermeable membranes. *Circulation* **21,** 845–854.

Mawe, R. C., and Hempling, H. G. (1965). The exchange of ^{14}C glucose across the membrane of the human erythrocyte. *J. Cellular Comp. Physiol.* **66,** 95–104.

Meyer, K. H., and Sievers, J-F. (1936a). La perméabilité des membranes. I. Théorie de la perméabilité ionique. *Helv. Chim. Acta* **19,** 649–664.

Meyer, K. H., and Sievers, J-F. (1936b). La perméabilité des membranes. II. Essais avec des membranes sélectives artificielles. *Helv. Chim. Acta* **19,** 665–677.

Miller, D. M. (1960). The osmotic pump theory of selective transport. *Biochim. Biophys. Acta* **37,** 448–462.

Miller, D. M. (1965a). The kinetics of selective biological transport. I. Determination of transport constants for sugar movements in human erythrocytes. *Biophys. J.* **5,** 407–415.

Miller, D. M. (1965b). The kinetics of selective biological transport. II. Equations for induced uphill transport of sugars in human erythrocytes. *Biophys. J.* **5,** 417–423.

Miller, D. M. (1966). A re-examination of Stein's dimer theory of sugar transport in human erythrocytes. *Biochim. Biophys. Acta* **120,** 156–158.

Miner, C. S., and Dalton, N. N. (1953). "Glycerol," p. 279. Reinhold, New York.

Mishima, S., and Hedbys, B. O. (1966). Permeability of the corneal epithelium and endothelium to water. *Abstr. Am. Biophys. Soc. 10th Ann. Meeting, Boston, 1966* p. 146 [*Biophys. J.* **6** (1966)].

Mitchell, C. D., and Hanahan, D. J. (1966). Solubilisation of certain proteins from the human erythrocyte stroma. *Biochemistry* **5,** 51–57.

Mitchell, C. D., Mitchell, W. B., and Hanahan, D. J. (1965). Enzyme and hemoglobin retention in human erythrocyte stroma. *Biochim. Biophys. Acta* **104,** 348–358.

Mitchell, P. (1954). Transport of phosphate through an osmotic barrier. *Symp. Soc. Exptl. Biol.* **8,** 254–261.

Mitchell, P. (1957). A general theory of membrane transport from studies of bacteria. *Nature* **180,** 134–136.

Mitchell, P. (1959). Structure and function in microorganisms. *In* "Structure and function of subcellular components." *Proc. 16th Biochem. Soc. Symp., London, 1957* pp. 73–93. Cambridge Univ. Press, London and New York.

Mitchell, P. (1961). Biological transport phenomena and the spatially anisotropic characteristics of enzyme systems causing a vector component of metabolism. *In* "Membrane Transport and Metabolism" (A. Kleinzeller and A. Kotyk, eds.), pp. 22–34. Academic Press, New York.

Mitchell, P., and Moyle, J. (1956). Permeation mechanisms in bacterial membranes. *Discussions Faraday Soc.* **21**, 258–265.

Mizobuchi, K., Demerec, M., and Gillespie, D. H. (1962). Cysteine mutants of *Salmonella typhimurium. Genetics* **47**, 1617.

Monod, J., Wyman, J., and Changeux, J-P. (1965). On the nature of allosteric transitions; a plausible model. *J. Mol. Biol.* **12**, 88–118.

Morgan, H. E., and Park, C. R. (1957). The effect of insulin, alloxan diabetes and phlorizin on sugar transport across the muscle cell membrane. *J. Clin. Invest.* **36**, 916.

Morgan, H. E., Randle, P. J., and Regen, D. M. (1959). Regulation of glucose uptake by muscle. 3. The effects of insulin, anoxia, salicylate and 2 : 4 dinitrophenol on membrane transport and intracellular phosphorylation of glucose in the isolated rat heart. *Biochem. J.* **73**, 573–579.

Morgan, H. E., Regen, D. M., and Park, C. R. (1964). Identification of a mobile carrier-mediated sugar transport system in muscle. *J. Biol. Chem.* **239**, 369–374.

Morgan, T. E., and Hanahan, D. J. (1966). Solubilisation and characterisation of a lipoprotein from erythrocyte stroma. *Biochemistry* **5**, 1050–1059.

Moskowitz, M., and Calvin, M. (1952). Components and structure of the human red cell membrane. *Exptl. Cell Res.* **3**, 33–46.

Najjar, V. A. (1962). Phosphoglucomutase. *In* "The Enzymes" (P. D. Boyer, H. Lardy, and K. Myrbäck, eds.), 2nd ed., Vol. 6, pp. 161–178. Academic Press, New York.

Narahara, H. T., and Özand, P. (1963). Studies of tissue permeability. IX. The effect of insulin on the penetration of 3-methylglucose-H^3 in frog muscle. *J. Biol. Chem.* **238**, 40–49.

Newey, H., and Smyth, D. H. (1964). The transfer system for neutral amino acids in the rat small intestine. *J. Physiol.* (*London*) **170**, 328–343.

Nirenberg, M. W., and Hogg, J. F. (1958). Hexose transport in ascites tumor cells. *J. Am. Chem. Soc.* **80**, 4407–4412.

Nixon, D. A. (1963). The transplacental passage of fructose, urea and meso-inositol in the direction from foetus to mother, as demonstrated by perfusion studies in the sheep. *J. Physiol.* (*London*) **166**, 351–362.

Ogilvie, J. T., McIntosh, J. R., and Curran, P. F. (1963). Volume flow in a series-membrane system. *Biochim. Biophys. Acta* **66**, 441–444.

Ohnishi, T. (1962). Extraction of actin- and myosin-like proteins from erythrocyte membrane. *J. Biochem.* (*Tokyo*) **52**, 307–308.

Oken, D. E., Whittembury, G., Windhager, E. E., and Solomon, A. K. (1963). Single proximal tubules of Necturus kidney. V. Unidirectional sodium movement. *Am. J. Physiol.* **204**, 372–376.

Oosterbaan, R. A., and Cohen, J. A. (1964). The active site of esterases. *In* "The Structure and Activity of Enzymes" (T. W. Goodwin, B. S. Hartley, and J. I. Harris, eds.), pp. 87–95. Academic Press, New York.

Orloff, J., and Handler, J. S. (1964). Mechanism of action of antidiuretic hormones on epithelial structures. *In* "The Cellular Functions of Membrane Transport" (J. F. Hoffman, ed.), pp. 251–268. Prentice-Hall, Englewood Cliffs, New Jersey.

Ørskov, S. L. (1935). Eine Methode zur fortlaufenden photographischen Aufzeichnung von Volumanderunge der roten Blutkörperchen. *Biochem. Z.* **279**, 241–249.

Osborn, M. J., McLellan, W. L., and Horecker, B. L. (1961). Galactose transport in *Escherichia coli.* III. The effect of 2,4-dinitrophenol on entry and accumulation. *J. Biol. Chem.* **236**, 2585–2589.

Osterhout, W. J. V. (1935). How do electrolytes enter the cell? *Proc. Natl. Acad. Sci. U.S.* **21**, 125–132.

Oxender, D. L. (1962). Personal communication to the author.

Oxender, D. L., and Christensen, H. N. (1963a). Evidence for two types of mediation of neutral amino acid transport in Ehrlich cells. *Nature* **197**, 765–767.

Oxender, D. L., and Christensen, H. N. (1963b). Distinct mediating systems for the transport of neutral amino acids by the Ehrlich cell. *J. Biol. Chem.* **238**, 3686–3699.

Oxender, D. L., and Whitmore, B. (1966). Separation of uptake from exchange processes for amino acids in Ehrlich cells. *Federation Proc.* **25**, 592 (abstr.).

Paganelli, C. V. (1962). *Meeting Red Cell Club, Atlantic City, 1962* Oral report.

Paganelli, C. V., and Solomon, A. K. (1957). The rate of exchange of tritiated water across the human red cell membrane. *J. Gen. Physiol.* **41**, 259–277.

Pappenheimer, J. R., Renkin, E. M., and Borrero, L. M. (1951). Filtration, diffusion and molecular sieving through peripheral capillary membranes. *Am. J. Physiol.* **167**, 13–46.

Pardee, A. B., and Prestidge, L. S. (1966). Cell-free activity of a sulfate binding site involved in active transport. *Proc. Natl. Acad. Sci. U.S.* **55**, 189–191.

Park, C. R. (1961). Discussion. *In* "Membrane Transport and Metabolism" (A. Kleinzeller and A. Kotyk, eds.), pp. 453–454. Academic Press, New York.

Park, C. R., Morgan, H. E., Henderson, M. J., Regen, D. M., Cadenas, E., and Post, R. L. (1961). Hormones and organic metabolism. V. The regulation of glucose uptake in muscle as studied in the perfused rat heart. *Recent Progr. Hormone Res.* **17**, 493–538.

Parpart, A. K., and Dziemian, A. J. (1940). The chemical composition of the red cell membrane. *Cold Spring Harbor Symp. Quant. Biol.* **8**, 17.

Parrish, J. E., and Kipnis, D. M. (1964). Effects of Na^+ on sugar and amino acid transport in striated muscle. *J. Clin. Invest.* **43**, 1994–2002.

Parsons, D. S., and Wingate, D. L. (1961). Changes in the fluid content of rat intestine segments during fluid absorption *in vitro.* *Biochim. Biophys. Acta* **46**, 184–186.

Partington, J. R. (1951). "An Advanced Treatise on Physical Chemistry," Vol. 2 ("The Properties of Liquids"). Longmans, Green, New York.

Passow, H. (1964). Ion and water permeability of the red blood cell. *In* "The Red Blood Cell" (C. Bishop and D. M. Surgenor, eds.), Chapter 3. Academic Press, New York.

Passow, H., Rothstein, A., and Clarkson, T. W. (1961). The general pharmacology of the heavy metals. *Pharmacol. Rev.* **13**, 185–224.

Patlak, C. S. (1956). Contributions to the theory of active transport. *Bull. Math. Biophys.* **18**, 271–315.

Patlak, C. S. (1957). Contributions to the theory of active transport. II. The gate type non-carrier mechanism and generalisations concerning tracer flow, efficiency and measurement of energy expenditure. *Bull. Math. Biophys.* **19**, 209–235.

Patlak, C. S., Goldstein, D. A., and Hoffman, J. F. (1963). Flow of solute and solvent across a two-membrane system. *J. Theoret. Biol.* **5**, 426–442.

Pauling, L. (1948). "The Nature of the Chemical Bond," 2nd ed. Cornell Univ. Press, Ithaca, New York.

Pauling, L. (1960). "The Nature of the Chemical Bond," 3rd ed. Cornell Univ. Press, Ithaca, New York.

Phillips, R. A., Love, A. H. G., Mitchell, T. G., and Neptune, E. M. (1965). Cathartics and the sodium pump. *Nature* **206**, 1367.

Pimentel, G. C., and McClennan, A. L. (1960). "The Hydrogen Bond." Freeman, San Francisco, California.

Pitts, R. F. (1963). "Physiology of the Kidney and Body Fluids." Year Book Publ., Chicago, Illinois.

Ponder, E. (1949). Permeability of human red cells to cations after treatment with resorcinol, butyl alcohol and similar lysins. *J. Gen. Physiol.* **32**, 53–62.

Ponder, E. (1955). Red cell structure and its breakdown. "Protoplasmatologia," Vol. 10, Part 2. Springer, Vienna.

Ponder, E. (1961). The cell membrane and its properties. *In* "The Cell" (J. Brachet and A. E. Mirsky, eds.), Vol. 2, Chapter 1, pp. 1–84. Academic Press, New York.

Post, R. L. (1957). Substitution of lithium for potassium in active cation transport across the human erythrocyte membrane. *Federation Proc.* **16**, 102.

Post, R. L., and Albright, C. D. (1961). Membrane adenosine triphosphatase as part of a system for active sodium and potassium transport. *In* "Membrane Transport and Metabolism" (A. Kleinzeller and A. Kotyk, eds.), pp. 219–229. Academic Press, New York.

Post, R. L., and Jolly, P. C. (1957). The linkage of sodium, potassium, and ammonium active transport across the human erythrocyte membrane. *Biochim. Biophys. Acta* **25**, 118–128.

Post, R. L., Merritt, C. R., Kinsolving, C. R., and Albright, C. D. (1960). Membrane adenosine triphosphatase as a participant in the active transport of sodium and potassium in the human erythrocyte. *J. Biol. Chem.* **235**, 1796–1802.

Post, R. L., Morgan, H. E., and Park, C. R. (1961). Regulation of glucose uptake in muscle. III. The interaction of membrane transport and phosphorylation in the control of glucose uptake. *J. Biol. Chem.* **236**, 269–272.

Post, R. L., Sen, A. K., and Rosenthal, A. S. (1965). A phosphorylated intermediate in adenosine triphosphate-dependent sodium and potassium transport across kidney membranes. *J. Biol. Chem.* **240**, 1437–1445.

Poulik, M. D., and Lauf, P. K. (1965). Heterogeneity of water-soluble structural components of human red cell membrane. *Nature* **208**, 874–876.

Prescott, D. M., and Zeuthen, E. (1953). Comparison of water diffusion and water filtration across cell surfaces. *Acta Physiol. Scand.* **28**, 77–94.

Prestidge, L. S., and Pardee, A. B. (1965). A second permease for methyl-thio-β-D-galactoside in *Escherichia coli*. *Biochim. Biophys. Acta* **100**, 591–593.

Ray, W. J., and Koshland, D. E. (1961). A method for characterising the type and number of groups involved in enzyme action. *J. Biol. Chem.* **236**, 1973.

Razin, S., Argaman, M., and Avigan, J. (1963). Chemical composition of mycoplasma cells and membranes. *J. Gen. Microbiol.* **33**, 477–487.

Reed, C. F., Swisher, S. N., Marinetti, G. V., and Eden, E. G. (1960). Studies of the lipids of the erythrocyte. I. Quantitative analysis of the lipids of normal human red blood cells. *J. Lab. Clin. Med.* **56**, 281–289.

Reeves, R. E. (1951). Cuprammonium-glycoside complexes. *Advan. Carbohydrate Chem.* **6**, 107–134.

Regen, D. M., and Morgan, H. E. (1964). Studies of the glucose transport system in the rabbit erythrocyte. *Biochim. Biophys. Acta* **79**, 151–166.

Reinwein, D. (1961). Quoted by Park (1961).

Renkin, E. M. (1954). Filtration, diffusion and molecular sieving through porous cellulose membranes. *J. Gen. Physiol.* **38**, 225–243.

Rickenberg, H. V., and Maio, J. J. (1961). The accumulation of galactose by mammalian tissue culture cells. *In* "Membrane Transport and Metabolism" (A. Kleinzeller and A. Kotyk, eds.), pp. 409–422. Academic Press, New York.

Rickenberg, H. V., Cohen, G. N., Buttin, G., and Monod, J. (1956). La galactoside-perméase d'*Escherichia coli*. *Ann. Inst. Pasteur* **91**, 829–857.

Riklis, E., and Quastel, J. H. (1958). Effects of cations on sugar absorption by isolated surviving guinea pig intestine. *Can. J. Biochem. Physiol.* **36**, 347–362.

Robbins, E., and Mauro, A. (1960). Experimental study of the independence of diffusion and hydrodynamic permeability coefficients in collodion membranes. *J. Gen. Physiol.* **43**, 523–532.

Robertson, J. D. (1960). The molecular structure and contact relationships of cell membranes. *Progr. Biophys. Biophys. Chem.* **10**, 343–418.

Robertson, J. D. (1964). Unit membranes: A review with recent new studies of experimental alterations and a new subunit structure in synaptic membranes. *In* "Cellular Membranes in Development" (M. Locke, ed.), p. 1–81. Academic Press, New York.

Robertson, J. J., and Halvorson, H. O. (1957). The components of maltozymase in yeast, and their behaviour during deadaptation. *J. Bacteriol.* **73**, 188–198.

Robertson, R. N. (1960). Ion transport and respiration. *Biol. Rev.* **35**, 231–264.

Robinson, R. A., and Stokes, R. H. (1959). "Electrolyte Solutions," 2nd ed. Butterworth, London and Washington, D.C.

Rosenberg, L. E., Blair, A., and Segal, S. (1961). Transport of amino acids by slices of rat-kidney cortex. *Biochim. Biophys. Acta* **54**, 479–488.

Rosenberg, L. E., Downing, S. J., and Segal, S. (1962). Competitive inhibition of dibasic amino acid transport in rat kidney. *J. Biol. Chem.* **237**, 2265–2270.

Rosenberg, T., and Wilbrandt, W. (1955). The kinetics of membrane transports involving chemical reactions. *Exptl. Cell Res.* **9**, 49–67.

Rosenberg, T., and Wilbrandt, W. (1957a). Strukturabhängigkeit der Hemmwirkung von Phlorizin und anderen Phloretinderivaten auf den Glukosetransport durch die Erythrocytenmembran. *Helv. Physiol. Acta* **15**, 168–176.

Rosenberg, T., and Wilbrandt, W. (1957b). Uphill transport induced by counterflow. *J. Gen. Physiol.* **41**, 289–296.

Rothstein, A. (1954). Enzyme systems of the cell surface involved in the uptake of sugars by yeast. *Symp. Soc. Exptl. Biol.* **8**, 165–201.

Rothstein, A., and Bruce, M. (1958). The potassium efflux and influx in yeast at different potassium concentrations. *J. Cellular Comp. Physiol.* **51**, 145–160.

Rotman, B. (1959). Separate permeases for the accumulation of methyl-β-D-galactoside and methyl-β-D-thiogalactoside in *Escherichia coli*. *Biochim. Biophys. Acta* **32**, 599–601.

Salton, M. R. J., and Freer, J. H. (1965). Composition of the membranes isolated from several Gram-positive bacteria. *Biochim. Biophys. Acta* **107**, 531–538.

Saunders, S. J., and Isselbacher, K. J. (1965). Inhibition of intestinal amino acid transport by hexoses. *Biochim. Biophys. Acta* **102**, 397–409.

Schatzmann, H. J. (1953). Cardioactive glycosides as inhibitors for the active transport of potassium and sodium through the erythrocyte membrane. (In German.) *Helv. Physiol. Pharmacol. Acta* **11,** 346–354.

Schatzmann, H. J. (1962). Lipoprotein nature of red cell adenosine triphosphatase. *Nature* **196,** 677.

Schneider, P. B., and Wolff, J. (1965). Thyroidal iodide transport. VI. On a possible role for iodide-binding phospholipids. *Biochim. Biophys. Acta* **94,** 114–123.

Schniedermann, L. J. (1965). Solubilisation and electrophoresis of human red cell stroma. *Biochem. Biophys. Res. Commun.* **20,** 763–767.

Schoellmann, G., and Shaw, E. (1963). Direct evidence for the presence of histidine in the active center of chymotrypsin. *Biochemistry* **2,** 252–255.

Schultz, S. G., and Solomon, A. K. (1961). Determination of the effective hydrodynamic radii of small molecules by viscometry. *J. Gen. Physiol.* **44,** 1189–1199.

Schultz, S. G., and Zalusky, R. (1964a). Ion transport in isolated rabbit ileum. I. Short-circuit current and Na fluxes. *J. Gen. Physiol.* **47,** 567–584.

Schultz, S. G., and Zalusky, R. (1964b). Ion transport in isolated rabbit ileum. II. The interaction between active sodium and active sugar transport. *J. Gen. Physiol.* **47,** 1043–1059.

Schultz, S. G., and Zalusky, R. (1965). Interactions between active sodium transport and active amino-acid transport in isolated rabbit ileum. *Nature* **205,** 292–294.

Schwartz, J. H., Maas, W. K., and Simon, E. J. (1959). An impaired concentrating mechanism for amino acids in mutants of *Escherichia coli* resistant to L-canavanine and D-serine. *Biochim. Biophys. Acta* **32,** 582–583.

Seaman, G. V. F., and Heard, D. H. (1960). The surface of the washed human erythrocyte as a polyanion. *J. Gen. Physiol.* **44,** 251–268.

Sen, A. K., and Post, R. L. (1964). Stoichiometry and localization of adenosine triphosphate-dependent sodium and potassium transport in the erythrocyte. *J. Biol. Chem.* **239,** 345–352.

Sen, A. K., and Post, R. L. (1966). A second phosphorylated intermediate in active sodium and potassium transport. *Abstr. Am. Biophys. Soc. 10th Ann. Meeting, Boston, 1966* p. 152 [*Biophys. J.* **6** (1966)].

Sen, A. K., and Widdas, W. F. (1962a). Determination of the temperature- and pH-dependence of glucose transfer across the human erythrocyte membrane measured by glucose exit. *J. Physiol.* (*London*) **160,** 392–403.

Sen, A. K., and Widdas, W. F. (1962b). Variations of the parameters of glucose transfer across the human erythrocyte membrane in the presence of inhibitors of transfer. *J. Physiol.* (*London*) **160,** 404–416.

Shaw, F. H., Simon, S. E., Johnstone, B. M., and Holman, M. E. (1956). The effect of changes of environment on the electrical and ionic pattern of muscle. *J. Gen. Physiol.* **40,** 263–288.

Shaw, T. I. (1955). Potassium movements in washed erythrocytes. *J. Physiol.* (*London*) **129,** 464–475.

Sheppard, C. W., and Martin, W. R. (1950). Cation exchange between cells and plasma of mammalian blood. I. Methods applied to potassium exchange in human blood. *J. Gen. Physiol.* **33,** 703–722.

Sidel, V. W., and Hoffman, J. F. (1962). Water transport across membrane analogues. *Federation Proc.* **20,** 137.

Sidel, V. W., and Solomon, A. K. (1957). Entrance of water into human red cells under an osmotic pressure gradient. *J. Gen. Physiol.* **41,** 243–257.

Sistrom, W. R. (1958). On the physical state of the intra-cellularly accumulated substrates of β-galactoside-permease in *Escherichia coli*. *Biochim. Biophys. Acta* **29**, 579–587.

Sjölin, S. (1954). Resistance of red cells *in vitro*. *Acta Paediat.* **43**, Suppl., 98.

Sjöstrand, F. S. (1962). Critical evaluation of ultrastructural patterns with respect to fixation. *In* "Symposia of the International Society for Cell Biology," Vol. I (R. J. C. Harris, ed.), pp. 47–68. Academic Press, New York.

Sjöstrand, F. S. (1963). A new ultrastructural element of the membranes in mitochondria and of some cytoplasmic membranes. *J. Ultrastruct. Res.* **9**, 340–361.

Skála, I., and Hrubá, F. (1964). Accumulation of vitamin A by small intestine of the rat *in vitro*. *Am. J. Physiol.* **206**, 458–460.

Skou, J. C. (1957). The influence of some cations on an adenosine triphosphatase from peripheral nerves. *Biochim. Biophys. Acta* **23**, 394–401.

Skou, J. C. (1961). The relationship of a Mg^{2+}- and Na^{+}-activated, K^{+}-stimulated enzyme or enzyme system to the active linked transport of Na^{+} and K^{+} across the cell membrane. *In* "Membrane Transport and Metabolism" (A. Kleinzeller and A. Kotyk, eds.), pp. 228–236. Academic Press, New York.

Skou, J. C. (1964). Enzymatic aspects of active linked transport of Na^{+} and K^{+} through the cell membrane. *Progr. Biophys. Biophys. Chem.* **14**, 131–166.

Skou, J. C. (1966). *Protoplasma* (to be published).

Smith, H. M. (1958). Effects of sulfhydryl blockade on axonal function. *J. Cellular Comp. Physiol.* **51**, 161–171.

Smith, H. W. (1951). "The Kidney: Structure and Function in Health and Disease." Oxford Univ. Press, London and New York.

Smith, H. W. (1956). "Principles of Renal Physiology." Oxford Univ. Press, London and New York.

Smyth, D. H. (1961). Studies on the transport of amino acids and glucose by the intestine. *In* "Membrane Transport and Metabolism" (A. Kleinzeller and A. Kotyk, eds.), pp. 488–499. Academic Press, New York.

Snell, F. M., and Chowdhury, T. K. (1965). Contralateral effects of sodium and potassium on the electrical potential in frog skin and toad bladder. *Nature* **207**, 45.

Soldano, B. A. (1953). The kinetics of ion exchange processes. *Ann. N.Y. Acad. Sci.* **57**, 116–124.

Solomon, A. K. (1952). Permeability of human erythrocytes to sodium and potassium. *J. Gen. Physiol.* **36**, 57–110.

Solomon, A. K. (1960). Pores in the living cell. *Sci. Am.* **203**, 146–156.

Solomon, A. K. (1963). Single proximal tubules of *Necturus* kidney. VII. Ion fluxes across individual faces of cell. *Am. J. Physiol.* **204**, 381–386.

Solomon, A. K., Lionetti, F., and Curran, P. F. (1956). Possible cation-carrier substances in blood. *Nature* **178**, 582–583.

Sols, A. (1956a). The hexokinase activity of the intestinal mucosa. *Biochim. Biophys. Acta* **19**, 144–152.

Sols, A. (1956b). Selective fermentation and phosphorylation of sugars by Sauternes yeast. *Biochim. Biophys. Acta* **20**, 62–68.

Sols, A., and de la Fuente, G. (1961). Transport and hydrolysis in the utilization of oligosaccharides by yeasts. *In* "Membrane Transport and Metabolism" (A. Kleinzeller and A. Kotyk, eds.), pp. 361–377. Academic Press, New York.

Sols, A., de la Fuente, G., Villar-Palasi, C., and Asensio, C. (1958). Substrate specificity and some other properties of baker's yeast hexokinase. *Biochim. Biophys. Acta* **30**, 92–101.

Spencer, R. P., and Brody, K. R. (1964). Biotin transport by small intestine of rat, hamster and other species. *Am. J. Physiol.* **206,** 653–657.

Spencer, R. P., Brody, K. R., and Mautner, H. G. (1965). Intestinal transport of cystine analogues. *Nature* **207,** 418–419.

Squires, R. F. (1965). On the interactions of Na^+, K^+, Mg^{++} and ATP with the Na^+ plus K^+ activated ATPase from rat brain. *Biochem. Biophys. Res. Commun.* **19,** 27.

Stadelmann, E. J. (1956). Zur Versuchsmethodik und Berechnung der Permeabilität pflanzlicher Protoplasten. *Protoplasma* **46,** 692–710.

Stadelmann, E. J. (1963). Vergleich und Umrechnung von Permeabilitätskonstanten für Wasser. *Protoplasma* **57,** 660–718.

Stadelmann, E. J. (1964). Zu Plasmolyse und Deplasmolyse von *Allium*-Epidermen. *Protoplasma* **59,** 14–68.

Staverman, A. J. (1952). Non-equilibrium thermodynamics of membrane processes. *Trans. Faraday Soc.* **48,** 176–185.

Stein, W. D. (1956). The permeability of erythrocyte ghosts. *Exptl. Cell Res.* **11,** 232–234.

Stein, W. D. (1958a). N-terminal histidine at the active centre of a permeability mechanism. *Nature* **181,** 1662–1663.

Stein, W. D. (1958b). The mechanism of the penetration of glycerol through the red cell membrane. Ph.D. Thesis, University of London.

Stein, W. D. (1962a). Diffusion and osmosis. *In* "Comprehensive Biochemistry" (M. Florkin and E. H. Stotz, eds.), Vol. 2, Chapter 3. Elsevier, Amsterdam.

Stein, W. D. (1962b). Spontaneous and enzyme-induced dimer formation and its role in membrane permeability. I. The permeability of non-electrolytes at high concentration. *Biochim. Biophys. Acta* **59,** 35–46.

Stein, W. D. (1962c). Spontaneous and enzyme-induced dimer formation and its role in membrane permeability. II. The mechanism of movement of glycerol across the human erythrocyte membrane. *Biochim. Biophys. Acta* **59,** 47–65.

Stein, W. D. (1962d). Spontaneous and enzyme-induced dimer formation and its role in membrane permeability. III. The mechanism of movement of glucose across the human erythrocyte membrane. *Biochim. Biophys. Acta* **59,** 66–77.

Stein, W. D. (1962e). Unpublished results.

Stein, W. D. (1964a). Facilitated diffusion. *Recent Progr. Surface Sci.* **1,** 300–337.

Stein, W. D. (1964b). A procedure which labels the active centre of the glucose transport system of the human erythrocyte. *In* "The Structure and Activity of Enzymes" (T. W. Goodwin, B. S. Hartley, and J. I. Harris, eds.), pp. 133–137. Academic Press, New York.

Stein, W. D. (1964c). Unpublished results.

Stein, W. D., and Danielli, J. F. (1956). Structure and function in red cell permeability. *Discussions Faraday Soc.* **21,** 238–251.

Steinbrecht, I., and Hofmann, E. (1964). Die Permeabilität von Kaninchenerythrocyten für 2-Desoxy-D-ribose. *Z. Physiol. Chem.* **339,** 194–201.

Stewart, D. R., and Jacobs, M. H. (1936). Further studies on the permeability of the egg of *Arbacia punctulata* to certain solutes and to water. *J. Cellular Comp. Physiol.* **7,** 333–358.

Stoeber, F. (1961). Ph.D. Thesis, reported in Kepes and Cohen (1962).

Stoeckenius, W. (1962). The molecular structure of lipid-water systems and cell membrane models studied with the electron microscope. *In* "Symposia of the International Society for Cell Biology," Vol. I (R. J. C. Harris, ed.), pp. 349–368. Academic Press, New York.

Stricks, W., and Kolthoff, I. M. (1953). Reactions between mercuric mercury and cysteine and glutathione. Apparent dissociation constants, heats and entropies of formation of various forms of mercuric mercapto-cysteine and glutathione. *J. Am. Chem. Soc.* **75**, 5673–5681.

Swift, H. (1962). Nucleoprotein localization in electron micrographs. Metal binding and autoradiography. *In* "Symposia of the International Society for Cell Biology," Vol. I (R. J. C. Harris, ed.), pp. 213–232. Academic Press, New York.

Taylor, H. S. (1938). The temperature variation of diffusion processes. *J. Chem. Phys.* **6**, 331–334.

Teorell, T. (1935). An attempt to formulate a quantitative theory of membrane permeability. *Proc. Soc. Exptl. Biol. Med.* **33**, 282–285.

Teorell, T. (1952). Permeability properties of erythrocyte ghosts. *J. Gen. Physiol.* **35**, 669.

Teorell, T. (1956). Transport phenomena in membranes. *Discussions Faraday Soc.* **21**, 9–26.

Thompson, T. E. (1964). The properties of bimolecular phospholipid membranes. *In* "Cellular Membranes in Development" (M. Locke, ed.), pp. 83–96. Academic Press, New York.

Thompson, T. E. (1966). *Protoplasma* (to be published).

Tosteson, D. C. (1959). Halide transport in red blood cells. *Acta Physiol. Scand.* **46**, 19–41.

Tosteson, D. C. (1964). Regulation of cell volume by sodium and potassium transport. *In* "The Cellular Functions of Membrane Transport" (J. F. Hoffman, ed.), pp. 3–22. Prentice-Hall, Englewood Cliffs, New Jersey.

Tosteson, D. C., and Hoffman, J. F. (1960). Regulation of cell volume by active cation transport in high and low potassium sheep red cells. *J. Gen. Physiol.* **44**, 169–194.

Tosteson, D. C., Blount, R., and Cook, P. (1964). Separation of ATPase from sheep red cell membranes. *Federation Proc.* **23**, 114.

Troshin, A. S. (1961). Sorption properties of protoplasm and their role in cell permeability. *In* "Membrane Transport and Metabolism" (A. Kleinzeller and A. Kotyk, eds.), pp. 45–53. Academic Press, New York.

Ullrich, K. J., Rumrich, G., and Fuchs, G. (1964). Wasserpermeabilität und transtubulärer Wasserfluss corticaler Nephronabschnitte bei verschiedenen Diuresezustande. *Arch. Ges. Physiol.* **280**, 99–119.

Ussing, H. H. (1949a). The active transport through the isolated frog skin in the light of tracer studies. *Acta Physiol. Scand.* **17**, 1–37.

Ussing, H. H. (1949b). Distinction by means of tracers between active transport and diffusion. The transfer of iodide across the isolated frog skin. *Acta Physiol. Scand.* **19**, 43–56.

Ussing, H. H. (1954). Active transport of inorganic ions. *Symp. Soc. Exptl. Biol.* **8**, 407–422.

Ussing, H. H. (1959). Ionic movements in cell membranes in relation to the activity of the nervous system. *Proc. 4th Intern. Congr. Biochem. Vienna, 1958* Vol. 3, pp. 1–17. Pergamon Press, Oxford.

Ussing, H. H., and Zerahn, K. (1951). Active transport of sodium as the source of the electric current in the short-circuited isolated frog skin. *Acta Physiol. Scand.* **23**, Suppl. 80, 110–127.

Vaidhyanathan, V. S., and Perkins, W. H. (1964). On the permeability of non-electrolytes through biological membranes. *J. Theoret. Biol.* **7**, 329–333.

van Deenen, L. L. M., and de Gier, J. (1964). Chemical composition and metabolism of lipids in red cells of various animal species. *In* "The Red Blood Cell" (C. Bishop and D. M. Surgenor, eds.), pp. 243–308. Academic Press, New York.

van Gastel, C., *et al.* (1964). Quoted in van Deenen and de Gier (1964).

van Steveninck, J., Weed, R. I., and Rothstein, A. (1965). Localization of erythrocyte membrane sulphydryl groups essential for glucose transport. *J. Gen. Physiol.* **48**, 617–632.

Vidaver, G. A. (1964a). Transport of glycine by pigeon red cells. *Biochemistry* **3**, 662–667.

Vidaver, G. A. (1964b). Glycine transport by hemolyzed and restored pigeon red cells. *Biochemistry* **3**, 795–799.

Vidaver, G. A. (1964c). Some tests of the hypothesis that the sodium-ion gradient furnishes the energy for glycine-active transport by pigeon red cells. *Biochemistry* **3**, 803–808.

Vidaver, G. A. (1966). Inhibition of parallel flux and augmentation of counter flux shown by transport models not involving a mobile carrier. *J. Theoret. Biol.* **10**, 301–306.

Villegas, L. (1963). Equivalent pore radius in the frog gastric mucosa. *Biochim. Biophys. Acta* **75**, 131–134.

Villegas, R., and Barnola, F. V. (1961). Characterisation of the resting axolemma in the giant axon of the squid. *J. Gen. Physiol.* **44**, 963–977.

Villegas, R., and Villegas, G. M. (1960). Characterisation of the membranes in the giant nerve fiber of the squid. *J. Gen. Physiol.* **43**, Suppl. 5, 73.

Villegas, R., and Villegas, G. M. (1962). The endoneurium cells of the squid giant nerve and their permeability to ^{14}C glycerol. *Biochim. Biophys. Acta* **60**, 202–204.

Villegas, R., Barton, T. C., and Solomon, A. K. (1958). The entrance of water into beef and dog red cells. *J. Gen. Physiol.* **42**, 355–369.

Villegas, R., Caputo, C., and Villegas, L. (1962). Diffusion barriers in the squid nerve fiber. The axolemma and the Schwann layer. *J. Gen. Physiol.* **46**, 245–255.

Visscher, M. B., Varco, R. H., Carr, C. W., Dean, R. B., and Erickson, D. (1944). Sodium ion movement between the intestinal lumen and the blood. *Am. J. Physiol.* **141**, 488–505.

Walker, D. G. (1960). The transmission of sugars across the goat placenta. *Biochem. J.* **74**, 287–297.

Wallach, D. F. H., and Zahler, P. H. (1966). Protein conformations in cellular membranes. *Proc. Natl. Acad. Sci.* **56**, 1552–1559.

Wallenfels, K., and Malhotra, O. P. (1960). β-Galactosidase. *In* "The Enzymes" (P. D. Boyer, H. Lardy, and K. Myrbäck, eds.), 2nd ed., Vol. 4, pp. 409–430. Academic Press, New York.

Wang, J. H. (1951). Self-diffusion and structure of liquid water. I. Measurement of self-diffusion of liquid water with deuterium as a tracer. *J. Am. Chem. Soc.* **73**, 510–513.

Wang, J. H., Robinson, C. V., and Edelman, I. S. (1953). Self-diffusion and structure of liquid water. III. Measurement of the self-diffusion of liquid water with H^2, H^3, and O^{18} as tracers. *J. Am. Chem. Soc.* **75**, 466–470.

Wartiovaara, V. (1942). Über die Temperaturabhängigkeit der Protoplasmapermeabilität. *Ann. Botan. Soc. Zool. Botan. Fennicae "Vanamo"* **16**, 1.

Wartiovaara, V. (1949). The permeability of the plasma membranes of *Nitella* to normal primary alcohols at low and intermediate temperatures. *Physiol. Plantarum.* **2,** 184–196.

Wartiovaara, V., and Collander, R. (1960). Permeabilitätstheorien. *In* "Protoplasmatologia: Handbuch der Protoplasmaforschung," Vol. II, C. 8d. Springer, Vienna.

Ways, P., and Hanahan, D. J. (1964). Characterization and quantification of red cell lipids in normal man. *J. Lipid Res.* **5,** 318–328.

Weed, R. I., Eber, J. V., and Rothstein, A. (1962). Interaction of mercury with human erythrocyte. *J. Gen. Physiol.* **45,** 395–410.

Weed, R. I., Reed, C. F., and Berg, G. (1963). Is hemoglobin an essential structural component of human erythrocyte membranes? *J. Clin. Invest.* **42,** 581–588.

Weed, R. I., van Steveninck, J., and Rothstein, A. (1964). See van Steveninck *et al.* (1965).

Welch, K., Sadler, K., and Gold, G. (1966). Volume flow across choroidal ependyma of the rabbit. *Am. J. Physiol.* **210,** 232–236.

Wheeler, K. P., and Whittam, R. (1964). Structural and enzymic aspects of the hydrolysis of adenosine triphosphate by membranes of kidney cortex and erythrocytes. *Biochem. J.* **93,** 349–363.

Wheeler, K. P., Inui, Y., Hollenberg, P. F., Eavenson, E., and Christensen, H. N. (1965). Relation of amino acid transport to sodium-ion concentration. *Biochim. Biophys. Acta* **109,** 620–622.

Whittam, R. (1958). Potassium movements and ATP in human red cells. *J. Physiol. (London)* **140,** 479–497.

Whittam, R., and Ager, M. E. (1964). Vectorial aspects of adenosine-triphosphatase activity in erythrocyte membranes. *Biochem. J.* **93,** 337–348.

Whittam, R., and Ager, M. E. (1965). The connexion between active cation transport and metabolism in erythrocytes. *Biochem. J.* **97,** 214–227.

Whittembury, G., Sugino, N., and Solomon, A. K. (1960). Effect of anti-diuretic hormone and calcium on the equivalent pore radius of kidney slices from *Necturus*. *Nature* **187,** 699–701.

Widdas, W. F. (1952). Inability of diffusion to account for placental glucose transfer in the sheep and consideration of the kinetics of a possible carrier transfer. *J. Physiol. (London)* **118,** 23–39.

Widdas, W. F. (1953). An apparatus for recording erythrocyte volume changes in permeability studies. *J. Physiol. (London)* **120,** 20P.

Widdas, W. F. (1954a). Difference of cation concentrations in foetal and adult sheep erythrocytes. *J. Physiol. (London)* **125,** 18–19.

Widdas, W. F. (1954b). Facilitated transfer of hexoses across the human erythrocyte membrane. *J. Physiol. (London)* **125,** 163–180.

Widdas, W. F. (1955). Hexose permeability of foetal erythrocytes. *J. Physiol. (London)* **127,** 318–327.

Wilbrandt, W. (1954). Secretion and transport of non-electrolytes. *Symp. Soc. Exptl. Biol.* **8,** 136–161.

Wilbrandt, W. (1959). Transportsysteme für Zucker. *Mod. Probl. Pediat.* **4,** 30–49.

Wilbrandt, W., and Kotyk, A. (1964). Transport of sugar mono- and di-complexes in human erythrocytes. *Arch. Exptl. Pathol. Pharmakol.* **249,** 279–287.

Wilbrandt, W., and Rosenberg, T. (1961). The concept of carrier transport and its corollaries in pharmacology. *Pharmacol. Rev.* **13,** 109–183.

Wilbrandt, W., Mislin, H., and Strauss, F. (1955). Spezifitäten der Zellpermeabilität und stammesgeschichtliche Verwandtschaft. *Z. Vergleich. Physiol.* **37**, 211–220.

Winkler, H. H., and Wilson, T. H. (1966). The role of energy coupling in the transport of β-galactosides by *Escherichia coli*. *J. Biol. Chem.* **241**, 2200–2211.

Winter, C. G., and Christensen, H. N. (1964). Migration of amino acids across the membrane of the human erythrocyte. *J. Biol. Chem.* **239**, 872.

Winter, C. G., and Christensen, H. N. (1965). Contrasts in neutral amino acid transport by rabbit erythrocytes and reticulocytes. *J. Biol. Chem.* **240**, 3594–3600.

Witzel, H., and Barnard, E. A. (1962). Mechanism and binding sites in the ribonuclease reaction. I. Kinetic studies on the second step of the reaction. *Biochem. Biophys. Res. Commun.* **7**, 289–294.

Wolff, J. (1960). Thyroidal iodide transport. I. Cardiac glycosides and the role of potassium. *Biochim. Biophys. Acta* **38**, 316.

Wolff, J. (1964). Transport of iodide and other anions in the thyroid gland. *Physiol. Rev.* **44**, 45–90.

Wolpert, L., Thompson, C. M., and O'Neill, C. H. (1965). Studies on the isolated membrane and cytoplasm of *Amoeba proteus* in relation to ameboid movement. *In* "Primitive Motile Systems in Cell Biology" (R. D. Allen and N. Kamiya, eds.), pp. 143–172. Academic Press, New York.

Wong, J. T. (1965). The possible role of polyvalent carriers in cellular transports. *Biochim. Biophys. Acta* **94**, 102–113.

Yudkin, M. (1965). Isolation and analysis of the protoplast membrane of *Bacillus megaterium*. *Biochem. J.* **98**, 923–928.

Yunis, A. A., Arimura, G., and Kipnis, D. M. (1962). Amino acid transport and the sodium-potassium pump in human leukocytes. *J. Lab. Clin. Med.* **60**, 1028 (abstr.).

Zabin, I. (1963). Crystalline thiogalactoside transacetylase. *J. Biol. Chem.* **238**, 3300–3306.

Zabin, I., Kepes, A., and Monod, J. (1962). Thiogalactoside transacetylase. *J. Biol. Chem.* **237**, 253–257.

Zadunaisky, J. A., Parisi, M. N., and Montoreano, R. (1963). Effect of antidiuretic hormone on permeability of single muscle fibres. *Nature* **200**, 365–366.

Zerahn, K. (1961). Active sodium transport across the isolated frog skin in relation to metabolism. *In* "Membrane Transport and Metabolism" (A. Kleinzeller and A. Kotyk, eds.), pp. 237–246. Academic Press, New York.

Zierler, K. L. (1961). A model of a poorly-permeable membrane as an alternative to the carrier hypothesis of cell membrane penetration. *Bull. Johns Hopkins Hosp.* **109**, 35–48.

Zwolinski, B. J., Eyring, H., and Reese, C. E. (1949). Diffusion and membrane permeability. I. *J. Phys. & Colloid Chem.* **53**, 1426–1453.

Author Index

Numbers in italics refer to pages on which the complete references are listed.

A

Abrams, A., 130, *324*
Adair, G. S., 315, *324*
Adamson, A. W., 32, 34, *324*
Adrian, R. H., 93, *324*
Ager, M. E., 186, 216, 218, 219, 222, 228, *350*
Ahmed, K., *324, 327, 336*
Akedo, H., 171, 235, 277, 278, 279, *324, 327*
Alberty, R. A., 60, *329*
Albright, C. D., 304, 305, 306, *343*
Alvarado, F., 171, 186, 282, *324*
Andersen, B., 260, *324*
Anfinsen, C. B., 289, *324*
Annegers, J. H., 186, 276, *324*
Anton, J., 130, 171, *327*
Argaman, M., 13, *343*
Arimura, G., 132, *351*
Asensio, C., 267, *346*
Ashmore, J., 275, *334*
Aspen, A. J., 279, *327*
Austin, G., 111, *324*
Autilio, L. A., 13, *325*
Avigad, G., 224, *325*
Avigan, J., 13, *343*
Avria, I., 8, *331*

B

Baker, P. F., 214, *325*
Bangham, A. D., 101, 102, 104, *325*
Barlund, H., 58, 73, 76, 82, 83, 85, *328*
Barnard, E. A., 289, *325, 351*
Barnola, F. V., 57, *349*
Barron, D. H., 276, *325*
Barton, T. C., 111, *325, 349*
Battaglia, F. C., 276, *325*
Beament, J. W. L., 245, *325*
Bégin, N., 133, 171, 307, *325*
Bender, M. L., 281, *325*
Benedetti, E. L., 13, *331*
Berg, G., 7, 8, *350*
Bernstein, S. S., 8, *331*
Biber, T., 237, *326*
Bihler, I., 178, 179, 181, 273, *325, 328*
Blair, A., 133, 171, 277, *344*
Blank, M., 97, *325*
Blecher, M., 283, 294, *325*
Blount, R., 306, *348*
Bobinski, H., 11, 297, *325*
Bockris, J. O'M., 92, *325*
Boezi, J. A., 280, *325*
Bonsall, R. B., 11, 297, *326, 335*
Bonting, S. L., 306, *326*
Borrero, L. M., 57, 108, *342*
Bos, C. J., 13, *331*
Bower, B. F., 228, 262, *326*
Bowyer, F., 127, 131, 137, 149, 154, 171, 283, 290, 291, 292, 275, *326*
Boyd, G. E., 104, *326*
Bradbury, M. W. B., 133, *326*
Branton, D., 3, 31, *326*
Bricker, N. S., 216, 237, *326, 337*
Britten, J. S., 97, *325*
Britton, H. G., 147, 153, 154, 163, 171, 172, 308, *326*
Brody, K. R., 225, 276, *347*
Brown, D. A. J., 111, *325*
Bruce, M., 225, *344*
Budreau, A., *339*
Burgen, A. S. V., 315, 316, *326*
Burger, M., 58, 130, 143, 144, 152, 275, *326, 337*
Butler, J. A. V., 61, 71, 91, *326*
Buttin, G., 214, 233, 300, *344*

C

Cadenas, E., 132, *342*
Caldwell, P. C., 216, *326*
Calvin, M., 9, *341*
Caputo, C., 94, *349*
Carr, C. W., 210, *349*
Carter, J. R., *332*
Castleman, H., 268, *331*
Cecil, R., 291, *326*
Cereijido, M., 131, 171, 239, 240, *326*

Chain, E. B., 133, *329*
Chan, S. S., 283, *327*
Chance, B., 221, *327*
Changeux, J-P., 308, 321, *341*
Chen, L., 275, *327*
Chinard, F. P., 45, 109, *327*
Chirigos, M. A., 133, *327*
Chowdhury, T. K., 238, 239, *346*
Christensen, H. N., 131, 132, 171, 177, 178, 189, 194, 203, 235, 276, 277, 278, 279, 321, *324, 327, 335, 342, 350, 351*
Christie, G. S., *327, 336*
Cirillo, V. P., 130, 171, 276, 275, *327*
Clarkson, E. M., 104, *328*
Clarkson, T. W., 294, *342*
Coggins, C. H., 216, *339*
Cohen, B. L., 250, *331*
Cohen, G. N., 214, 223, 224, 233, 234, 275, 276, 280, 300, *328, 336, 344*
Cohen, J. A., 11, 281, *328, 341*
Cohen, W., 268, *331*
Collander, R., 58, 65, 73, 74, 76, 82, 83, 85, *328, 350*
Colowick, S. P., 13, *337*
Conway, E. J., 93, *328*
Cook, G. M. W., 9, *328*
Cook, P., 306, *348*
Cooper, A. G., 268, *331*
Cori, C. F., 132, 171, 275, 276, 283, *328*
Crane, R. K., 132, 133, 171, 178, 180, 181, 186, 195, 272, 273, 274, 275, 276, 281, 282, 283, 317, 318, *324, 325, 328, 337*
Crestfield, A. M., 289, *328*
Crofford, O. B., 133, *328*
Csáky, T. Z., 131, 178, 179, 186, 306, *328, 329*
Curie, P., *329*
Curran, P. F., 56, 90, 131, 171, 210, 239, 240, 255, 256, 261, 262, 264, 295, 296, *326, 329, 334, 341, 346*

D

Dainty, J., 53, 54, 56, 57, 58, 59, 77, 118, 123, 124, *329*
Dalton, N. N., 112, *340*
Danielli, J. F., 19, 32, 34, 39, 52, 97, 100, 108, 126, 127, 131, 163, 287, 288, 315, 322, *329, 347*
Daniels, F., 60, *329*
Davies, D. R., 281, *329*
Davson, H., 19, 32, 34, 52, 97, 100, 126, 133, *326, 329*
Dawson, A. C., 171, 175, 290, 292, 293, *329*
Dawson, R. M. C., 308, *332*
De Almeida, D. F., 133, *329*
Dean, R. B., 210, *349*
Deavin, A., 289, *340*
de Gier, J., 8, 9, *329, 349*
De Groot, S. R., 41, 42, 44, 208, *330*
de la Fuente, G., 130, 267, *346*
Demerec, M., 298, *341*
de Moss, R. D., 280, *325*
Dempsey, E. F., 131, 171, *332*
Derviehian, D. G., 1, 22, 25, *330*
Diamond, J. M., 53, 210, 251, 253, 255, 257, 258, 259, 264, *330*
Dick, D. A. T., 53, 55, 116, *330*
Diedrich, D. F., 284, *330*
Dixon, M., 139, 175, 266, 267, 268, 270, *330*
Dodge, J. T., 7, 8, *330*
Downing, S. J., 171, 277, 279, *344*
Drapeau, G. R., 130, *330*
Dreyfuss, J., 298, *330*
Dunham, E. T., 304, 306, *330*
Durbin, R. P., 45, 46, 57, 111, *330*
Dziemian, A. J., 8, *342*

E

Eavenson, E., 188, 194, *350*
Eber, J. V., 149, *350*
Eddy, A. A., 190, 198, 203, 205, *330*
Edelman, I. S., 115, *349*
Eden, E. G., 8, *343*
Edwards, C., *330*
Egan, J. B., 223, 303, *330, 331*
Eichberg, J., 308, *332*
Eichholz, A., 195, *328*
Einstein, A., 67, *331*
Eisen, H. N., 132, 275, 276, 283, *334*
Emmelot, P., 13, *331*
Erickson, B. N., 8, *331*
Erickson, D., 210, *349*
Erlanger, B. F., 268, *331*
Evans, J. V., 250, *331*
Ewell, R. H., 71, *331*

Eyring, H., 65, 66, 67, 69, 70, 71, 72, 87, 88, 89, *331, 332, 351*

F

Fernald, G. W., 131, 178, 179, 186, *328, 329*
Fernández-Morán, H., 15, *331*
Fick, A., 37, *331*
Field, R. A., 132, 171, 275, 276, 283, *328*
Finch, L. R., 171, *331*
Finean, J. B., 10, 21, 31, *331*
Fischer, H., 177, *327*
Fisher, R. B., 171, 186, *331*
Fishman, R. A., 133, *331*
Flanigan, W. J., 131, 210, 261, 262, *327, 329*
Fleischer, S., 22, 24, 33, *333*
Fleming, W. E., 211, *331*
Folch, J., 319, *331*
Forstner, G., 195, *328*
Foulkes, E. C., 132, 294, *331*
Fox, C. F., 299, 302, *332*
Fox, M., 134, *332*
Frank, H., 111, *330*
Frazier, H. S., 131, 171, *332*
Freer, J. H., 13, *344*
Frenkel, J. I., 66, *332*
Frey-Wyssling, A., 16, 32, 33, *332*
Fuchs, G., 111, *348*
Fuhrman, F. A., 241, *332*
Fuhrman, G. J., 241, *332*

G

Gaines, G. L., 315, *332*
Gale, E. F., *332*
Gardos, G., 217, *332*
Garrahan, P. J., 228, *332*
George, J., 108, 131, 290, 294, *335*
Geren, B. B., 19, *332*
Giebel, O., 101, *332*
Gilbert, J. C., 133, 171, *332*
Gillespie, D. H., 298, *341*
Ginzburg, B. Z., 49, 50, 51, 53, 54, 56, 57, 77, 111, 118, 123, 124, *329, 332*
Giraldez, A., 284, *337*
Glassman, H. N., 78, 79, 81, 106, 107, *335*
Glasstone, S., 66, 67, 69, 70, 72, 87, 88, 89, *332*
Glauert, A. M., 23, 26, 27, 28, 31, *339*
Glynn, I. M., 104, 209, 222, 225, 226, 227, 228, 304, 306, 308, 314, *330, 332*
Gold, G., 111, *350*
Goldstein, D. A., 55, 57, 112, 113, 116, 117, 118, 119, 255, 256, *332, 342*
Goodman, J., 225, *332*
Goresky, C. A., 133, *333*
Gorman, J., 130, 224, 231, *333*
Green, A. L., 281, *329*
Green, D. E., 22, 24, 33, *333*
Green, K., 210, *335*
Greengard, P., 133, *327*
Grobecker, H., *337*
Gross, J. D., *339*
Guroff, G., 132, *333*

H

Habich, K. I., 296, *338*
Hagihira, H., 171, 277, *333, 339*
Halvorson, H. O., 130, 224, 231, *333, 344*
Hanahan, D. J., 7, 8, 9, 12, *330, 340, 341, 350*
Hanai, T., 111, *333*
Handler, J. S., 240, 264, *341*
Handovsky, H., 73, *333*
Harned, H. S., 91, *333*
Harris, E. J., 45, 92, 93, 99, 100, 101, 109, 161, 163, 171, 209, 314, *330, 333*
Harris, H., 250, *331*
Hartley, B. S., 300, *333*
Hartzog, H. G., 178, 179, *329*
Hatch, F. T., 319, *333*
Hauser, G., 133, *333*
Hawkins, K. A., 180, 306, *325*
Hawkins, N. M., *326*
Haydon, D. A., 22, 32, 84, 111, *333*
Hays, R. M., 57, 111, 123, *333, 338*
Heard, D. H., 9, *328, 333, 345*
Heckmann, K. D., 242, 243, *334*
Hedbys, B. O., 57, *340*
Heilbrunn, L. V., 261, *334*
Heinz, E., 171, 306, *334, 337*
Heisey, S. R., 111, *334*
Hejmová, L., 130, 275, *326*
Held, D., 111, *334*
Hellegers, A. E., 276, *325*

Helmreich, E., 132, 275, 276, 283, *334*
Hempling, H. G., 53, 58, 81, 104, 123, 124, 150, 158, 171, *334, 340*
Henderson, M. J., 132, *334, 342*
Herrera, F. C., 131, 171, 210, 261, 262, *327, 329, 334*
Hess, H., 296, *338*
Hillman, R. S., 275, *334*
Hird, F. J. R., 171, *331*
Hirs, C. H. W., 289, *334*
Hodgkin, A. L., 93, 94, 95, 96, 97, 104, 214, 216, 222, *325, 326, 334*
Hodgman, C. D., *334*
Höber, R., 106, 107, *334*
Höfer, M., 224, *337*
Hoffman, J. F., 97, 98, 114, 187, 217, 227, 248, 250, 251, 252, 253, 255, 256, 306, *334, 335, 342, 345, 348*
Hofmann, E., 171, *347*
Hogg, J. F., 132, *341*
Hokin, L. E., 300, 308, *335*
Hokin, M. R., 308, *335*
Hollenberg, P. F., 188, 194, *350*
Holman, M. E., 314, *345*
Holmstedt, B., 281, *335*
Horecker, B. L., 130, 223, *335, 342*
Hoskin, F. G. G., 131, *335*
House, C. R., 210, *335*
Hrubá, F., 225, *346*
Hudson, M. R., 296, *338*
Hunt, S., 11, 297, *326, 335*
Hunter, F. R., 108, 131, 290, 294 *335*

I

Inui, Y., 171, 188, 194, *335, 350*
Isselbacher, K. J., 187, *344*

J

Jacob, F., 300, *335*
Jacobs, M. H., 39, 53, 78, 79, 81, 82, 98, 106, 107, *335, 347*
Jacquez, J. A., 132, 153, 180, 283, 307, *335*
Jardetzky, O., 208, *335*
Jenerick, H. P., *335*
Jervis, E. L., 283, *336*
Johns, D. G., 133, *333*
Johnson, F. R., 283, *336*
Johnstone, B. M., 314, *345*
Jolly, P. C., 213, 222, 225, 226, 227, 228, 245, 246, *343*
Jones, G., 112, *336*
Jones, R. L., 8, *331*
Jost, W., 87, 88, 89, *336*
Judah, J. D., *324, 327, 336*

K

Kaback, H. R., 187, 215, *336*
Katchalsky, A., 40, 42, 48, 49, 50, 51, 55, 57, 109, 111, 117, *332, 336*
Katz, B., 93, *334*
Kedem, O., 40, 42, 48, 49, 50, 55, 57, 63, 109, 117, 208, *336*
Kendrew, J. C., 319, *336*
Kennedy, E. P., 299, 302, *332*
Kepes, A., 130, 214, 216, 220, 221, 223, 224, 234, 275, 276, 277, 280, 294, 300, *336, 351*
Kessel, D. H., *339*
Keston, A. S., 319, *336*
Keynes, R. D., 58, 94, 95, 96, 97, 104, 209, 216, 217, 218, 222, 307, *326, 334, 336*
Kézdy, F. J., 281, *325*
King, J. W. B., 250, *331*
Kinsolving, C. R., 304, 305, *343*
Kinter, W. B., 191, *336*
Kipnis, D. M., 132, 275, *337, 342, 351*
Kirschner, L. B., 295, *337*
Klahr, S., 216, *337*
Kleinzeller, A., 130, 209, 275, 314, 319, *326, 337*
Koch, A. L., 130, 223, 229, 294, 302, *337*
Koefoed-Johnsen, V., 108, 235, 259, 260, *337*
Kolber, A. R., 159, 171, 301, 302, *337*
Kolber, J., 214, 215, 223, *337*
Kolthoff, I. M., 291, *348*
Kono, T., 13, *337*
Koshland, D. E., 289, 320, *337, 343*
Kotyk, A., 162, 209, 224, 307, 314, 319, *337, 350*
Krane, S. M., 133, 171, 272, 274, *328, 337*
Kromphardt, H., *337*
Kupchan, S. M., 300, *335*

L

Lacko, L., 58, 143, 144, 152, 275, *337*
Laidler, K. J., 66, 67, 69, 70, 72, 87, 88, 89, *332*
Landau, B. R., 275, *334*
Larralde, J., 284, *337*
Larsen, P. R., 186, 277, *337*
Lassen, U. V., 131, 171, *337*
Laster, L., 171, 186, 277, *340*
Lauf, P. K., 11, *343*
Leaf, A., 53, 57, 111, 123, 131, 171, 210, 216, 220, 240, 244, 257, 259, 260, 261, *332, 333, 338*
Lee, J. S., 257, *338*
Lees, M., 319, *331*
LeFevre, P. G., 3, 53, 108, 131, 137, 140, 141, 142, 152, 158, 159, 163, 171, 176, 187, 242, 268, 269, 270, 271, 275, 283, 284, 285, 286, 287, 290, 292, 295, 296, 297, 307, *327, 337, 338*
Lehninger, A. L., 232, *338*
Lenard, J., 34, *338*
Leoncini, G., 132, 176, 275, 276, *339*
Levine, M., 150, 151, 158, 161, 163, 171, 172, 173, *338, 339*
Lin, E. C. C., 171, 277, *333, 339*
Lindemann, B., 57, 111, *339*
Linderholm, H., 294, *339*
Ling, G. N., 209, 314, *339*
Lionetti, F., 295, 296, *346*
Longsworth, L. G., 104, *339*
Longuet-Higgens, H. C., 111, *324*
Lotspeich, W. D., 283, *327*
Love, A. H. G., 210, 241, *339, 343*
Love, W. E., 78, 80, 82, *339*
Lovelock, J. E., 7, *339*
Lubin, M., *339*
Lucké, B., 123, *339*
Lucy, J. A., 23, 26, 27, 28, 31, *339*
Lumry, R., 87, *339*
Luzzato, L., 132, 176, 275, 276, *339*

M

Maas, W. K., *345*
McClennan, A. L., 86, *343*
McCutcheon, M., 123, *339*
McGinniss, G. F., 140, 141, 163, 171, *338*
McIntosh, J. R., 255, 256, 329, *341*
McLean, A. E. M., *327, 337*
McLellan, W. L., *342*
MacLeod, R. A., 130, *330*
MacRobbie, E. A. C., 220, *339*
Macy, J. G., 8, *331*
Maddy, A. H., 11, *339*
Maffly, R. H., 216, *339*
Maio, J. J., 132, 171, 275, 276, 283, *344*
Maizels, M., 99, 104, 214, 218, 228, *328 339, 340*
Mandelstam, J., *340*
Mares-Guia, M., 299, *340*
Marinetti, G. V., 8, *343*
Marshall, J. K., 141, 270, 287, *338*
Martin, W. R., 104, *345*
Martins-Ferreira, H., 93, *333*
Mathias, A. P., 289, *340*
Matthews, D. M., 171, 186, 277, *340*
Mauro, A., 110, *340, 344*
Mautner, H. G., 277, *347*
Mawe, R. C., 58, 150, 171, *340*
Merrit, C. R., 304, 305, *343*
Meschia, G., 276, *325*
Meyer, K. H., 100, *340*
Miller, D., 178, 180, 181, 273, *328*
Miller, D. M., 158, 161, 163, 171, 172, 176, 314, *340*
Miner, C. S., 112, *340*
Mishima, S., 57, *340*
Mislin, H., 290, *351*
Mitchell, C. D., 7, 8, 9, 12, *330, 340*
Mitchell, P., 13, 130, 151, 209, 225, 319, 320, *340, 341*
Mitchell, T. G., 210, 241, *339, 343*
Mitchell, W. B., 7, 9, *340*
Mizobuchi, K., 298, *341*
Mokotoff, M., 300, *335*
Monod, J., 130, 214, 223, 233, 300, 308, 321, *335, 341, 344, 351*
Montoreano, R., 57, 111, *351*
Moore, S., 289, *328*
Morgan, H. E., 132, 145, 148, 153, 162, 163, 171, 275, 276, *341, 342, 343, 344*
Morgan, T. E., 12, 283, *341*
Morse, M. L., 223, 303, *330, 331*
Moskowitz, M., 9, *341*
Mothon, S., 279, *327*
Moyle, J., 12, *341*

Mulcahy, M., 190, 198, 203, 205, *330*
Malhotra, O. P., 300, *349*

N

Najjar, V. A., 306, *341*
Narahara, H. T., 131, 171, *341*
Neptune, E. M., 210, 241, *339*, *343*
Newey, H., 277, *341*
Nirenberg, M. W., 132, *341*
Nixon, D. A., 133, 276, *341*
Norton, W. T., 13, *325*

O

Ørskov, S. L., 53, 106, 107, *334*, *342*
Özand, P., 131, 171, *341*
Ogilvie, J. T., 255, *341*
Ohnishi, T., 11, *341*
Okada, H., 130, 224, 231, *333*
Oken, D. E., 210, *341*
O'Neill, C. H., 13, *351*
Oosterbaan, R. A., 281, *341*
Orloff, J., 240, 264, *341*
Osborn, M. J., *342*
Ospina, B., 108, 131, 290, 294, *335*
Osterhout, W. J. V., 148, *342*
Overgaard-Hansen, K., 131, 171, *337*
Owen, B. B., 91, *333*
Oxender, D. L., 132, 150, 151, 158, 163, 171, 172, 203, 277, 278, 279, 294, *327*, *339*, *342*

P

Paganelli, C. V., 58, 107, 110, 111, 123, *342*
Paine, C. M., 132, 294, *331*
Palatine, I. M., 177, *327*
Pappenheimer, J. R., 57, 108, 111, *334*, *342*
Pardee, A. B., 276, 298, *330*, *342*, *343*
Parisi, M. N., 57, 111, *351*
Park, C. R., 132, 145, 148, 163, 283, 296, *341*, *342*, *343*
Parpart, A. K., 8, 78, 79, 81, 106, 107, *335*, *342*
Parrish, J. E., 132, *342*
Parsons, D. S., 186, 242, 243, 257, *331*, *334*, *342*
Partington, J. R., 71, *342*
Passow, H., 99, 100, 101, 294, *332*, *342*
Patlak, C. S., 255, 256, *342*
Pauling, L., 72, 75, 86, *343*
Perkins, W. H., 49, *349*
Pfeiffer, B., 171, *334*
Phillips, R. A., 241, *343*
Pichler, A. G., 171, *334*
Pimentel, G. C., 86, *343*
Pitts, R. F., 241, *343*
Pocchiari, F., 133, *329*
Ponder, E., 1, 3, 7, 246, *343*
Post, R. L., 132, 171, 213, 216, 222, 225, 226, 227, 228, 245, 246, 304, 305, 306, 308, *342*, *343*, *345*
Poulik, M. D., 11, *343*
Prankerd, T. A. J., 209, 314, *333*
Prescott, D. M., 111, *343*
Prestidge, L. S., 276, 298, *342*, *343*

Q

Quastel, J. H., 186, *344*

R

Rabin, B. R., 289, *340*
Randle, P. J., 275, 276, *341*
Ray, N. E., 177, 178, *327*
Ray, W. J., 289, *343*
Razin, S., 13, *343*
Reed, C. F., 7, 8, *343*, *350*
Reese, C. E., 65, *351*
Reeves, R. E., 269, *343*
Regen, D. M., 132, 145, 148, 153, 162, 163, 171, 275, 276, *341*, *342*, *344*
Reinwein, D., 295, 296, *344*
Remington, M., 104, *340*
Renkin, E. M., 57, 108, 112, 113, 116, *342*, *344*
Renold, A. E., 133, *328*
Rickenberg, H. V., 132, 171, 214, 224, 233, 275, 276, 283, 300, *328*, *344*
Riggs, T. E., 177, 178, 279, *327*
Riklis, E., 186, *344*
Ring, K., *337*
Robbins, E., *344*
Robertson, J. D., 2, 5, 20, 21, 26, 28, 29, *344*
Robertson, J. J., 224, *344*
Robertson, R. N., 216, 220, *344*
Robinson, C. V., 115, *349*
Robinson, R. A., 91, *344*
Ron-Noya, J., 284, *337*

Rosenberg, L. E., 133, 171, 277, 279, *332, 344*
Rosenberg, P., 131, *335*
Rosenberg, T., 127, 137, 145, 147, 148, 152, 204, 282, 283, 284, 287, 314, 322, 344, *350*
Rosenthal, A. S., 306, *343*
Ross, J. E., 171, 186, 277, *337*
Rothstein, A., 130, 149, 225, 290, 294, 295, *332, 342, 344, 349, 350*
Rotman, B., 223, 276, *344*
Rümke, P. L., 13, *331*
Rumrich, G., 111, *348*

S

Sadler, K., 111, *350*
Salton, M. R. J., 13, *344*
Sato, M., 111, *324*
Saunders, S. J., 187, *344*
Schatzmann, H. J., 228, 306, *345*
Schmitt, F. O., 19, *332*
Schneider, P. B., 295, 296, *345*
Schneidermann, L. J., 11, *345*
Schoellmann, G., 299, *345*
Scholefield, P. G., 133, 171, 176, 307, *325*
Schultz, S. G., 112, 133, 183, 184, 185, 186, 187, 194, 197, 210, *345*
Schwartz, G. F., *329*
Schwartz, J. H., *345*
Seaman, G. V. F., 9, *328, 333, 345*
Segal, S., 133, 171, 277, 279, *332, 344*
Sen, A. K., 143, 150, 160, 161, 162, 163, 174, 216, 228, 290, 291, 306, 308, *343, 345*
Shaw, E., *340, 345*
Shaw, F. H., 314, *345*
Shaw, T. I., 214, 216, 225, 226, 299, *325, 326*
Sheff, M. F., 283, *336*
Sheppard, C. W., 104, *345*
Sherman, J. H., 180, *335*
Sidel, V. W., 54, 110, 111, 114, *345*
Sievers, J-F., 100, *340*
Simon, E. J., *345*
Simon, K. A., 306, *326*
Simon, S. E., 314, *345*
Singer, S. J., 34, *338*
Sistrom, W. R., 298, *346*
Sjölin, S., 111, *346*
Sjöstrand, F. S., 3, 6, 30, *346*
Skála, I., 225, *346*
Skou, J. C., 294, 304, 305, 306, 316, 321, *346*
Slayman, C. W., 308, *332*
Smith, H. M., 294, *346*
Smith, H. W., 264, 281, *346*
Smyth, D. H., 133, 277, 279, 281, 283, *336, 341, 346*
Snell, F. M., 238, 239, *346*
Soldano, B. A., 104, 105, *326, 346*
Solomon, A. K., 53, 54, 55, 57, 58, 101, 104, 107, 108, 110, 111, 112, 113, 116, 117, 118, 119, 210, 240, 264, 295, 296, *329, 330, 332, 339, 341, 342, 345, 346, 349, 350*
Sols, A., 130, 267, 272, *346*
Spencer, R. P., 225, 277, *347*
Squires, R. F., 227, 307, *347*
Stadelmann, E. J., 53, 55, *347*
Stadtman, E. R., 187, 215, *336*
Standish, M. M., 101, 102, 104, *325*
Staverman, A. J., 45, *347*
Stein, W. D., 3, 39, 48, 67, 68, 69, 107, 108, 121, 127, 134, 150, 151, 158, 160, 163, 171, 172, 173, 176, 187, 214, 215, 223, 287, 288, 289, 290, 291, 292, 293, 297, 299, 301, 302, 307, 315, 317, 322, *325, 337, 338, 339, 347*
Stein, W. H., 289, *328*
Steinbrecht, I., 171, *347*
Stewart, D. R., 78, 82, *335, 347*
Stoeber, F., 234, *347*
Stoeckenius, W., 15, 16, 18, *347*
Stokes, R. H., 91, *344*
Strauss, F., 290, *351*
Stricks, W., 291, *348*
Sugino, N., 57, *350*
Swift, H., 6, *348*
Swisher, S. N., 8, *343*

T

Talley, S. K., 112, *336*
Tapley, D. F., 171, 186, 277, *337*
Taylor, F. H., 22, 84, *333*
Taylor, H. S., 71, *348*
Taylor, J., 32, 111, *333*
Teorell, T., 3, 100, 187, *348*
Terry, R. D., 13, *325*

Thale, M., 131, 178, 179, *329*
Thier, S., *332*
Thomas, J., 130, 223, *335*
Thomas, M. A. W., 291, *326*
Thompson, C. M., 13, *351*
Thompson, T. E., 114, *348*
Tormey, J. McD., 258, 259, *330*
Tosteson, D. C., 97, 98, 225, 227, 244, 248, 250, 251, 252, 253, 306, *348*
Troshin, A. S., 209, 314, *348*
Truscoe, R., 104, *340*

U

Udenfriend, S., 132, 133, *327*, *333*
Ullrich, K. J., 111, *348*
Ussing, H. H., 62, 108, 131, 156, 209, 210, 211, 212, 220, 235, 236, 237, 238, 259, 260, 294, *324*, *326*, *337*, *348*

V

Vaidhyanathan, V. S., 49, *349*
van Deenan, L. L. M., 8, 9, *329*, *349*
van Gastel, C., 8, *349*
van Steveninck, J., 290, 295, *349*, *350*
Varco, R. H., 210, *349*
Vidaver, G. A., 131, 171, 176, 187, 188, 189, 195, 199, 200, 235, 307, 308, *349*
Villar-Palasi, C., 267, *346*
Villegas, G. M., 94, 111, 131, 294, *349*
Villegas, L., 94, *349*
Villegas, R., 57, 94, 111, 131, 294, *349*
Visscher, M. B., 210, *349*
Vratsanos, S. M., 268, *331*

W

Walker, D. G., 275, 276, *349*
Wallach, D. F. H., 34, *349*
Wallenfels, K., 300, *349*
Walsh, P. M., 151, *334*
Wang, J. H., 115, 123, *349*
Warren, F. L., 250, *331*
Warringa, M. G. P. J., 11, *328*
Wartiovaara, V., 65, 78, 80, *349*, *350*
Watanabe, H., 133, *333*
Watkins, J. C., 101, 102, 104, *325*
Ways, P., 8, *350*
Webb, E. C., 139, 175, 266, 267, 268, 270, *330*
Weed, R. I., 7, 8, 149, 290, 295, 349, *350*
Welch, K., 111, *350*
Wheeler, K. P., 188, 306, *350*
White, F. H., 289, *324*
Whitmore, B., 294, *342*
Whittam, R., 187, 216, 217, 218, 219, 222, 228, 306, *350*
Whittembury, G., 57, 93, 210, *341*, *350*
Widdas, W. F., 53, 133, 137, 143, 148, 149, 150, 152, 158, 160, 161, 162, 163, 171, 174, 175, 290, 291, 292, 293, *326*, *329*, *345*, *350*
Wilbrandt, W., 127, 137, 145, 147, 148, 149, 152, 162, 171, 176, 205, 282, 283, 284, 287, 289, 307, 314, 322, *344*, *350*, *351*
Wilkins, P. O., 130, 171, *327*
Williams, H. H., 8, *331*
Wilson, T. H., 130, 171, 190, 191, 229, 230, 277, *333*, *336*, *339*, *351*,
Windhager, E. E., 210, *341*
Wingate, D. L., 257, *342*
Winkler, H. H., 130, 229, 230, *351*
Winter, C. G., 131, 171, 276, 278, 279, *327*, *351*
Witzel, H., 289, *351*
Wolff, J., 133, 171, 295, 296, *345*, *351*
Wolpert, L., 13, *351*
Wong, J. T., 308, *351*
Wyman, J., 308, 321, *341*

Y

Yudkin, M., 13, *351*
Yunis, A. A., 132, *351*

Z

Zabin, I., 300, *351*
Zachariah, P., 171, *331*
Zadunaisky, J. A., 57, 111, *351*
Zahler, P. H., 34, *349*
Zalusky, R., 133, 183, 184, 185, 186, 187, 194, 197, 210, *345*
Zerahn, K., 131, 209, 216, 220, 294, *348*, *351*
Zeuthen, E., 111, *343*
Zierler, K. L., 315, *351*
Zollicoffer, L., 178, *329*
Zwolinski, B. J., 65, *351*

Subject Index

Page numbers in boldface type are those which contain data given in tables or figures.

A

Accelerative exchange diffusion, 150–152, 156
Activation energy, 70–71
  ion transfer, **104**
  water permeability, **123**
Active transport
  amino acids, 203, **186, 224**
  ascites tumor cells, 203
  cations, 209–215, 248–251
  coupled with facilitated diffusion, 177–206
  erythrocytes, **188**, 213, **216, 225**
  inhibitors, 179
  irreversible thermodynamics, 62–64
  kinetics, **186**
  link to chemical reaction, 215–221
  primary, 201, 207–241, 309–311
    erythrocytes, 218, **225**
    kinetics, 221–228, 231–235
    stoichiometry, **216**
    uncoupled to energy output, 228–231
    vitamin A, **225**
  secondary, 178, **186**, 188, 207–209
  sodium, 209–215, **216, 225**
  sodium-potassium pump, 248–**251**
  specificity, 272–277
  sugars, **186, 223**, 272, 273
  sulfate, 298
  tertiary, 203–206
Adenosine diphosphate, in *Escherichia coli*, 220
Adenosine triphosphatase, 306, 319, 321
  in brain, 307
  in erythrocytes, 218, 226, 229
  in *Escherichia coli*, 230
  isolation, 304–306, 308
Adenosine triphosphate
  in erythrocytes, 217, 218, 228
  in nerve, 216, 217, 218
ADP, *see* Adenosine diphosphate
Algae, *see also* names of individual algae
  cell volume changes, 53, 56
  reflection coefficients, **56**
  solute diffusional coefficient, 58–59
Allosteric enzymes, 321
Allosteric model, 308
Amino acid transport
  active, 177–178
    primary, **224**
    secondary, **186**, 188
    tertiary, 203
  ascites tumor cells, 294, 307
  co-transport, 189, 194, 200
  erythrocytes, 307
  facilitated diffusion, **167–170**, 177–178, **283**
  inhibition, 294
  pancreas, 307
  reflection coefficients, **56**–57
  specificity requirements, 276–277
Anion transport, active, primary, **225**
Antidiuretic hormone, *see* Hormone, antidiuretic
Arbacia punctulata
  nonelectrolyte diffusion, **78, 81, 82**
  permeability data, **78, 82, 123**
  volume changes, 53
Arginine phosphate, 216, 218

Arrhenius expression, 70
Artificial membranes, *see* Synthetic membranes
Ascites tumor cells
amino acid transport, 194, 203, 277, 294, 307
facilitated diffusion, 132, **167, 169,** 283
ions, **104**
nonelectrolyte diffusion, **81**
nucleosides, **283**
permeability data, **123**
proteins in membrane, 34
sugar transport, 159, 276, **283**
Associated liquids, 71–72
ATP, *see* Adenosine triphosphate
ATPase, *see* Adenosine triphosphatase
Axolemma, *see* Myelin sheath
Axon
ion transfer, 94–97, **104**
reflection coefficients, **56**

B

Bacteria, *see also* names of individual bacteria
amino acid transport, 277, 280
chemical composition of membrane, **13**
facilitated diffusion, 130
permeases, 302
Bimolecular lipid leaflet, 16, 22–32, 72, 88, 90
ion transfer, passive, 106
physical stability of, 22–31
Biotin, **225**
Bivalency, 176, 288, 307–308, 321
glucose transport, 291–292
glycerol transport, 291–292
Bladder
active transport, primary, **210, 216**
permeability data, **111**
reflection coefficients, **56**
Brain
ATPase, 294, 307
facilitated diffusion, **167**
permeability data, **111**

C

Calcium, effect on permeability, 262
Capillaries, cat, reflection coefficients, **57**
Car gene, 303
Carbohydrates, active transport, primary, **223–224**
Cardiac glycosides, **98,** 181, 218, 262, 263, 305
Carrier model, 147, 148–157
accelerative exchange diffusion, 151
cation transport, 227
counterflow, 147
glucose transport, 172
kinetics, 153–157, 172–173
mobile site, 147, 148
Cation transport, *see also* Potassium, Sodium
active, primary, 213, 227
epithelial cell layers, 235–241
erythrocytes, 305–306
in amino acid transport, 188
in co-transport, 178–182, 188
in sugar transport, 178–182
inhibitors, 305
passive movement, 97–101
temperature coefficients, **104**
Cation transport system, isolation, 296
Cell membranes
chemical composition, 3–**13**
diffusion, 72–84
structure, 2–6
Cell volume, 242–251
changes, 52–57, 161
ion pumping and, 250, 252
Channels in membrane, 116
Chara, nonelectrolyte diffusion, 58, **82, 83**
Chara australis, nonelectrolyte diffusion, 77
Chara ceratophylla, nonelectrolyte diffusion, **74**
Cholesterol, in membranes, **8,** 9, 34
Co-transport
distribution ratios, 198–202
kinetics, 168–171, 192–198
model, 181–191
sugar transport, 181–182, 195
Column chromatography, 297, 301
Competitive exchange diffusion, 144, 146–147, 156
Competitive inhibition, **287–8**
Compulsory exchange diffusion, 156
Conformational stability, 269–**271**
Copper, as inhibitor, 290–292, 307

Cornea, reflection coefficients, **57**
Counter-transport, *see* Counterflow
Counterflow, 128
  sugar transport, 145–148
Cross-coefficients, 42–44
Cross-phenomena, *see* Cross-coefficients
Curie's theorem, 208

D

Diaphragm, facilitated diffusion, 132
Diffusion
  across cell membrane, 72–84
  coefficients, 36–37
    as function of molecular weight, 68
    nonelectrolytes, **88**
    water, 58–59, **110–111**
  flow, 108–116
  gases, temperature dependence, **71**
  in liquids, 65–72
    temperature dependence, 69–70, **71**
Dimerizer, 317
Dimers
  facilitated diffusion, 176
  water permeability, 121
Dinitrofluorobenzene, 290, 291, 297, 299, 307
Distribution ratio, 61
  co-transport, 198–202
DNFB, *see* Dinitrofluorobenzene
Donnan distribution, 62, 246, 248
Donnan potential, 62
Double-labeling, 299, 301
Double-membrane model, 256–7
Drugs, 281–289, *see also* Cardiac glycosides

E

Ehrlich ascites tumor cells, *see* Ascites tumor cells
Electrical potential, 210–211
  ion transfer, passive, 102
  skin, 238–240
  sodium transport, 184
Electrochemical potential, 62–63
Electrolytes, *see* Ions
Electron microscopy, 2–3, 18
  lecithin/cholesterol/water mixtures, 27
  myelin forms, 14, 15
  nerve, 5, 21
  plant cell membranes, 3
  synthetic membranes, 102
Electroneutrality condition, 61
Enzymes, 267–268, 300, *see also* Adenosine triphosphatase, Allosteric enzymes
  in whole cells, 9
  inhibition, 299–300
  mechanism of action, 281, 289
  membrane-bound, 9
  specificity, 267–268, 299–300
  transport, 317–321
Epithelial tissue, *see also* Skin
  active transport, primary, **210**
  ion pump, 235–241, 253–255
  volume changes, 53
  water transport, 253–265
Erythrocyte ghosts, protein, 34
Erythrocyte membrane
  chemical composition, 7–12
  lattice parameters, **89**
Erythrocytes
  accelerative exchange diffusion, 150
  active transport, 213, **216**, 218, **225**, 229, 248–252
  adenosine triphosphatase activity, 218
  amino acid transport, 178, **188**, 200, 307
  cation transport, 213, **216**, 218, **225–226**, 229, 248–251, 305–306
  cell volume, 55, 242
  competitive exchange diffusion, 144
  facilitated diffusion, 108, 131, 134, 136, 140–141, 161, **164–169**, 283, 288, 290
  glucose transport, 134, 140–141, 144, 149, 151, 161, **164–166**, **174**, 268–272, **283**, 285, 287, 290, 307
  glycerol transport, 283, **288**, 290–292, 294, 307
  ion transfer, passive, 97–101, **104**
  nonelectrolyte diffusion, **79**, **81**, **107**, 113, 121
  permeability data, **110**, **111**, **118**, **123**
  pore model, 116
  reflection coefficients, **57**
  solute diffusional coefficient, 58
  sugar transport, 136, 142, **164–166**, 172, **174**, 176, 275, 294, 296–297
  unit membrane, 4, 6

urea transport, 294
water transport, 54, **110**
Escherichia coli
active transport, primary, **216**, 223, 224
cation transport, 214–215
facilitated diffusion, 130
glucuronide transport, 234
sugar transport, 214–**216**, 220–221, 229, 233, 276, 294, 300–303
N Ethyl maleimide, 292, 302
Exocrine pancreas cells, unit membrane, 6

F

Facilitated diffusion, 126–176
amino acids, **167–170**
co-transport systems, **168–171**
coupled with active transport, 177–206
glucuronides, 234
inhibition, 148–150, 283, 290
ions, **168, 171,** 236
kinetic parameters, **164–171**
kinetics, 126–176, 234–5
model, 320–321
nonelectrolytes, 108
specificity requirements, 275
sugars, 136, 142, **164–167, 171,** 174–175, 181, 268–272, 275
Facilitated diffusion systems
characteristics, 127–128
distribution, 130–133
properties, 309–311
Fick's law, 40
Fixed charge model, 100–101, 103, 105
Fixed site model, 148
Frictional coefficients, ion transfer, passive, 96
Flux, 40
Flux ratio, 95
Flux ratio test, 62, 142–143
Frictional coefficients, 48–49, **51**
pore model, 114–115
Frog skin, *see* Skin

G

Galactoside transport, *E. coli*, 214–**216**, 220–221, **223**, 229, **233**, 294, 300–303
Gall bladder
volume changes, 53
water transport, 258, 262, 264
Gases, diffusion, **71**
Ghosts
active transport, 217
chemical composition, 7–12
co-transport, 187
sugar-binding, 297
Glucose transport
accelerative exchange diffusion, 150–151, 172
active, 179, **224**
ascites tumor cells, 283
competitive exchange diffusion, 144
co-transport, 195
counterflow, 145
erythrocytes, 268–272, 285, **287**, 294, 299, 307
facilitated diffusion, 134, 140–141, 147, 149, 161, **283**
frictional coefficient, **51**
ghosts, 297
inhibition, 184, 285, **287**, 290–294, 297, 299
intestine, 179, 181, 184–185, 195, **283**, 294, 317
kidney, **283**, 294
muscle, **283**
Glucose transport system, isolation, 297
Glucuronide permease, 234
Glycerol transport
erythrocytes, 121, **288**, 291, 292, 299, 307
facilitated diffusion, **283**, 299
inhibition, **288**, 290–292, 299
nerve, 294
permeability data, **118**
reflection coefficients, **56–57**
Glycerol transport system, isolation, 299
Glycine, co-transport, 188, 200
Glycols, reflection coefficients, **56–57**

H

Half-equilibrium time, 158–159
Half-times, 37
Heart
counterflow, 145
sugar transport, 145
Hemolysis times, 37
Homologous pairs, **83**
Hormone, antidiuretic, 260–262, 265

Hydraulic permeability coefficients, **110–111**
Hydrodynamic flow, 108–116
Hydrogen bonds, **76, 79–80,** 84
  thermodynamic parameters, **86**
  water permeability, 120–122
Hydrostatic flow, **46**

I

Ileum, *see* Intestine
Inhibition, 137, **287–288**
  glycerol transport, **288,** 290–292, 299
  potassium transport, 218, 228
  sodium transport, 217–218, 262, 263
  sugar transport, 179, 181, 184, 284–**287,** 294
  urea transport, 294
Inhibitors, 290, 307
  cardiac glycosides, 98, 181, 218, 262–263, 305
  chemical reagents, 289–307
  drugs, 149, 181, 184, 281–289, 297, 307
  nonpermeating, 148–150
Integrated rate equation, facilitated diffusion, 157, 159
Intestine
  active transport, **186, 225,** 272–274
  amino acid transport, 190–191, **283**
  facilitated diffusion, **169–171, 283**
  monocarboxylic acids, 294
  permeability data, **111, 210**
  reflection coefficients, **57**
  sugar transport, 179–181, 184–185, **186,** 190–191, 195, 273, 274, 276, **283,** 294, 317
Ion transport, 90–106, *see also* Anion transport, Cation transport
  cell volume and, 243
  diffusion, 59–62
  epithelial tissue, 235–241, 294
  facilitated diffusion, 168, **171**
  inhibition, 294
  long pore model, 96
  passive, 92–101, 102–3
  permeability data, **93**
  water permeability and, 245, 247–248, 253
Ions, hydration numbers, **91**
Irreversible thermodynamics, 41–52
  active transport, 63–64
  passive transport, 63–64
  pore model, 114–115, 118–119
Isolation of transport systems, 295–308

K

Kidney
  amino acid transport, 277, 294
  facilitated diffusion, 133, **169, 170**
  ion transport, **93**
  permeability data, **111, 210**
  reflection coefficients, **56**
  sugar transport, **283,** 294
  water permeability, 264
Kinetic parameters
  active transport systems
    primary, **223–225**
    secondary, **186**
  facilitated diffusion systems, **164–171**
Kinetics
  active transport systems, primary, 221–228, 231–235
  carrier model, 153–157
  co-transport, 192–198
  facilitated diffusion, 137, 157, 161

L

Lac operon, 300–301
Lattice model, 66–85
  diffusion across cell membrane, 72–84
    in liquids, 65–72
  ion transfer, 90–92, 95–97, 99, 105
  water permeability, 122
Lattice parameters, **79, 89**
"Leaks," kinetics, 231–235
Light-scattering, cell volume changes, 53–55
Lipids
  chemical structure, 10
  in membranes, 7, **8, 13**
Lipoproteins in membrane, 9–12
Liquids, diffusion, **71**
Liver, **13,** 243
Long pore model, 96

M

M-Protein, 302
Mediated transfer
  flux ratio test, 142
  properties, 309–311

Mercurials, 290, 291
Metabolism and ion transport, 215–217, 220–221
Micelles, 23, 28, 30–31
Michaelis-Menten kinetics, 134, 187–188
Mitochondrion, 30
Mobile carrier, 314–315
Mobile site, *see* Carrier model
Models
  cell membrane, 18, 19, 23, 32, 73
    fixed charge model, 100–101, 103, 105
    pore model, 106–120
  cell volume regulation, 244, 246
  epithelia
    ion transport, 236–240
    water transport, 255, 257, 259
  transport systems, 96, 115, 148, 176, 308, 311–321, *see also* Carrier model, Pore model, Pump model
Muscle, *see also* Diaphragm, Heart
  chemical composition of membrane, **13**
  facilitated diffusion, 131, **166, 283**
  ion transport, **93,** 307
  permeability data, **111**
  reflection coefficients, **57**
  sugar transport, 276, **283**
  water transport, **111**
Mutants
  E. coli, 301–302
  permeaseless, 230, **231**
Myelin forms, 12–22, 29
Myelin sheath, **13,** 19–21, 29, 31

N

Nerve, *see also* Axon, Myelin sheath, Schwann cell
  active transport, primary, 214, **216**
  cation transport, 304–305
  facilitated diffusion, 131
  ion transfer, 94–97, **104,** 215–**216,** 304–305
  permeability data, **111**
  solute diffusional coefficient, 58
  unit membrane, 5
  water permeability, **111**
Nerve myelin, **13,** *see also* Myelin sheath
Net flux, 135, 137–143
Neurolemma, *see* Myelin sheath
Nitella
  nonelectrolyte diffusion, **82**
  permeability data, **123**
Nonelectrolytes
  diffusion coefficients, **88**
  molecular radii, 113
  permeability data, **77, 78, 80, 81, 83, 107, 118**
  simple diffusion, **74**
Nonequilibrium kinetics, facilitated diffusion, 175

O

Optical birefringence
  myelin forms, 16
  myelin sheath, 20, 29
Optical rotatory dispersion spectra, 34
Osmosis
  irreversible thermodynamics, 43
  local, 255, 257
  water transport, 242–243, 245, 247
Osmotic flow, **46,** 108–116
Osmotic permeability coefficient, 39, **110–111**
Ouabain, 218

P

Pancreas
  amino acid transport, 307
  facilitated diffusion, 133, **169**
Passive transfer
  ions, 92–103
  irreversible thermodynamics, 62–64
"Paucimolecular" model, 19, 32
Peltier effect, 43
Permeability
  coefficients, 36–40, 47
    interconversion of units, 38–39
    units of measurement, 38–39
  data, *see also* Kinetic parameters, reflection coefficients
    Arbacia punctulata, **78, 82, 123**
    ascites tumor cells, **123**
    bladder, **111, 210**
    brain, **111**
    Chara australis, **77**
    Chara ceratophylla, **74, 82**
    egg cells, **110**
    epithelial tissue, **210**
    erythrocytes, **79, 98, 107, 110, 111, 117, 118, 123**

intestine, **111, 210**
ions, **93, 98, 104, 210**
kidney, **111, 210**
muscle, **111**
nerve, **111**
Nitella, **82, 123**
nonelectrolytes, **74, 77–81, 83, 107, 118**
Phascolosoma, **80, 82**
skin, **210**
synthetic membranes, **46, 111**
Tollypellopsis, **80**
water, **46, 110, 111**
methods of measuring, 52–59
Permeases, 300–303
adenosine triphosphate level and, 220
Escherichia coli, 229, **233**, 234
isolation, 300–303
specificity requirements, 275, 278
Phascolosoma, nonelectrolyte diffusion, **80, 82**
Phloretin, 149, 282–285, 297, 307
Phlorizin, 181, 184, 281–284
Phospholipids
as possible carriers, 296
in membrane, **8, 13**
model membranes, 16–18
Placenta, sugar transport, 276
Plant cells
salt transport, **216**
volume changes, 53
unit membrane, 3
Poiseuille's law, 109, 114, 255
Polar liquids, 72
Polysaccharides, in human erythrocyte membrane, 7–9
Pore model, 106–120, 315–316
erythrocytes, 116
irreversible thermodynamics, 114–115, 118–119
water transport, 115
Pores, size, 56–57, 90, 110, 112–113, 116
Potassium transport
active, 178–182, 213–214, **216**–218, **225**, 250–**251**
erythrocytes, 218, 250–**251**, 304
in co-transport, 178–182
inhibition, 218, 228
muscle, **93**
nerve, 94–97, 304–305
passive transfer, 95, **98**, 102–103
skin, 236–238, 239
synthetic membranes, 102–103
Pressure-filtration coefficient, 42, 45
measurement, 53–55
Primary active transport, *see* Active transport, primary
Proteins
in membrane, **8**, 9, 11, **13**
model membranes, 16–18
solubilization, 9
Pump model
epithelial cell layers, 235–241
kinetics, 231–235
water permeability, 264

R

Red blood cells, *see* Erythrocytes
Reflection coefficients, **45–46, 55–57, 117–118**
measurement, 55–**57**
pore model, 116–120
Resins, ion transfer, **104**
Ribose, facilitated diffusion, 142
RNA, in isolated membrane, **13**

S

Salmonella typhimurium, **224**, 298
Salts, *see also* Ions
hydration numbers, **91**
Schiff's base, 279
Schwann cell
glycerol transport, 294
membrane structure, 20
Secondary active transport, *see* Active transport, secondary
Semipermeable membranes, irreversible thermodynamics, 42, 45, 46
Sialomucopeptides, in human erythrocyte membrane, 9
Sieve model, 73
Simple diffusion, 68, 69
cells, 73–85
Sites, number, glucose transport, 293, 295
Size of permeant, 113, 117
Skin
active transport, 212, **216**
cation transport, 212, **216**, 239–240, 294

chloride transport, 294
facilitated diffusion, 131
permeability data, **210**
pump model, 236–240
water permeability, 258–9, 262, 263
Sodium chloride, diffusion, 59–61
Sodium pump, 181–182, 236, 239–240, 250, **251**, 252, 253, 255, *see also* Pump
Sodium transport, *see also* Sodium pump
active transport, primary, 209, 213–214, **216**, **225**, 229
erythrocytes, **225**–226, 229, 304
facilitated diffusion, **168**, 236
in amino acid transport, 189, 194, 200
in co-transport, 178–182, **186**, 274
in sugar transport, 180, 195
inhibition, 217–218, 262, 263
kinetic constants, **186**
muscle, **93**, 307
nerve, 94–97, 215–**216**, 217–218, 304
passive transfer, 95, **98**
Solute diffusional coefficient, 58–59, **110**–119
Solute-membrane drag, 49
Solute-solvent drag, 49
Solvent-membrane drag, 48
Sorbose binding, stroma, 297
Sorbose transport
facilitated diffusion, 136
inhibition, 297
Specificity studies, 266–281
sugars, 268–278
Square-dependence, sugar transport, 176
Staphylococcus aureus, **223**, **225**, 303
Sterols, 286
Stokes-Einstein relation, 67
Stroma, *see* Ghosts
Strophanthidin, 98, 262, 263
Substrate-binding, 295–298, 307
Sucrose, frictional coefficients, **51**
Sugar-binding, ghosts, 297
Sugar transport, 283–288
active, 179–180, **186**, 214–215, 220–221, **223**, 230–231
bacteria, 214–215, 220–221, 229, 294, 300–303
competitive exchange diffusion, 144
co-transport, 181–182
erythrocytes, 294, 296–297
facilitated diffusion, 159, **164–167**, 171, **174**, 176, 181–182
inhibitors, 149, 179, 181, 184, 284–287, 294
intestine, 179–181, 184–**186**, 190–191, 195, 273–274, 276, **283**, 294, 317
kidney, **283**, 294
muscle, 276
reflection coefficients, **56**–57
specificity requirements, 268–278
yeast cells, 276, 294
Sugar transport systems, isolation, 296–297
Sulfate transport system, isolation, 298
Surface tension of cell, 31–34
Synthetic membranes
diffusion coefficients (nonelectrolytes), **88**
frictional coefficients, **51**
ion transfer, 101–**104**
permeability data, **111**
phenomenological coefficients, **51**
reflection coefficients, **57**
water flow, **46**, **111**, 256

T

Temperature coefficients
cation transfer, **104**
water permeability, **123**
Tertiary active transport, 203–206
Thiol reagents, 295
Tollypellopsis stelligera, nonelectrolyte diffusion, **80**
Transition state, 79
Translocase, 319
Transport systems, molecular properties, 266–308

U

Unidirectional flux, 162–163
carrier model, 155
co-transport, 193
facilitated diffusion, 135, 140–143
Unit membrane, 3–6, 20, 30
Urea transport, **51**, 294

V

van der Waal's interaction, 115
van't Hoff's law, 47
Vesicle, 28

Viscosity, **88**
  temperature dependence, **71**
Vitamin A, active transport, primary, **225**
Volume, *see* Cell volume

W

Water
  diffusional coefficients, **110–111**
  heavy, frictional coefficients, **51**
Water balance, 243, 245–246, 250, 264
Water permeability, 108–116, 120–124, 242–265
  activation energies, **123**
  affected by hormone, 260, 261
  epithelial cell layer, 253–265
  gall bladder, 257, 258, 262, 264
  model systems, 115, 244, 246, 255–257
  permeability data, **110, 111**
  skin, 258–259, 262, 263
  synthetic membranes, **46, 111,** 256
  temperature coefficients, **123**

X

X-Ray crystallography
  myelin forms, 16, 17
  myelin sheath, 20

Y

Yeasts
  active transport, primary, **224, 225, 231**
  facilitated diffusion, 130, **167**
  sugar transport, 276, 294
  unit membrane, 3

5
6
7
D 8
E 9
F 0
G 1
H 2
I 3
J 4